Algonquin Area Public Library
2600 Harnish Dr.
Algonquin, IL 60102
www.aapld.org

Food Chemistry

Scrivener Publishing
100 Cummings Center, Suite 541J
Beverly, MA 01915-6106

Publishers at Scrivener
Martin Scrivener (martin@scrivenerpublishing.com)
Phillip Carmical (pcarmical@scrivenerpublishing.com)

Food Chemistry

The Role of Additives, Preservatives and Adulteration

Edited by

Mousumi Sen

Amity Institute of Applied Sciences, Department of Chemistry, Amity University, Uttar Pradesh, India

WILEY

This edition first published 2022 by John Wiley & Sons, Inc., 111 River Street, Hoboken, NJ 07030, USA and Scrivener Publishing LLC, 100 Cummings Center, Suite 541J, Beverly, MA 01915, USA
© 2022 Scrivener Publishing LLC
For more information about Scrivener publications please visit www.scrivenerpublishing.com.

All rights reserved. No part of this publication may be reproduced, stored in a retrieval system, or transmitted, in any form or by any means, electronic, mechanical, photocopying, recording, or otherwise, except as permitted by law. Advice on how to obtain permission to reuse material from this title is available at http://www.wiley.com/go/permissions.

Wiley Global Headquarters
111 River Street, Hoboken, NJ 07030, USA

For details of our global editorial offices, customer services, and more information about Wiley products visit us at www.wiley.com.

Limit of Liability/Disclaimer of Warranty
While the publisher and authors have used their best efforts in preparing this work, they make no representations or warranties with respect to the accuracy or completeness of the contents of this work and specifically disclaim all warranties, including without limitation any implied warranties of merchantability or fitness for a particular purpose. No warranty may be created or extended by sales representatives, written sales materials, or promotional statements for this work. The fact that an organization, website, or product is referred to in this work as a citation and/or potential source of further information does not mean that the publisher and authors endorse the information or services the organization, website, or product may provide or recommendations it may make. This work is sold with the understanding that the publisher is not engaged in rendering professional services. The advice and strategies contained herein may not be suitable for your situation. You should consult with a specialist where appropriate. Neither the publisher nor authors shall be liable for any loss of profit or any other commercial damages, including but not limited to special, incidental, consequential, or other damages. Further, readers should be aware that websites listed in this work may have changed or disappeared between when this work was written and when it is read.

Library of Congress Cataloging-in-Publication Data

ISBN 978-1-119-79161-4

Cover image: Pixabay.Com
Cover design by Russell Richardson

Set in size of 11pt and Minion Pro by Manila Typesetting Company, Makati, Philippines

Printed in the USA

Contents

Preface   xix

1 Food Chemistry: Role of Additives, Preservatives, and Adulteration   1
Mousumi Sen
- 1.1 Introduction   2
- 1.2 Categories of Food Colors   2
- 1.3 Natural Colors Are Best Over Artificial Colors   3
- 1.4 Classification of Food Colorants   3
    - 1.4.1 Natural Colorants   3
    - 1.4.2 Synthetic Colorants   7
        - 1.4.2.1 Water Soluble Synthetic Colors   7
        - 1.4.2.2 Fat Soluble Synthetic Colorants   8
        - 1.4.2.3 Lake Colorants   8
- 1.5 Classification of Food Additives   9
    - 1.5.1 Why Food Colors Are Preferred   12
    - 1.5.2 E-Numbering   13
- 1.6 Food Spoilage and Preservation   14
    - 1.6.1 Causes of Spoilage   14
    - 1.6.2 Principle of Food Preservation   15
- 1.7 Preservatives   15
    - 1.7.1 Factors Affecting Preservative Efficiency   15
        - 1.7.1.1 Interaction With Formulation Components   15
        - 1.7.1.2 Properties of Preservatives   16
        - 1.7.1.3 Effect of Containers   16
        - 1.7.1.4 Types of Micro-Organisms   16
        - 1.7.1.5 Influence of pH   16
    - 1.7.2 Factors Affecting Chemical Preservation   17
    - 1.7.3 Classification of Chemical Preservatives   17
    - 1.7.4 Types of Chemical Preservatives   17
    - 1.7.5 Natural Chemical Preservatives   18

		1.7.6 Methods of Food Preservation	19
1.8	Antioxidants		20
1.9	Oils and Spices		21
1.10	Introduction to Hurdle Technology		22
	1.10.1	Advantages of Food Additives and Preservatives	23
	1.10.2	Disadvantages of Food Additives and Preservatives	23
	1.10.3	Effects of Food Additives and Food Preservatives	24
	1.10.4	Safety of Food Additives and Preservatives	25
1.11	Adulteration		26
	1.11.1	History of Food Adulteration	27
	1.11.2	Types of Food Adulteration	27
		1.11.2.1 Intentional Adulteration	27
		1.11.2.2 Incidental Adulteration	28
		1.11.2.3 Metallic Adulteration	28
	1.11.3	A Food Is Considered Adulterated if It Has the Following Factors	28
	1.11.4	Effects of Adulterated Food on Human Health	28
	1.11.5	Reasons for Food Adulteration	29
	1.11.6	Methods of Food Adulteration	29
	1.11.7	Trends of Food Adulteration in Developing Countries	30
1.12	Food Safety and Standards Act		30
	1.12.1	Few Steps to Avoid Adulteration	32
	1.12.2	Detection Methods of Adulteration	33
	1.12.3	Technique to Check Food Adulteration	33
1.13	Conclusion		33
	References		35

2 Additives and Preservatives Used in Food Processing and Preservation, and Their Health Implication — 43
Sunita Adhikari (Nee Pramanik)

Abbreviations			44
2.1	Introduction		44
2.2	Merits and Demerits of Food Additives and Preservatives		47
	2.2.1	Merits of Food Additives and Preservatives	47
	2.2.2	Demerits of Food Additives and Preservatives	47
2.3	Types of Food Additives and Preservatives		48
	2.3.1	Preservatives	48
	2.3.2	Nutritional Additives	51
	2.3.3	Flavoring Agent	51

	2.3.4	Coloring Agent	51
	2.3.5	Texturizing Agent	53
	2.3.6	Miscellaneous Additives	54
		2.3.6.1 Acidity Regulator	54
		2.3.6.2 Anti-Caking Agent	54
		2.3.6.3 Antifoaming Agent	54
		2.3.6.4 Flour Treatment Agents	54
		2.3.6.5 Fat Replacers	56
		2.3.6.6 Sweeteners	56
		2.3.6.7 Leavening Agent	61
		2.3.6.8 Firming Agent	62
		2.3.6.9 Glazing Agent	62
		2.3.6.10 Humectant	62
		2.3.6.11 Sequestering Agent	62
		2.3.6.12 Gelling Agent	63
		2.3.6.13 Propellants	63
		2.3.6.14 Foaming Agent	63
		2.3.6.15 Seasoning	63
		2.3.6.16 Curing Agents	64
		2.3.6.17 Probiotics	64
		2.3.6.18 Other Food Additives	65
		2.3.6.19 Indirect Food Additives	65
2.4	Health Effect of Food Additives and Preservatives		66
2.5	Conclusion		70
	References		71

3 Role of Packaging in Food Processing — 73
Bhasha Sharma, Susmita dey Sadhu, Rajni Chopra and Meenakshi Garg

3.1	Introduction		74
3.2	State-of-the-Art		76
3.3	Raw Materials Used in Food Packaging		78
	3.3.1	Metals	79
	3.3.2	Glass	80
	3.3.3	Plastics	80
	3.3.4	Paper and Cardboard	80
3.4	Packaging Footprints on Quality, Shelf Life, and Safety of Food		81
3.5	Prolegomenon on Active and Smart Packaging Systems		81
	3.5.1	Active Packaging	82
	3.5.2	Intelligent Packaging System	83
3.6	Aseptic Packaging in Food Processing		84

viii Contents

  3.7 The Paradigm in Strategies for Improvement of Food
       Packaging   85
       3.7.1 Bequest of Packaging Into the Cycle of Food Chain
              Sustainability   85
       3.7.2 Selection of Materials With the Objective
              of Recyclability   86
       3.7.3 Escalating Protective Role of Packaging   87
       3.7.4 How Biodegradable Polymers can Mitigate the Plight
              of Packaging in Food Processing   88
  3.8 Integration of Nanotechnology to Ameliorate Food Packaging   89
  3.9 Life Cycle Assessment (LCA)   90
  3.10 Deciphering the Challenges for Sustainable Food Packaging   91
  3.11 Conclusion and the Way Forward   92
       Acknowledgement   92
       References   92

4 **Laws Impacting Chemicals Added to Food**   97
  Preeti Khanna, Rajni Chopra and Meenakshi Garg
  4.1 Introduction   98
  4.2 Functions of Food Additives   98
       4.2.1 Sustain or Enhance the Shelf Life and Freshness
              of a Product   99
       4.2.2 Sustain or Enhance the Nutritional Quality of a Product 99
       4.2.3 Improve the Aesthetic Appeal and Sensory
              Attributes of a Product   99
  4.3 Classification of Food Additives   99
       4.3.1 Classification Based on Functionality   99
              4.3.1.1 Flavoring Agents   100
              4.3.1.2 Enzyme Preparations   100
              4.3.1.3 Other Additives   100
  4.4 Classification Based on Primary and Secondary
       Technological Roles—Direct and Indirect Additives   101
  4.5 Evaluating the Health Risk of Food Additives   101
  4.6 International Regulations for the Efficacy of Food Additives 102
  4.7 International Laws   102
       4.7.1 US Food and Drug Administration   102
  4.8 Indian Regulations—Food Safety and Standards Authority
       of India (FSSAI), Additives Regulations (Regulation 3.1)   103
  4.9 Safety Assessment: Redbook's Principles of Safety Evaluation 105
  4.10 Levels of Concern for Direct Food Additives   106

4.11	Threshold Regulation Exemption for Indirect Food Additives	107
4.12	Estimated Daily Intakes	108
4.13	Human Data and Clinical Studies	109
4.14	GRAS Substances	109
4.15	European Union Legislation	110
4.16	Categorization of Food Additives	110
	4.16.1 Additives Can Be Used for the Following Purposes	110
4.17	Safety Assessment of Food Additives	111
4.18	Safety Evaluation Process and Authorization	112
4.19	Use of Food Additives in Food Products	113
	4.19.1 Traditional Foods	113
	4.19.2 Restricted Provisions	114
4.20	Labeling Regulations and Guidelines	114
4.21	Conclusion	114
	References	114

5 Detection of Food Adulterants in Different Foodstuff 117

Aditi Negi, P. Lakshmi Praba K., R. Meenatchi and Akash Pare

5.1	Introduction	118
5.2	Types of Adulteration	118
5.3	Impact of Adulteration on Health	120
5.4	Approaches for Adulterant Authentication in Food Materials	121
5.5	Physical Authentication Techniques	122
5.6	Application of Biochemical and Analytical Methods in Adulterant Authentication	124
	5.6.1 Adulterant Authentication Through HPLC	124
	5.6.2 Adulterant Authentication Through GCMS	127
	5.6.3 Adulterant Authentication Through Spectroscopic Method	127
	5.6.4 Adulterant Authentication Through Ambient Mass Spectroscopy Techniques	128
	5.6.5 Adulterant Authentication Through Nuclear Magnetic Resonance Technique	128
5.7	Adulterant Identification by Molecular Techniques	136
	5.7.1 Polymerase Chain Reaction–Based Techniques for Adulterant Identification	137
	5.7.2 Application of Real-Time PCR in Adulterant Authentication	142
	5.7.3 Isothermal Amplification Methods for Adulterant Identification	142

		5.7.4	Sequencing and Hybridization-Based Methods in Adulterant Identification	147
	5.8	Limitation in Use of Molecular-Based Methods for Adulterant Authentication		148
	5.9	Conclusion		152
		References		152

6 Trends of Food Adulteration in Developing Countries and Its Remedies 165

Satyam Chachan, Anand Kishore, Khushbu Kumari and Arun Sharma

6.1	Introduction		166
6.2	Food Fraud in Developing Countries		166
	6.2.1	Impact of Adulteration	168
6.3	Classification of Food Adulteration		169
	6.3.1	Intentional Adulteration	171
6.4	Common Food Adulterants		172
6.5	Adulteration Remedy Strategies		177
	6.5.1	Government and Regulatory Agency Initiative	177
	6.5.2	Loopholes in Existing Method of Eliminating Adulteration	179
	6.5.3	Process and Product Verification	181
	6.5.4	Higher Levels of Transparency/Traceability in Supply Chain	182
	6.5.5	Use of Novel Technology	182
	6.5.6	Training	183
	6.5.7	Awareness	183
6.6	Conclusion		185
	References		186

7 Food Adulteration and Its Impacts on Our Health/Balanced Nutrition 189

Suka Thangaraju, Nikitha Modupalli and Venkatachalapathy Natarajan

7.1	Introduction		190
7.2	Types of Adulteration		192
	7.2.1	Intentional Adulteration	192
	7.2.2	Incidental Adulteration	193
	7.2.3	Other Types of Adulteration	193
		7.2.3.1 Natural Contamination	193

		7.2.3.2	Metallic Contamination	193

7.2.3.2 Metallic Contamination — 193
7.2.3.3 Microbial Contamination — 194
7.2.3.4 Adulteration in Organic Foods — 195
7.2.3.5 Adulteration During Irradiation of Foods — 195
7.2.3.6 Genetically Modified Foods — 195
7.3 Adulteration in Foods — 196
    7.3.1 Global Food Environment — 197
7.4 Effects of Food Adulteration — 201
    7.4.1 Health Effects — 201
    7.4.2 Balanced Nutrition — 205
7.5 Measures to Mitigate Food Adulteration — 206
    7.5.1 Producer's or Manufacturer's End — 206
    7.5.2 Consumer's End — 206
    7.5.3 Government and Regulatory Agencies — 207
References — 209

8 Natural Food Toxins as Anti-Nutritional Factors in Plants and Their Reduction Strategies — 217
Naman Kaur, Aparna Agarwal, Manisha Sabharwal and Nidhi Jaiswal

Abbreviations — 218
8.1 Introduction — 218
8.2 Anti-Nutritional Factor — 221
    8.2.1 Tannins — 221
        8.2.1.1 Types — 222
        8.2.1.2 Adverse Effects — 222
    8.2.2 Saponins — 223
        8.2.2.1 Saponins — 223
        8.2.2.2 Adverse Effects — 224
    8.2.3 Lectins and Hemagglutinin — 225
        8.2.3.1 Adverse Effects — 226
    8.2.4 Alkaloids — 227
        8.2.4.1 Adverse Health Effects — 227
    8.2.5 Oxalates — 228
        8.2.5.1 Adverse Effects — 229
    8.2.6 Cyanogenic Glycosides — 230
        8.2.6.1 Adverse Effects — 231
    8.2.7 Goitrogens — 231
        8.2.7.1 Adverse Effects — 232
8.3 Methods to Reduce Levels of Anti-Nutritional Factors in Foods — 234

		8.3.1	Soaking	234
		8.3.2	Fermentation	235
		8.3.3	Germination	236
		8.3.4	Milling	237
		8.3.5	Extrusion	237
		8.3.6	Heating-Autoclaving (Wet Heating) and Roasting (Dry Heating)	238
		8.3.7	Gamma Radiation	239
		8.3.8	Genomic Technology	239
	8.4	Conclusion		240
		References		240

9 Feeding the Future—Challenges and Limitations — 249
Baishakhi De and Tridib Kumar Goswami

- 9.1 Introduction — 250
- 9.2 Early Life Nutrition and Healthy Future — 252
  - 9.2.1 Choice of Food and "Nutrition Transition" — 253
- 9.3 Challenges and Opportunities in Developing the Future Food Systems — 255
- 9.4 Sustainable Diet for the Future — 257
- 9.5 Research Trends and Green Food Technologies — 259
  - 9.5.1 Green Technologies in Food Processing — 260
  - 9.5.2 Nanotechnology in Food Processing and Food Safety — 262
  - 9.5.3 CRISPR-Based Technologies — 262
  - 9.5.4 Future Directives — 265
    - 9.5.4.1 3D Food Printing and Mass Customization of Diet — 266
- 9.6 Regulations and Trade — 270
- 9.7 Conclusion — 270
- References — 271

10 Alternate Food Preservation Technology — 275
Pratik S. Gaikwad, Chayanika Sarma, Aditi Negi and Akash Pare

- 10.1 Introduction — 276
- 10.2 Non-Thermal Preservation Technique — 277
  - 10.2.1 Packaging Technology — 277
    - 10.2.1.1 Challenges and Future Scope of MAP Processing — 282
  - 10.2.2 Ozone (O_3) Treatment — 282
    - 10.2.2.1 Properties of O_3 — 291

		10.2.2.2	Principle of O_3 Generation	291
		10.2.2.3	Challenges and Future Scope of O_3 Processing	292
	10.2.3	High Hydrostatic Pressure Treatment		292
		10.2.3.1	Principles of HPP Treatment	294
		10.2.3.2	HPP Time	294
		10.2.3.3	Challenges and Future Scope of HPP Treatment	295
	10.2.4	Ultrasound Treatment		295
		10.2.4.1	Principle of Ultrasound Treatment	296
		10.2.4.2	Challenges and Future Scope of Ultrasound Treatment	296
	10.2.5	Pulsed Electric Field Treatment		296
		10.2.5.1	Principle of PEF Treatment	297
		10.2.5.2	Challenges and Future Scope of PEF Treatment	298
	10.2.6	Cold Plasma Treatment		298
		10.2.6.1	Generation of CP Treatment	298
		10.2.6.2	Challenges and Future Scope of CP	300
	10.2.7	Oscillating Magnetic Field		300
		10.2.7.1	Challenges and Future Scope of OMF	301
	10.2.8	Membrane Filtration Process		301
		10.2.8.1	Principle of the Membrane Filtration Process	301
		10.2.8.2	Microfiltration	301
		10.2.8.3	Ultrafiltration	302
		10.2.8.4	Nanofiltration	303
		10.2.8.5	Reverse Osmosis	303
		10.2.8.6	Challenges and Future Scope of the Membrane Filtration Process	303
10.3	Novel-Thermal Preservation Technique			303
	10.3.1	Ohmic Heating Treatment		303
		10.3.1.1	Application of OH Treatment	304
		10.3.1.2	Challenges and Future Scope of OH Treatment	311
	10.3.2	Microwave Heating		311
		10.3.2.1	Principle of MW Heating	311
		10.3.2.2	Applications of MW Heating	312
		10.3.2.3	Challenges and Future Scope of MW Heating	312
	10.3.3	Infrared Heating (IRH)		312

		10.3.3.1	Application of IRH	313
		10.3.3.2	Challenges and Future Scope of IRH	313
	10.3.4	Radio Frequency Heating		313
		10.3.4.1	Principle of RF Heating	314
		10.3.4.2	Factor Influencing of RF Heating	314
		10.3.4.3	Challenges and Future Scope of RF Heating	314
10.4	Other Alternate Preservation Techniques			315
	10.4.1	Freezing		315
		10.4.1.1	Challenges and Future Scope of Freezing	316
	10.4.2	Dehydration		316
	10.4.3	Frying		317
	10.4.4	Chilling		318
	10.4.5	Extrusion		318
	10.4.6	Three-Dimensional (3-D) Printing		319
		10.4.6.1	Principle of 3-D Printing	319
		10.4.6.2	Factor Influencing 3-D Printing	319
	10.4.7	Blanching		320
10.5	Hurdle Technology for Preservation of Food			320
10.6	Irradiation Process for Preservation of Food			321
	10.6.1	Electron Beam		327
	10.6.2	X-Radiation (X-Ray)		327
	10.6.3	Gamma Rays		327
10.7	Food Additives for the Preservation of Food			328
	10.7.1	Natural Additives		328
	10.7.2	Synthetic Additives		328
	10.7.3	Challenges and Future Scope of Additives		331
10.8	Conclusion			332
	References			332

11 Green Solvents for Food Processing Applications — 341
A. Surendra Babu, A. Sangeetha and R. Jaganmohan

11.1	Introduction		342
11.2	Green Solvents		344
	11.2.1	Water as Green Solvent	345
	11.2.2	Subcritical Water Extraction	346
	11.2.3	Supercritical Fluids as Green Solvent	346
	11.2.4	Gas Expanded Liquids as Green Solvent	347
	11.2.5	Ionic Liquids as Green Solvent	348
		11.2.5.1 Classification of Ionic Liquids	348
	11.2.6	Solvents Derived From Biomass as Green Solvent	349

		11.2.7	Deep Eutectic Solvents as Green Solvents	352
	11.3	Synthesis of NADES		353
		11.3.1	NADES for Extraction of Phenolic Compounds	355
		11.3.2	NADES for Extraction of Flavonoids	360
		11.3.3	NADES for Extraction of Other Polar Compounds	364
			11.3.3.1 Ferulic Acid Extraction From *Ligusticum Chuanxiong* Hort and NADES	364
		11.3.4	NADES for Extraction of Food Samples	364
			11.3.4.1 Extraction of Vanillin With NADES	364
			11.3.4.2 Extraction of Anthocyanins With NADES	364
			11.3.4.3 Extraction of Phenolic Compounds With NADES	364
		11.3.5	General Considerations Using NADES as Extraction Solvents	365
	11.4	Conclusion and Future Trends		366
		References		366

12 Technological Advancement in Food Additives and Preservatives 375
Shikha Pandhi, Arvind Kumar and Akansha Gupta

	Abbreviations		376
12.1	Introduction		377
12.2	Food Additives and Preservatives		378
	12.2.1	Classes of Food Additives	379
	12.2.2	Significance in Food Processing and Preservation	381
	12.2.3	Mechanism of Action of Food Preservatives	381
12.3	Regulatory Aspects of Food Additives and Preservatives		382
	12.3.1	Generally Recognized as Safe	383
	12.3.2	FSSAI Regulations on Permissible Limits of Food Additives	383
12.4	Health Concerns of Conventional Food Additives		383
12.5	Technological Advancements in Food Additives and Preservatives		384
	12.5.1	Novel Food Additives	384
		12.5.1.1 Essential Oils/Phytochemicals	385
		12.5.1.2 Metallic Nanoparticles as Antimicrobial (Green Route)	386
12.6	Novel Technological Approaches for Enhanced Functionality		386
	12.6.1	Nanoencapsulation	386

| | | 12.6.1.1 | Fundamentals and Techniques | 387 |
| | | 12.6.1.2 | Types of Encapsulating Material | 388 |

- 12.7 Methods for Food Additives Determination — 389
  - 12.7.1 Analytical Methods — 389
    - 12.7.1.1 Spectroscopy Techniques — 389
    - 12.7.1.2 Chromatographic Techniques — 390
    - 12.7.1.3 Electroanalytical Techniques — 391
- 12.8 Future Prospects — 391
- 12.9 Conclusion — 392
- References — 393

13 Sensors for Non-Destructive Quality Evaluation of Food — 397
Krishna Gopalakrishnan, Arun Sharma, Neela Emanuel, Pramod K. Prabhakar and Ritesh Kumar

- 13.1 Introduction — 398
- 13.2 Different Types of Non-Destructive Methods — 400
  - 13.2.1 Mechanical Method — 400
    - 13.2.1.1 Mechanical Thumb Method — 400
    - 13.2.1.2 Sinclair IQ™–Firmness Tester (SIQ-FT) — 401
    - 13.2.1.3 Laser Air-Puff — 401
  - 13.2.2 Chemical Method — 401
    - 13.2.2.1 Electronic Nose — 401
  - 13.2.3 Electromagnetic Method — 404
    - 13.2.3.1 Nuclear Magnetic Resonance (NMR) — 404
    - 13.2.3.2 Magnetic Resonance Imaging — 406
  - 13.2.4 Optical Method — 408
    - 13.2.4.1 NIR Spectroscopy — 408
    - 13.2.4.2 Image Analysis Techniques — 410
    - 13.2.4.3 Time-Resolved Reflectance Spectrometry — 415
  - 13.2.5 Dynamic Method — 418
    - 13.2.5.1 X-Rays — 418
    - 13.2.5.2 Computed Tomography — 419
    - 13.2.5.3 Ultrasonic — 420
    - 13.2.5.4 Acoustic Techniques — 421
  - 13.2.6 Sensor Fusion — 423
- 13.3 Non-Destructive Quality Testing in Various Food Commodities — 425
  - 13.3.1 Staple Foods — 425
    - 13.3.1.1 Sensory Aspect — 426
    - 13.3.1.2 Adulteration Aspects — 427

		13.3.1.3	Chemical Aspects	427
	13.3.2	Fruits		427
		13.3.2.1	Fruit Quality Inspection Using Electronic Nose	429
		13.3.2.2	Fruit Quality Inspection Using UV-VIS-NIR Spectroscopy	429
		13.3.2.3	Fruit Quality Inspection Using Ultrasound Sensing Technique	430
		13.3.2.4	Fruit Quality Inspection Using Machine Vision Sensing Technique	430
		13.3.2.5	Fruit Quality Inspection Using Acoustic Impulse Technique	431
	13.3.3	Vegetables		431
		13.3.3.1	Spectroscopic Techniques	431
		13.3.3.2	Sound Waves Techniques	434
		13.3.3.3	Imaging Analysis Techniques	435
13.4	Conclusion			436
	References			437

Index   **451**

Preface

Food processing is no longer as simple and straightforward as it was in the past and has now become more of a highly interdisciplinary science than an art, and although the basic principles of food chemistry remain the same, much additional research has been carried out in recent years that has extended and deepened our knowledge. In light of the fact that each of the processes involved in food production—including storage, preparation, and distribution—influences the different qualities of food in either a beneficial or harmful way, it is crucial to have a basic understanding of every safeguarding strategy and how it impacts different food systems. Thus, it is of utmost importance to acquire the requisite knowledge about the technology, methods, and the science of the mode of action. Because of the importance of this complex subject, *Food Chemistry: Role of Additives, Preservatives, and Adulteration* is designed to present basic information on the composition of foods and the chemical and physical changes that their characteristics undergo during processing, storage, and handling. Details concerning recent developments and insights into the future of food chemical risk analysis are also presented, along with topics such as food chemistry, the role of additives, preservatives and food adulteration, food safety objectives, risk assessment, and quality assurance and control. Moreover, good manufacturing practices, food processing systems, design and control, and rapid methods of analysis and detection are covered, as well as sensor technology, environmental control, and safety.

Today, the variety of food items—mainly fast foods—is greatly increasing for food lovers. Due to tremendous advancements, synthetic and natural supplements are added to these food items and beverages in huge amounts, sometimes with serious consequences. For example, when color preservatives overshoot the sanctioned limit, they gradually cause hyperactivity in children. Therefore, it is mandatory to ensure that the food quality and supplements being added to food for consumption are safe. Consequently, this book contains detailed information about the chemistry of each major class of food additive and their multiple functionalities. In addition, numerous recent findings are covered, along with an explanation of how their quality is ascertained and consumer safety ensured.

A brief description of the topics presented in each chapter of the book follows, starting with the first chapter on health-related aspects such as the limited use of additives to perform and fulfill a certain function within the legal framework. The various rules for consumer safety that should be followed for food to remain healthy and nutritious and the importance of spreading awareness about food adulteration are discussed. The next chapter presents various facts about additives and preservatives used in food processing and preservation, and their health implications. For example, the effect of additives on the population, especially children, as consumption of additives is related to hyperactivity and hypersensitivity. Synthetic additives should especially be avoided due to their potential toxic effect.

The third chapter includes the zealous objective to elevate sustainability or recyclability of plastic packaging via enhanced recovery and the choice of packaging design that takes its end of life into consideration. Progress toward the efficacy of packaging materials can also enhance the removal of food pathogens and alleviate environmental waste by maintaining food quality. Because laws regulating food additives are extremely important for safe use and application, the national and international laws associated with food additives used in the food industry are highlighted in the fourth chapter.

Since screening and detecting adulterants in food commodities is a step that ensures the quality of the product, the fifth chapter highlights detection techniques such as physical, chemical, analytical, and immuno-based techniques. The objective of the sixth chapter is to identify common cases of food adulteration malpractice in developing countries. A preventive and remedial approach to counter food adulteration by evaluating existing laws and technology and looking for gaps in the law is discussed.

Food adulteration, why it is done, and its effect on human health and balanced nutrition are covered in the seventh chapter. Regulations that severely punish illegal traders and dishonest producers that adulterate different food products in various ways should be enforced. Next, the eighth chapter attempts to elucidate different types of antinutrients, their structures and adverse effects, as well as strategies for reducing their levels to improve the quality of food. Also discussed is the extensive research on strategies to reduce the levels of anti-nutritional factors (ANFs) in foods, which is essential to produce foods with few or no ANFs without altering their nutritional or organoleptic value. The ninth chapter is a detailed discussion of the challenges and constraints in future food security. Green technologies and associated research trends in food processing are covered, with a major focus on topics like the search for alternative protein sources, novel technologies in food processing, nutrigenomics, biomimicry, DNA barcoding, lab-on-chip, and nanosensors in food safety technology as remedial solutions and associated regulations in attaining a future sustainable food supply.

Alternative food preservation technologies are advantageous over conventional technologies for beverages and solid foods as they can retain nutritional qualities, sensory attributes, lower the chances of cross-contamination, and increase shelf life. The tenth chapter discusses these alternative food preservation techniques for extending and improving the shelf life of food products with shorter treatment duration and higher product safety. The next chapter discusses the tractability of natural deep eutectic solvents (NADES) in relation to environmental impact, stability, solute recovery, and toxicity, taking into consideration the costs.

The twelfth chapter provides a succinct overview of various conventional food additives and preservatives with a major emphasis on their type, functionality, mode of action, and significance for food applications. Furthermore, it stipulates various technological advancements made over the years with novel alternative ingredients and discusses the opportunities shaped by the advent of technologies like nanoencapsulation. The last chapter discusses the broader concept of using nondestructive methods in quality assessment of food products and their applications for food evaluation and quality. The overall quality measurement in the food industry is attributed to the sensor fusion technique, which is a collaborative multi-sensor approach to enhance the quality assessment of agro-food products. These methods, which help to ensure customer satisfaction of products by providing good quality products without rupturing the food product, are discussed in this chapter.

In conclusion, I am very grateful for all the hard work and efforts of the many contributors to this book. I would like to thank all the authors for sharing their insightful research and information with us. I am also very much thankful to Aarushi Sen for her unending encouragement and support throughout the creation of this book. Her help was much appreciated. I am also most grateful for the efforts put in by Martin Scrivener of Scrivener Publishing, whose help made this book possible. I thank him for his patience and consistent support throughout the journey. Finally, I express my sincere thanks to the Department of Chemistry, Amity University, Uttar Pradesh for all the help and support I have received.

Dr. Mousumi Sen
October 2021

1

Food Chemistry: Role of Additives, Preservatives, and Adulteration

Mousumi Sen

Amity Institute of Applied Sciences, Department of Chemistry, Amity University, Uttar Pradesh, India

Abstract

Food is one of the most important substances for survival, it contains all the components that are required for nutrition to maintain a healthy life. Nowadays, due to tremendous advancement, synthetic and natural supplements are added to food items and beverages in huge amount. The synthetic or artificial colors need a strong accreditation from the administrative bodies for their consumption [1–4].

When color preservatives overshoot the sanctioned limit it gradually causes hyperactivity in children. It is mandatory to oversee food quality and the supplements being added to the food for consumption. Also, natural color is rapidly being replaced with artificial or synthetic colors [5].

It is very crucial and critical to understand the effects of each preservation methods and their handling procedure on various foods because each step of storage, handling, processing, and distribution affects the various characteristics of food which could be desirable/undesirable.

Adulteration of food is a demonstration of deliberately corrupting the nature of nourishment offered available to be purchased either by the admixture or substitution of second rate substances and which antagonistically influence nature of food sources, yet in addition their coincidental pollution during the time of development, reaping, capacity, preparing, transport, and dispersion [6–8].

Keywords: Synthetic and natural supplements, artificial colors, food colors, chemical food preservation, sophisticated techniques, adulteration of food, health hazards, pollution

Email: mousumi1976@gmail.com

1.1 Introduction

Food colors are basically the supplements that are called as pigments or dyes. When these supplements are added to the food, these improve color of the eatables. These type of color supplements are basically increased within the food just to overcome the color loss that is caused due to the prolong storage conditions. These additives also provide color to the colorless food that makes them more attractive [9]. Food colorants are being added to food in one or another form since many years. Food color plays an eminent role in today's scenario.

Colorant are different from one another on the basis of their chemical properties and physical properties such as structure, sources, and usage purpose. Colorants are usually added to processed food such as candies, snacks, margarine, cheese, soft drinks, jam/jelly, gelatin, pudding, and pastry fillings [10, 14]. It is known that in medieval ages nitrate was used to enhance the color of meat and to prevent botulism apart from salt and smoke that were used as preservatives.

Food colorants were used by Egyptians in 400 BC to regulate the color of wine and confectionery products. Around the mid-1800 some of the natural colorants added to food were vegetable-derived products such as saffron, carrot, mulberry and flower, various animal originated pigments, and minerals from copper and iron. Around the end of 19th century, the first synthetic dye obtained from organic coal tar was used in butter and cheese [11, 13]. It has been reported that food colorants were used in Europe during the Bronze age.

1.2 Categories of Food Colors

Food colors are classified into three categories:

- *Natural colors*: Natural colors are the pigments that are made by living organisms. For example: beetroot extract, lutein, and annatto.
- *Nature-identical colors*: Nature-identical colors are the man-made pigments which are basically found in nature. For example: beta-carotene and canthaxanthin.
- *Artificial colors*: Artificial colors are purely man-made colors. For example: allura red, and brilliant blue.

1.3 Natural Colors Are Best Over Artificial Colors

Artificial food coloring causes many ill effects mainly to the health. Some of them are mentioned below:

- ➢ The major effect caused is the behavioral problems especially in children.
- ➢ Depression in youth is observed on a large scale.
- ➢ Food allergies and food poisoning are quite common.
- ➢ Headaches and migraines in people are also seen.

1.4 Classification of Food Colorants

1.4.1 Natural Colorants

Natural food colorants are used worldwide, are known to have significant benefits when consumed, and are demanded by people for their reliability, functionality, biological potential, and health effects [12]. Many consumers associate good and natural looking food and drinks with high quality while they think the other way around when it comes to faded and artificial shining products. In addition, the production of colorants from known sources such as beetroot, grape, cabbage, and paprika makes the consumer feel safe and makes it easier to familiarize and accept the product. Natural colorants are less stable to heat, light, or pH, and their production is inadequate to meet industrial demand. Chemical classification of natural colorants is shown in Table 1.1 [17, 18].

They quickly fade when exposed to light and show low resistance to acidity and high temperature. For example, annatto turns to pink from yellow at low pH and chlorophyll turns to brown from green. This makes natural origin colorants more expensive. For example, natural red and yellow colorants may cost 100 times more than synthetic products with the same effect. Natural coloring matters are synthesized by plant and animal organisms or microorganisms and they naturally exist in them. The most notable colorants obtained from animal sources are Natural Sepia (Cuttle fish), Crimson (Kermes Lotus), and Tyrian purple (Murexshellfish).

We all need food to survive and to maintain our health and as the population is increasing day by day, demand for food is also increasing.

Table 1.1 Chemical classification of natural colorants.

Sl. no	Color	Chemical classification	Plant sources
1	Orange-yellow	Flavone dyes, isoquinoline dyes, polyene dyes, pyran dyes, chromene dyes	Marigold, β-carotene, lycopene, gentism, turmeric, saffron, Sanguinaria canadensis
2	Brown	Naphthoquinone dyes	Camellia thea, Lawsonia inermis
3	Red	Quinone dyes, anthraquinone dyes, chromene dyes	Annatto, Beta vulgaris, paprica, grapes vitacea, Alkanna tinctoria
4	Purple-blue	Benzopyrone dyes	Centaurea cyanus, Indigofera inctoria, Vaccinium myrtillus

In modern days, consumer wants good food with appropriate quality, taste, and safety. For these improvements and increasing demand of food, food additives are required in our daily food for good quality to fulfil the demands of consumers. Ingredients or substance that added in a food for processing, preserve, or storage is called food additives. Food additives are of little or no nutritive values and it includes food coloring agents, flavoring substance, and food preservatives. Food additives are also used for enhancing the taste and appearance.

As we all need food for living and for good health, but the demand of food is increases with increasing population day by day. In modern days, consumer wants food with increasing standard of quality, taste, and safety. For this improvement, we added food additives in our daily food for good quality. So, when food is manufacture, it has to face many challenges which can be technologically fascinating.

Food additives are used to prevent from spoiling and it also prevents the growth of bacteria, yeast, in food. The most commonly used additives are salt and sugar. Other examples of additives are ammonium carbonates, sodium nitrates, white sugar, potassium bromate, sodium benzoate, etc.

As with the time, people's lifestyle changes, and nowadays, most of the people depends on the factory made food (also called processed food) rather than the homemade food. So, the factory made food should be balance its taste, appearance, texture, and its safety. The processed food should remain safe for long time after it has been made [15, 16]. The important roles of additives in our food are as follows:

1. For improving quality of food, for example, preservatives.
2. For giving the texture and consistency, for example, gelling and emulsifier.
3. For improving the nutritional value in food, for example, sweeteners.
4. For improving the flavor and taste, for example, acidulants and flavoring.
5. For preserving the food from different types of bacteria or fungi, etc.

All additives are assigned by European wide basis. European wide regulates and controls the use of additives in food. All food additives are tested and must be shown to be safe before food manufacture able to use them. Food preservatives are a part of food additives as they come under the category of food additives. For example, chips packet are not completely filled up with chips but also contains almost half the packet of gas which is nitrogen gas used for preserving the chips. Table 1.2 shows the common color and associated food [20–25].

Food additives can be natural or artificial. Natural food additives are extracted from fruits, plants, etc., for example, beetroot powder which is used to as coloring agent, whereas artificial food additives are made synthetically, for example, vanillin which is used as flavoring agent in desserts [20].

With time, the lifestyles of people are changing, and nowadays, most people depend on factory made food also known as processed food rather than using homemade food, for example, people prefer packed dairy products like paneer instead of making at home on their own. So, machine made food should have appropriate taste, appearance, texture, and its safety. Processed food should remain safe for longer time after it has been made.

The important functions of food additives are as follows:

- to prevent food from spoiling, for example, adding preservatives to food;
- to give texture and consistency, for example, gelling and emulsifiers;

Table 1.2 Common color and associated food.

Sl. no.	color	Chromophore	Plant sources	Nutrients
1	Purple-blue	Anthocyanins	Eggplant, blackberry, purple cabbage, plum, blueberry, raisins, prunes, purple grapes, figs	Lutein, zeaxanthin, resveratrol, vitamin C, flavonoid, ellagic acid, quecertin
2	Green	Chlorophyll	Avocado, cucumber, spinach, kale, broccoli, snow pea, zucchini, artichoke, lettuce, kiwi	Lutein, zeaxanthin, resveratrol, vitamin C, calcium, folate, β-carotene
3	White-tan	Anthoxanthins	Cauliflower, mushrooms, parsnip, potato, ginger, onions, jicama, banana, garlic	Ancilin, potassium, selenium
4	Yellow-orange	Carotenoids	Papaya, pineapple, apricot, pumpkin, peach, carrot, orange, corn	β-carotene, zeaxanthin, flavonoid, vitamin C, potassium
5	Red	Lycopene or anthocyanins	Cranberry, beet, watermelon, tomato, strawberry, pomegranate	Ellagic acid, quecertin, hesperidin

- to improve flavors, taste, for example, adding flavoring agents to food;
- to enhance the nutritional values, for example, sweeteners;
- to make food attractive and appealing;
- to garnish food;

- to expand affordability and convenience;
- to maintain the freshness of food.

1.4.2 Synthetic Colorants

Synthetic colorants are the colorants which are not found in nature due to their chemical structure and obtained by chemical synthesis. In 1856, William Henry Perkin discovered the first synthetic organic color which is purplish lilac color obtained from coaltar. Synthetic food colorants have their high coloring ability, various color tone, homogeneous color distribution, brightness, stability, and ease of application [19, 26].

Synthetic are divided into three groups according to their solubility:

- ➤ Water soluble synthetic colors
- ➤ Fat soluble synthetic colors
- ➤ Lake colors

1.4.2.1 Water Soluble Synthetic Colors

Allura Red AC: This synthetic colorant, generally known to be derived from insects, is actually produced from coal tar. Allura Red AC is used in the production of food like carbonated drinks, gums, snacks, sauces, soups, wine, and especially apple wine [27].

Sunset Yellow: It is orange red color, and it is usually used for food such as bread, drinks, cereals, sweet powders, ice cream, and snacks.

Brilliant Blue FCF and Brilliant Black BN: Available in blue and black colors, it exists in powder and granular form. It is easily soluble in water while being less soluble in ethanol. Brilliant black is used in the production of various cheese, wine, sauce, and beverages.

Tartrazine: Tartrazine is used to obtain lemon yellow color and is added to food products such as bread, beverages, cereals, peanuts, confectionery, cream, ice cream, and canned food.

Erythosine: Being a xanthen-class colorant in the structure of benzoate, erythosine exists the form of red powder or granules. It is added to flavored milk and puddings, ice products, chewing gum and candies, jelly and drink powders [29, 30].

Quinoline Yellow: Quinoline yellow is a synthetic substance used to obtain a greenish yellow color. It is used in soft drinks, jams and canned foods, edible ice, sweets, candies, pickles, sauces, and spices [28].

Brown FK and Brown HT: Brown FK is used in smoked and cured fish, meat, and chips, while Brown HT is used in various biscuits, chocolates, and cakes.

Other water soluble synthetic colorants are Green S, Indigotine, Patent Blue V, Litolurubin BK, Red 2G, Ponso 4R, and Azorubin.

1.4.2.2 Fat Soluble Synthetic Colorants

Artificial colorants soluble in oil or organic solvents are insoluble in water as they do not contain groups capable of forming salt form as in water-soluble colorants. This group of colorants are not allowed to be used for food coloring because of their toxic properties. For example, the use of oil-soluble Penso SX for the coloring of butter and margarine was banned in 1976. Oil Red XO, Yellow AB used in the coloring of orange peels and Yellow OB are not allowed to use because of their toxic properties [31–34].

1.4.2.3 Lake Colorants

Lake colorants are water-insoluble precipitation of aluminum hydrate substrate and are produced in the form of very fine powders. The dye content and particle size determine the color tone of the powder. As they are not soluble in water, oil, and other solvents, they are dispersed in food and produce color. They are used in cakes, biscuit fillings, confectionery, powder drinks, sweets, soups, and spice mixtures [36–38].

1.4.2.3.1 How to Minimize the Health Risk of Food Additives Consumption?

As we all know, food additives cause harmful health effects [39–44]. So, we should try to take precautions before consuming them. Some measures are as follows:

- ➢ The food containing danger preservatives and additives should be avoided to minimize the risks related to health.
- ➢ We should check the ingredients of packed food before purchasing them from markets.
- ➢ We should eat organic food because they do not have any artificial additives and preservatives in food.

- ➤ The freshly prepared food should be consumed to maximum extent and use of processed food should be discouraged.
- ➤ We should decrease the amount of consumptions of packed food items which contain lots of additives.
- ➤ We should buy certified food items which have been legally approved by FDA.
- ➤ We should read instructions given on food label before consuming them.

1.5 Classification of Food Additives

Food additives can be classified into many types [45–51], such as follows:

1. Acidity regulator:
   - Acidity regulators are those substances which are used for controlling the pH of food for the stability [35].
   - Types of acid regulators include acidity regulator, buffer, and acidifier.
   - Examples: sorbic acid, benzoic acid, and acetic acid.
2. Anti-caking agent:
   - These substances are used to prevent formation of lumps in dairy products.
   - Types of anti-caking agents are dusting agent, anti-sticking agent, and dusting agent.
   - Examples: baking powder, milk powder, table salt, icing sugar, and cake mixes.
3. Bulking agent:
   - These substances increase the bulk of food without changing its taste.
   - Bulking agents include fillers.
   - Examples: glycerin, polydextrose, and pectin.
4. Antioxidants:
   - These additives prevent spoiling of food by oxygen.
   - Types of antioxidants: antioxidant synergist and anti-browning agent.
   - Examples: vitamin C, vitamin E, polyphenols, and zinc.

5. Color retention agents:
   - These agents are used to maintain and retain the color of food.
   - Types of color retention agent are color stabilizer, color fixative, and color adjunct.
   - Examples: sodium nitrite, citric acid, EDTA, and sodium nitrate.
6. Emulsifiers:
   - These substances are used to form uniform emulsion of more than one phase in a food. They help oil and water to remain mixed together, like in mayonnaise and ice creams.
   - Types: plasticizer, dispersing agent, clouding agent, crystallization inhibitor, suspension agent, and surface active agent.
   - Examples: egg yolk, agar, albumin, and alignates.
7. Flavors:
   - These additives are used to give characteristic taste or smell to the food.
   - Flavors can be naturally derived ingredients or artificially synthesized.
   - Examples: citric acid, acetic acid, and ethylvanillin.
8. Stabilizers:
   - Stabilizers are used for maintaining uniform dispersion of more than one component in a food and it also gives suitable texture. They are not true emulsifiers but helps in stabilizing emulsions. These are used in jams, gellies, etc.
   - Types of stabilizers are foam stabilizers, colloidal stabilizers, and emulsion stabilizer.
   - Examples: polysaccharides, starch, and proteins like gelatin.
9. Preservatives:
   - Preservatives protect food from deterioration caused by microorganisms. It prevents food from spoiling and also stops the growth of microbes.
   - Types of preservatives: fungistatic agent, antimycotic agent, antimicrobial synergist, bacteriophages control agent, antimould, and antirope agent.

- Examples:
  - Benzoic acid: it increases the shelf life and protect from microorganism. Soft drink, beer, and acidic foods contain benzoic acid.
  - Sulfites: Fruit-based pie fillings and dried fruits are the examples of
  - where sulfites are found.
  - Nitrites: it helps in preserving the color in meat and dried fruits.

10. Sweeteners:
    - Sweeteners are added in food to give flavor. Sweeteners are unlike sugar, they are used to maintain low food calories. Sweeteners have fruitful effects on diabetes mellitus, diarrhoea, and tooth decay.
    - Sweetener can be naturally derived or artificially synthesized.
      i. Natural sweeteners are obtained from nature. For example: sugarcane, sugar beet, and fruits.
      ii. Artificial sweeteners are of two types:
         a) Non-caloric sweeteners: they do not add calories to foods like snack food and drinks. For example: saccharine and aspartame.
         b) Sugar alcohol: found in candies and chewing gums. They contain calories as sugar. For example: sorbitol and mannitol.
    - Some common sweeteners used are as follows:
      a) Saccharin: it is 300–500 times sweeter than sugar. Saccharin is most commonly used as sweeteners. They are harmless if used in limited quantity.
      b) Aspartame: it is 200% sweeter than sugar and it is low calorie sweetener. It can be added to all kind of food, beverages, and medicines. It usually found naturally in protein rich food.
      c) Acesulfame K: it is 130–200 times sweeter than sugar and 0-calorie sweetener. It helps to reduce the calorie of product. It can be found in food preserve, dairy product, and in beverages.

11. Fortifying agents:
    - Fortifying agents are used to increase the nutritional value.

- Examples: vitamins, lipids, carbohydrates, minerals, and dietary supplements.
12. Flavor enhancers:
    - These substances are used to enhance the existing flavor of food.
    - Types of flavor enhancer includes natural and artificial enhancers.
    - Examples: monosodium glutamate and salt.
13. Humectants:
    - These prevent food from drying.
    - Examples: aloe vera gel, honey, egg yolk, and lithium chloride.

1.5.1 Why Food Colors Are Preferred

Food supplements are quantized to modify the natural or original color of foods. For example, manufacturing of the strawberry jam that looks much brighter adds certain value to the foods. Sunlight can effectively affect the flavors in the foods [55–57]. Hence, dyes, pigments, or colors are added to the foods as they help to conserve the foods from detaching such nutrients. Food colors are also added to foods so they look attractive like in cakes. Foods that seems to look more appealing seem more appetizing and tend to be sold more faster.

Nowadays, food additives are more firmly studied, managed, and supervised than in past due to increasing use and demands. This responsibility of studying and determining the safe use of food additives is done by FDA (Food and Drug Administration). A manufacturer, producer, or sponsor should petition FDA for approval of food additives. After approval, they should be in agreement with specific food additive regulations [52].

FDA considers the following factors for studying the safety of a substance for its approval:

(i) Consumable amount
(ii) Properties and constitution of the substance
(iii) Long-term and instant effects of health
(iv) Evaluation of safety
(v) Environment assessment

After the approval, FDA issues regulations which includes the following: in which type of food, it must be used, highest amount to be consumed, and how it must be recognized on product labels. Since 1999, approval process is modified. Now, FDA may consult USDA (United States Department of Agriculture) during the review process of substances to be used in meats and poultry products. Regulations known as Good Manufacturing Practices (GMP) limit the quantity of food substances used in foods to the quantity necessary to attain the desired effect [53–55].

1.5.2 E-Numbering

- Additives are allotted a special number which is known as E-NUMBER. It is used in Europe for all the approved food additives. This process of giving characteristic number to food additives for easy identification is known as E-numbering [59].
- E-numbering is done to modulate these additives and to notify the consumers.
- The naming of all approved additives starts with the prefix "E" and followed by the number assigned for the approved additive [57].
- All countries other than Europe do not use the prefix "E". For example, in Europe, acetic acid is written as E260, whereas in other countries, it is simply written as additive 260.
- Some examples of E-numbers of food additive are as follows:

    - Tartrazine (E102)
    - Quinoline yellow (E104)
    - Amaranth (E123)
    - Saffron (E164)
    - Sorbic acid (E200)

    - Benzoic acid (E210)
    - Malic acid (E296)
    - Sodium alignate (E400)
    - Agar (E406)
    - Shellac (E904)

Food is defined as the complex mixture of carbohydrates, fats, proteins, minerals, and the vitamins. Because of these ingredients, the food undergoes spoilage mechanisms. Different microorganisms utilize these ingredients and grow and degradation of the product or any component causes spoilage of food. That is why we implement the food preservation technique.

There are two types of food preservation techniques: physical method and chemical method. Physical method includes manipulation of storage temperature, pasteurization, and sterilization to arrest the growth of microorganisms [56, 58, 60]. This type of method is although good in arresting the growth of microbes, they employ the economic investment as it requires a continuous supply of electricity which is not possible in developing countries like India. That is why we need to use chemical preservatives. Chemical preservatives are added in a very small quantity that they do not affect the organoleptic and physical properties of food.

Food preservatives contains a complex of substances having unsimilar molecular structures (organic and inorganic substances having different functional group and capacity to form ions). No methods are generally applied in the analysis of preservatives as a subgroup of food additive, the methods are specified to the preservative being examined.

1.6 Food Spoilage and Preservation

It is defined as an undesirable changes that occur in food due to the influence of factors like heat, light, air and moisture which favors the growth of various micro-organisms. Foods take different period of time to lose their natural form their spoilage [61–63]. The time for which a food can be kept fresh is called shelf life.

Food preservation is defined as a science which deals with the process of preservation of decay or spoilage of food to be stored in a fit condition for future use. It has been described as a state in which any food may be retained over a long period of time without being contaminated by pathogenic organisms or chemicals and losing optimum qualities of color, texture, flavor, and nutritive value. Preservation is important as it destroys pathogens and reduces the microbial, takes care of the excess produce, adds variety to our meals, and allows food items to be sent to places where they are not grown.

1.6.1 Causes of Spoilage

Germs are the major sources which infects food and water. The flies carry germs, they pass on these germs when they sit on our food and spoil it and that is when the food becomes infected. There are different factors that are responsible for food spoilage that are as follows:

Bacteria: There are variety of microorganisms present on the earth. They are small in size and has various shapes. Some bacteria are found to be useful like lactobacillus which converts milk to curd.

Protozoa: The single-celled microorganisms that are responsible for food poisoning, etc.

Fungi: They are the organisms that grows on the dead matter and found in the warm places.

Temperature: Too high temperature or too low temperature affects the food and becomes responsible for spoilage of food.

1.6.2 Principle of Food Preservation

Principle of food preservation includes killing the microorganisms or delaying the action of microorganisms and stopping the action of enzymes [64–67].

- Inhibit microbial growth is called as "microbistatic".
- Irreversibly inactive microorganisms (microbicidal).
- Mechanically remove microorganisms.
- Maintain asepsis.
- Prevent self-decomposition of food.
- Inactivate food enzymes.

1.7 Preservatives

There are different types of preservatives that are best for certain products and are useful and operative against particular chemical specific changes. In things like fruit juice, cheese, bread, and dried fruit, "antimycotics" stops the growth of molds which degrades the quality of those things. Some examples are sodium and calcium propionate and sorbic acid. Another one is "antioxidants" whose butylated hydroxytoluene or (BHT) slows down the occurrence of rancidity created by oxidation in margarine, shortening, and different sets of foods consisting of fats and oils [68]. The growth of harmful bacteria in poultry, seafood, and canned foods is stopped by "antibiotics" like tetracyclines, and to keep these products moisturized, humectants and substances are used. The preservatives may get reactive toward body [70]. It can cause reactions toward in sensitive people like skin allergies including itching and rashes, difficulty in breathing, sneezing, etc.

1.7.1 Factors Affecting Preservative Efficiency

1.7.1.1 Interaction With Formulation Components

Methyl cellulose, alignates, and tragacanth that are hydrocolloids can communicate with preservatives and stop their activity. In pharmaceutical preparations, different emulgents are utilized in preparation of many applications [69]. Preservatives, emulsified oil phase, and emulgent molecules react together. The concentration of preservatives is affected by multiple factors such as the nature of oil, type of concentration of emulgent, and water ratio.

1.7.1.2 Properties of Preservatives

The dispersal of preservative should be similar/homogeneous in nature and increased solubility should be present in the bulk phase is more suitable in multi-phase system. Different chemicals like chlorobutol may get hydrolyze on storage if pH do not favor it. Preservatives do get reactive with some compounds and results in the loss of their antimicrobial property.

1.7.1.3 Effect of Containers

The preparations packed in glass containers can be anticipated to keep their preservative content if the container is closed airtight [71, 72]. Preservatives may pass through the plastic container and reacts together. Rubber is the major things which sometimes reacts with some of the preservatives but it is still used for the closures. Containers may cause defilement of pathogens.

1.7.1.4 Types of Micro-Organisms

Clostridium species, *Bacillus anthracis*, are some of the plant products which may have some pathogenic microorganisms from the soil. The degradation of the pharmaceutical products is caused by these soil microorganisms. *Salmonella typhi* is the type of pathogen that is produced from the animal source [74].

1.7.1.5 Influence of pH

The chemical stability and the activity of the preservative may get affected by the adjustment of the pH. Mostly, preservatives are less determined by the pH.

1.7.2 Factors Affecting Chemical Preservation

Intrinsic	Extrinsic	Microbial	Processing
Food nutrients composition [73]	Processing temperature	Inherent resistance (vegetable cells/ spores/strain/ differences)	Changes in composition
pH/acidity [74]	atmosphere	Initial number	Changes in microbial composition
Buffering capacity [75]	Relative humidity	Growth rate	Changes in food microstructure

1.7.3 Classification of Chemical Preservatives

Many chemicals are used as a chemical preservative and are added into the food. According to Food Safety and Standards Authority of India (FSSAI), chemical preservatives are classified into two classes [76–78]:

Class 1: These are the preservatives that we can usually find them in our house in kitchens. They are the preservatives that are derived from the nature. They are not harmful in nature for our body. Example: common salt, sugar, spices, vinegar, vegetable oil, honey, and wood smoke.

Class 2: These are the man-made preservatives that are chemically prepared. There are more chances that they could react with the body and causes harmful effects to the body. Example: calcium or sodium propionate, sorbic acid or its sodium potassium and calcium salt, and acidulants.

1.7.4 Types of Chemical Preservatives

These preservatives are used in foods either as direct additives or are themselves developed during processes such as fermentation or decomposition [79, 80]. There are certain preservatives which can be used.

Natural	Sugar, Salt
Weak organic acid (Acidulants)	Acetic acid, lactic acid, sorbic acid and its salt, benzoic acid and its salt, propionic acid and its salt, malic, tartaric, citric, and ascorbic acid
Inorganic Salt	Nitrites and nitrates, chlorine, iodine, hydrogen peroxide phosphoric acid, and borax
Gaseous	Sulfur dioxide and sulfites, carbon dioxide and ethylene and propylene oxides
Other	Antioxidants, oils, and spices

1.7.5 Natural Chemical Preservatives

From ancient times, the salt and sugars are used as chemical preservatives. They are the regular part of our diet. The salt causes the food dehydration by drying out the water and tying up from the tissues of the food. It also ionizes yielding chlorine ion and interfere with the action of proteolytic enzymes which is harmful to the microorganisms. Similarly, sugar is also a part of our regular diet. It also produces the condition of high osmotic pressure that are unfavorable. Natural preservatives are the part of ancient method to save food from decaying. It delays the growth of unwanted substances like bacteria in the food for some time. They have different properties like antioxidative, antifungal, and antimicrobial which helps in the delay and prevention of the growth of unwanted substances [81–86].

1. Salt (Sodium Chloride, NaCl): It can be found in everyone's kitchen. It is considered as one of the best natural preservative. It helps in the preservation of non-vegetarian food like meat, chicken, and some vegetables, too. Different unwanted substances sometimes breed on the food in the aqueous medium like bacteria and yeast, and they feed and degrade the food making it unfit to eat so salt helps to prevent this spoilage by the help of the process osmosis in which it dehydrates the microbes.
2. Sugar (Sucrose, $C_{12}H_{22}O_{11}$): It is also a natural preservative which can be found in everyone's kitchen as the major role of it is to act as a sweetener. Sugar also helps to prevent the spoilage by the help of the process osmosis in which it dehydrates the microbes and bacteria, and yeast cannot breed and reproduce there.

3. Vinegar: Vinegar is also a natural preservative which everyone has in their kitchen helps in the preservation of meat and poultry food items. It also elevates the taste and antimicrobial property is also present in it. The fermentation of sugar is the process which helps in the formation of the vinegar. Vinegar has acetic acid in it which destroys the microbes and stops the food degradation. We have a very common example, i.e., "Pickling" which acts as a preservative.
4. Onion: Onion has some antioxidant and antimicrobial properties which makes it as another type of natural preservative. Onion has some components which has these properties. Basically, the onion extracts are considered as the natural preservatives. They stops the growth of microorganisms on the food.
5. Olive Oil: Olive oil acts as natural preservative as it segregates/separates the food from coming in contact with the air which detains oxidation, degradation and molding.
6. Cloves: It is the most traditional food preservative which we use from ancient times. It contains phenolic compounds in a large amount which consist of antioxidant properties which makes sure that it prevents the breeding of fungus and bacteria on the food. It is a very useful and effective method since the beginning.
7. Ginger: Ginger is also a natural preservative which helps in the preservation of the food due to its "antimicrobial" property.
8. Castor Oil: Castor oil has anti-fungal property which helps in the food preservation. It inhibits the growth of the fungus on the food. Many grains and pulses are coated with the castor oil so that they can be stored for the long time.

1.7.6 Methods of Food Preservation

1. Drying: It is an ancient method used for the food preservation. This method prevents hydrolysis makes the environment for bacteria unfit for the growth. It uncovers the food to the high temperature which helps the food to be demoisturized. Electric dehydrator is consider to be the best unit for process of drying [90].

2. Freezing: It is storing the food stuffs in the storages which are cold. Potatoes are being kept in the dark room but the potatoes should be frozen.
3. Smoking: It is the process in which food gets exposed to the smoke from the burning wood as smoke has a property of being antioxidant and antimicrobial, mostly meats and fish are burned and smoked. Hot smoking and cold smoking are the different methods of smoking but it has a disadvantage that it increases the probability of cancer [87].
4. Vacuum Packing: In this process, the bags and bottles are being air tight and makes it a place where no oxygen could be present and bacteria dies there.
5. Pickling: In this process, it helps to remove the moisture from the food making the environment unfit for microorganisms to grow. Chemical pickling and fermentation pickling are the types of it. EDTA is added in commercial pickles to increase its life span [90].
6. Sugar: It is also a natural preservative which can be found in everyone's kitchen as the major role of it is to act as a sweetener. Sugar also helps to prevent the spoilage by the help of the process osmosis in which it dehydrates the microbes and bacteria, and yeast cannot breed and reproduce there [89].
7. Lye: It helps to convert the food alkaline and helps to prevent the bacterial growth on it.
8. Canning: It is the process in which food is being sealed in cans and sterile bottles. Boiling is done in order to make sure that the bacteria is killed, and then, it is being sealed. When the seal is broken, then the chances of food spoilage increases again.

1.8 Antioxidants

Antioxidants usage in the field of food preservatives to save food from degradation has proved to be very effective. The oxidation of food is responsible by the two factors: oxygen and sunlight; so, the preservation procedure involves keeping food away from the contact of air and sun, keeping food in the dark and sealed container or maybe when required sealing it with the help of wax and cucumber where there is no chance for the food to get close or comes in the contact with the sunlight or oxygen but it does have one advantage as oxygen is the primary source for the respiration of plants so the plant products present there smells really bad and also the color of the products sometimes change. In addition, 8% oxygen is contained in the packaging of fresh fruits and vegetables. Antioxidants play an important part as preservatives, It helps to prevent the spoilage by fungi and bacteria [88].

Unsaturated fats are the most widely recognized atoms attacked by the process "oxidation" that makes them rancid. Oxidized lipids are regularly stained and result in displeasing tastes, for example, metallic or sulfurous flavors. It is essential to keep away the food from oxidation. These kinds of food are hardly preserved with the help of drying, and they are usually preserved with the help of salting/pickling, smoking, and fermenting. Before air drying the fruits which are less fatty foods are sprayed with antioxidants (sulfurous). Metals act as a catalysis in the process of oxidation, so it is better if fat oils like butter should not be kept in the Al foil or in any other metal container. Olive oil are halfway shielded from oxidation by their inherent antioxidant property but rest is sensitive toward photooxidation. Antioxidant additives are additionally added to fat based beautifying agents, for example, lipstick and lotions to forestall rancidity.

1.9 Oils and Spices

These are used as preservatives in pickles. Oil makes a protective covering and prevents contact of microorganisms and air with the food. Spices do not have bactericidal effect in concentrations usually used. Cinnamon and cloves containing cinnamic anhydride and eugenol, respectively, are more bacteriostatic than other spices. Extracts of garlic onion cabbage are inhibitory to *Bacillus subtilis* and *E. coli* [92].

The essential oils are the liquids that are extracted from the plant materials (such as fruits, flowers, barks, and peel), which are aromatic and volatile in nature. There are many uses of essential oils such as in medicine, perfumes, cosmetics, and as food spice and preservative. The initial use of essential oil was in medicine in the 19th century then they were used as odor and flavor ingredients and their business and usage increased in this field and became the priority. Out of 3,000 essential oils which are known yet, 300 out of them are used in the fragrances and flavor industry.

Essential oils has antimicrobial properties so they are regarded as secondary metabolites and they are essential for flora defense. Secondary metabolites "antimicrobial" property was tested by using vapors of essential oils by De la in 1881. From then, essential oils have been proved as the greatest examples that shows this property [94], and not only antimicrobial but also insecticidal, antiparasitic and antioxidant properties too. They act as the builders for the growth of animals.

In spite of the fact that the food business utilizes essential oils as flavoring agents, they also proved to be as the major source of antimicrobials in the process of food preservation. But, one should have all the knowledge about the properties of essential oils, i.e., the MIC (minimum inhibitory concentration), the mode of action, the range of target organisms, and the impact of food matrix components on their antimicrobial properties.

The diversity of compounds is created by plants containing antimicrobial activity. Some are constantly present while others are delivered because of the physical injury or microbial invasion. Essential oils are the blend of at least 45 various components so recognizing the most active component is not an easy task. There are various methods to extract the oil by mixing few components. Essential oils is widely spread family of organic compounds having low molecular weight with having huge contrasts in the antimicrobial activity [96]. There are four divisions in the active compounds as indicated by their chemical structures which are terpenoids, terpenes, phenylpropenes, and others.

1.10 Introduction to Hurdle Technology

Hurdle technology is a technique to control and eliminate pathogens by combining two or more factors. These factors are called hurdles. The pathogens have to overcome these hurdles to survive the right combination of the hurdles inactivates the pathogens and thereby makes the food product safe for consumption along with an extent shelf life [91, 93].

Hurdle technology is the technology where appropriate types of hurdles are selected and combined to provide microbial safety, stability, sensory, nutritional value, and economic viability to products.

The concept of combining several factors is known as hurdle effect. Food preservation can be attained by different types of hurdles. The different types of hurdles used for preservation are classified into three types.

1. Physical
2. Physicochemical
3. Microbial hurdles

These hurdles also influence the quality and safety of foods depending on the intensity of different types of hurdles used, and their effect on food quality may be either positive or negative. The main objective of the hurdles is to eliminate and inactivate or inhibit the growth of microorganisms present in food. Common salt used as hurdles to control the microbial load in foods [95]. Natural antimicrobials also called as bio preservatives like nisin, bacteriocins, and essential oils can also be used as one of the hurdles.

1.10.1 Advantages of Food Additives and Preservatives

Nowadays, food preservatives play an important role in our daily routine food. They give us safety from the bacteria and microorganisms that can damage the food as well as our health. We cannot take preservatives as for granted as they also play useful function in food. They improve the nutritional value of some food weakening unhealthy things from it and improving the quality of the product. Preservatives are play important role by increasing the life of a food and prevent from spoilage. Even though additives and preservatives are necessary for maintaining food safety, they are also danger for health [97–100]. The reasons of adding additives in food are as follows:

i) For maintaining consistency: by emulsifier, stabilizers and thickeners give food a good texture and prevent them from separating.
ii) For maintain quality: as they improve food texture, flavors, appearance, enrich in nutritional value, and maintain food consistency.
iii) For maintaining nutritional value: vitamins and minerals are adding in some product like milk cereals and flour for those likely to be lacking in a person's diet.
iv) For maintaining acidity or alkalinity: as we add baking soda for help cakes, biscuits, and other baked goods to rise during making.

1.10.2 Disadvantages of Food Additives and Preservatives

Even though additives and preservatives are necessary for food, they can cause many health problems. They can cause different kinds of allergies like hyperactivity and attention deficit disorder in some people who are sensitive to specific chemical [82, 91]. They also can cause asthma, hay fever, tight chest, headache, hives, and allergy like rashes. Some of the danger food additives are as follows:

- Benzoate: it causes allergies like rashes and asthma. It can also cause brain damage.
- Red dye 40: it can causes certain birth defects and possibly cancer.
- Caramel: it commonly used in flavoring and coloring agent. It causes vitamin B6 deficiencies, genetic defect and cancer.
- Saccharin: it causes bladder cancer, tumors, affecting skin, and toxic reaction.
- Caffeine: it is used as color and flavoring agent in food. It causes nervousness, occasionally heart defect.
- Sodium chloride: it can cause high blood pressure, kidney failure, stroke, and heart attack.

We should avoid the food which contains danger additives and preservatives for minimizing the risk of health problems. We should check the ingredients of sealed food before purchasing them from markets. We should eat organic food because they do not have any artificial additives and preservatives in food. We should try to eat freshly prepared food as much as possible rather than eating sealed food.

1.10.3 Effects of Food Additives and Food Preservatives

The first step toward increasing our health and lowering our risk of disease is to avoid toxins in our food. Food additives can cause immediate effect or harmful in a long run if you have constant exposure. Headache, change in energy level, and immune responses are the example of immediate effects [65]. Risks of cancer and cardiovascular diseases are the example of long-term effect.

Some food additives side effects are as follows:

i) Boric acids: it is a boron compound and it can be soluble and circulates in the plasma. It is a white powder water soluble solution which is usually used as pesticides to kill insects, fungi, bacteria, and cockroaches. In food, it is used as preservative. It is use for preserving meats, dairy products. Boric acid is toxic to cells. It harms more effectively if it takes in higher amount. It was reported by several times. It was tested on animals. Animals were treated with boric acid in higher amount [52–57]. It observed that boric acid suppressed the sperm release from testes, and hence, it reduces the fertility in male.

ii) Vinegar: it was reported for Esophageal injury. Different vinegar is varies in different content and pH. So, ingestion of heavy vinegar for long time could cause hyperreninemia, hypokalemia, and osteoporosis.
iii) Aspartame: it is a controversial artificial sweetener. One of the fact about aspartame is that it does not metabolized, so it causes metabolism disorder (phenylketonuria).
iv) Nitrites and nitrates: it causes blue color of skin because nitrites binds to hemoglobin, so it chemically altered hemoglobin which results in impair oxygen delivery to tissue. This disease is known as blue baby syndrome. Higher amount of nitrates and nitrites consuming can cause cancer and brain tumors.
v) Annatto: annatto is safe if it is consumes in small amount. Otherwise, it can cause allergies to people who are sensitive to it.

1.10.4 Safety of Food Additives and Preservatives

The limitation should be followed with due importance to factor. The SCF has introduced some guidelines of some test that must be carried out on food additives in order to their safety. It requires numbers of animals and other testes to access every conceivable risk to the consumer:

- Study for understanding how the body absorbs, metabolizes, and eliminates the substance.
- Study for potential of gene and chromosome damages.
- Life time studies and potential of fertility and birth defects.
- Study of causing cancer.

Every additive is safe if it takes in small amount, but if it take in high amount, they have their some causes [58–61]. The aim of doing test on animal with additives is to find their effects. The tests have been done both low and high doses. As observing that low doses are safe for health but some additives are harmful if it takes in high doses. Therefore, we should check the information about the additives that are labeled on packed food before consuming them. We should try to eat more organic food rather than eating processed food.

Nourishment is tainted if its quality is brought down or influenced by the expansion of substances which are damaging to well-being or by the evacuation of substances which are nutritious. It is characterized as the demonstration of deliberately degrading the nature of nourishment offered available to be purchased either by the admixture or substitution of sub-par substances or by the evacuation of some important fixing.

Nourishment is proclaimed debased if:

- A substance is included which deteriorates or harmfully influences it.
- Cheaper or substandard substances are subbed entirely or to a limited extent.
- Any important or fundamental constituent has been completely or to some degree disconnected.
- It is an impersonation.
- It is hued or in any case treated, to improve its appearance or in the event that it contains any additional substance damaging to well-being.
- For whatever reasons its quality is underneath the standard.

1.11   Adulteration

Health is the first priority of every human being and healthy body dwells with healthy mind. Many types of articles and programs related to health awareness are attracting millions of people. People have become health conscious nowadays. A substance which is found with other substances and make them unfit to use for other purposes known as adulterants, the process of addition of adulterants is known as adulteration [77–83]. So, food adulteration is a major issue to concentrate, for better and healthy life. When people consume adulterated food items, it adversely affects their health, i.e., it affects vital biological system which includes digestive system and nervous system in the human body. Moreover, to increase the yield of agricultural products, some hazardous chemicals which were banned due to their toxicity were used which resulted in many problems related to hormonal imbalance among children, increase in the usage of pesticides and preservatives has increased infection problems in skin, livers, kidneys, and eyes. Majority of consumers those who consume outside food is unhygienic which cause various health problems. Many farmers use harmful chemicals for growing crops by injecting them with such chemicals like oxytocin which is a human hormone, which is sometimes used by farmers for growing crop overnight as it speeds their growth. Many drugs are banned by the government but still they are sold by the chemists. Even the milk which we drink is adulterated, even if we are consuming products made from milk, they are adulterated and have adverse effect on our body.

1.11.1 History of Food Adulteration

The relationship goes back as far as history itself. In prehistoric time, people used to adulterate the food to increases its longevity and to enhance its taste. In the Victorian era, when adulterants were common, the use to mix lead with cheese similarly in United States such adulteration occurs but it was later banned by government [82, 88]. A German chemist (Frederick Accum) first investigated adulterant in 1820 and he identified toxic metal coloring in food and drink. His work was antagonized food suppliers, and he was ultimately discredited by a scandal over his alleged mutilation of books of the royal institution library. Then later, another studies can in late 1850s; then, in 1860, Food adulteration Act and other legislation can into existence. The only difference in the adulteration of food in ancient time and modern time is different methods of fraud. Today's sellers are using modern technology for adultering the food. A period of expeditious social change made it possible as a result of urbanization and industralization, and the distance between the consumers and producers and the market was anonymous. Whenever food product is highly commercialized, food sellers and food merchants succumb to the temptation of playing of playing trick in order to maximum profits. They use different flavors drug, color in food, fillers, and preservatives which is not easily detective.

1.11.2 Types of Food Adulteration

There are three major types of food adulteration:

- Intentional Adulterants
- Incidental Adulterants
- Metallic Adulterants

1.11.2.1 Intentional Adulteration

Intentional adulteration is an act of intentionally adding contaminants in food products by person or a group of people externally or internally to a food business [76]. On large scale, intentionally addition of the contaminants to the food is a threat to the consumers. People add substances which are similar to the food products hence making it difficult to recognize, for example, dyes, water, mud, marbles, sand, and chalk. These adulterants have adverse effect on human body.

1.11.2.2 Incidental Adulteration

Incidental adulteration is a type of adulteration in which adulterants are found in food substances due to ignorance, negligence or lack of proper facilities, or lack of hygiene during processes of making food items. This can come by spoilage of food by rodents, dust particles, stones, and residues from packaging. It is not a will full act on the part of an adulterer [89]. These types of adulterants are called unintentional, which accidently get added during the process of production of food including transportation, manufacturing, operation carried out in animal and crop husbandry, and as a result of environmental contamination.

1.11.2.3 Metallic Adulteration

When metallic substances are added intentionally or accidentally, it is known as metallic adulteration. Examples are arsenic, pesticides, lead from water, mercury from effluents, and tins from cans. It is one of the most poisonous type of adulteration as the metals which are added are toxic in nature.

1.11.3 A Food Is Considered Adulterated if It Has the Following Factors

- If a substance which is poisonous or unfit for consumption is added to the food, then it affects the quality of food.
- If cheaper quality substances are substituted in part or in wholly.
- If necessary constituents present in the substance are removed in part or wholly.
- It is a copy of original product.
- If the substance is colored, only for improving its physical appearance, then it can be adverse effects on health.
- If the quality is below standard for any reason that food can also be harmful for health.

1.11.4 Effects of Adulterated Food on Human Health

Adulterated food is a major threat to the consumers because just to maximize the profit vendors are using new techniques for adulteration of food which is difficult to find out [87–91]. Although there are many food safety acts launched by the government and there are many techniques through which the food undergoes examination before consumed by the consumer, the vendors are on step ahead from safety agencies because their techniques are increasing with the time. Vendors used poisonous chemical substances such as formalin which is applied on poultry products such as fish, meat, milk, and some fruits by many vendors which causes different types of skin diseases, cancer, and asthma to the consumer. Calcium carbide, coloring dyes, burnt engine oils, urea, and many more permitted preservatives are used in maximum amount which affects multiple organs of human body. In many cases, it becomes reason of cancer like colon, peptic ulcer, chronic level disease including liver failure, and cirrhosis. Bone marrow, blood disorder, and heart disease are also detected. There are also chances of malignancy increases and neurological impairment or brain functions are also compromised. Some food materials also caused different types of skin problems including allergic reactions. Many types of mineral oils which we consume often contains adulteration that is they are mixed with cheaper toxic oil which may cause heart problem, and if they are consumed by pregnant women, then it might lead to abortion or it can damage the brain of baby. Color which is added to food and vegetables can also lead to allergies, liver disease, and much more. $PbCrO_4$ when added to turmeric and species can causes various diseases such as paralysis, anemia, abortions and, brain damage.

1.11.5 Reasons for Food Adulteration

1. Rising population because of this the demand of food is more and supply is less.
2. Market competition is increasing day by day so as to cut of the products cost to beat the market competition.
3. To earn lots of profits.
4. Sometimes, authentic ingredients are not available at affordable price so shortage of ingredients leads to adulteration.
5. Lack of knowledge leading to food safety risk some people are not aware how much harmful is adulteration effect.
6. No update of processing, new techniques, and shortage of qualified personals.
7. Mostly seen that common people are not conscious enough.

1.11.6 Methods of Food Adulteration

Mixing: Mixing of stone, clay, sand, pebbles, marble chips in various pulses, and flour, for increasing its quantity.

Substitution: Inferior or cheap quality substances being substituted partially or entirely with pure substance.

Concealing Quality: By hiding the food quality by hiding the caption of low quality and labeling it with high-quality caption.

Decomposed Food: Mainly in fruits and vegetable this type of adulteration is favored in this decomposed food and vegetables are mixed with the fresh fruits and vegetable so that these can be sold during the sale and vendors get maximum profit.

Misbranding/False Labeling: This type of adulteration includes false labeling the products like changing their manufacturing or expiry dates or adding duplicate food stuffs.

Addition of Toxicants: Addition of non-edible toxic substances for altering the food quality like addition of argemone seed in mustard oil, adding food coloring, and poor or cheaper quality preservatives.

1.11.7 Trends of Food Adulteration in Developing Countries

1. The main part of our meal includes fruits and vegetables which are adulterated by farmers by self-applied fertilizers and pesticides for increasing the production and avoiding attach by the insects and disease [63–66].
2. Wax coating/dipping the fruits and vegetables in chemical water like copper sulfate solution so as to make them look fresh and attractive for increasing its sale.
3. In everyday lives, we consume vegetables and fruits which are produced on the fields which are irrigated by sewage water, which contain factory waste, human feces, detergents, and high amount of toxic heavy metals, e.g., Pb and As, which causes damage to liver and kidney, and may also cause cancer.
4. Red color dust is substituted in chili powder for increasing the quantity, $PbCrO_4$ and metanil yellow is substituted in turmeric and it depreciates human health adversely.
5. Salt and sugar are biased with calcium carbonate, like this only urea is mixed in brown rice.

1.12 Food Safety and Standards Act

The main objective of the food law is to provide safe, healthy, and nutritious food, to the citizens. The first ever food adulteration act was given in 1947 which was Vegetable Oil Products Order. This order was for control and improving the quality of vegetable oils which was further replaced by Edible Oil Products (Regulation), 1998. An organized manner was set up within the organization for production, transportation, and selling of goods. Many companies came into loss after the implementation of this act [78, 82, 87]. The head of these edible oil manufacturing company has the rights to form proper safety measures so that there should not be any fault and it is fit for consumption purpose, and set of rules should be set. (Prevention of Food Adulteration Act, 1954). It includes food quality, usual method of sampling, analysis of nutrients, powers of governing officials, proper punishments for fraudsters, those include additives, coloring, preservatives, labeling, packing stopping, and monitoring the sales, and another act was The Fruit Products Order (1955). This act is important and looks after the goods which are used daily; therefore, it comes under such section only. The main purpose of this is to look after the production vegetables and fruits, its hygiene, and control in the area of production only to as to maintain the quality of the goods and it comes under this act. All the producers of such products which make other products by using these as primary products should get a certified permit so that they can produce within the vigilance government (The Meat Food Products Order, 1973). Meat products are permitted and looked after during the production of complete meat products and enforcement of strict policies of government at every step of analysis, and it is to be kept in mind that during the storage, it should be frozen completely. In MFPO, various producers of meat products are engaged in producing, storing, and labeling, and the permit should be kept while selling the goods such as in restaurants and hotel (The Solvent Extracted Oil, De Oiled, and Edible Flour Order, 1967). All the oils should be refined before selling it to the consumers, and it should be labeling before selling. It should be mentioned fit at the time of sell. The following are its features:

1. It also looks after the production, quality, and transportation of the all types of oils and flours.
2. It provides quality assurance to the people for the nourishment products which are been purchased.
3. It eliminates the used of edible oils for other purposes.
4. The products which are unfit for consumptions are been expelled out and are labeled as unfit and are not for sale.

All these acts came under one act that is Food Safety and Standards Act (2006). This act illustrates the detailing and authorization of the safety act in India and was enacted by the FSSAI. The FSSAI works under the control of health department of India which works for the welfare of Indian families. The fundamental points of this act are as follows:

1. Set down science-based measures for articles of nourishment.
2. To organize work of distribution, manufacturing of products, storage in proper places, its import, and selling to proper authorized.
3. To ensure safety of food.

All above-mentioned laws were signed under the FSS 2006 Act. Certain guidelines have been formulated by the government for the research purposes, and the scientist those work for the food safety measure should work with some objectives and these were as follows:

1. They should come up with new ideas which helps in growing the country and provides some knowledge so that India should also compete with international institutes.
2. For building various laws, the researches which are carried out should be illustrated easily.

The organization which is working under the development of food safety measure should command various food safety measures for regulation of its standards, and these capacities not withstanding others incorporate setting down methodology and rules for notice of certify labs. There are some rules for permitting license under this government organization which are mainly three types:

1. Registration: for turnover under twelve lakh rupees.
2. State permit: for turnover between 12 lakhs to 20 crore.
3. Central permit: for turnover above 20 crore rupees.

There are various other criteria also which are needed at the time of assessing the permit.

1.12.1 Few Steps to Avoid Adulteration

Make Safe Choices: When you purchase food items, check FSSAI validated label. The FSSAI represents that they have FSSAI license number; also, check ingredient list, look up every ingredient mention there, and check manufacturing and expiry date.

Have Safe Water: The water you drink should be contamination free, and make sure you are using purified water by purifier or you can use boiled water.

Be Aware: Try to purchase food and vegetables either from organic markets or from farmers; do not purchase food items kept in open.

1.12.2 Detection Methods of Adulteration

We should know the identification of adulterants, so various types of methods come into progress based on anatomical/morphological characterization, color, texture, scent, and chemical testing.

The Following are the three basic strategies:

- By demonstrating the existence of foreign substances;
- By demonstrating the external appearance of the substance like color;
- By demonstrating the deviated components from its normal level.

1.12.3 Technique to Check Food Adulteration

Physical Method: In physical method, we can detect the adulteration by using microscopic and macroscopic visual methods [55–59]. Analysis of food adulteration by using physical methods includes bulk density, solubility, texture, and morphology. Microscopic examination is helpful to detect some powder species, namely, cloves, coriander, and cumin leads to easily detection of extraneous starch present in these powder species.

Chemical and Biochemical Methods: These types of techniques are based on chromatography, spectroscopy, immunology, and electrophoresis method. Biochemical methods are more helpful than physical methods because the analytical approach involves various steps as follows:

- Extraction of suitable solvents.
- Chromatographic separation.
- For removing interfering matrix components cleanup.
- Selective detection.

1.13 Conclusion

Food additives are being used since many years for preserving, thickening, blending, adding flavors and colors to food items. Artificial colors are found mainly in junk foods, which are generally high in calories and low in nutrients. Most of the artificial food dyes cause allergic reactions. Yellow 5 which is also known as tartrazine causes hives and asthma symptoms. In a survey conducted within people with chronic hives, many of them had an allergic reaction to the artificial or man-made food dyes. Some supplements may also contain certain contaminants that could even cause cancer like problems. Most of the food dyes does not cause any severe effects. Avoiding the processed foods that consists of the supplements can improve the overall health. Additives play important role in decreasing significant nutritional deficiencies within consumers. Preservatives have played a good role in food industry by preventing the spoilage of food, increasing its life and thus ensuring the safety for consumer's health. Food additives are capable of meeting the demands of manufacturer as well as of consumer's. Food additives also have harmful health effects if taken in higher amounts. Therefore, they are being strictly studied, monitored and regulated before approval by FDA. FDA ensures safety of food additives for consumer's health. Self-precautions must be taken to avoid health risks by checking the ingredients list and other mentioned points on food labels of packed food items before buying them and also replacing processed food items with natural and fresh food products. When evaluated in terms of health aspects, the use of additives must be performed within the limits and to fulfill a certain function within the legal framework.

Preservation has been used from centuries to ensure safety of different food stuff. It also delays the spoiling of food stuff and preventing any alteration in their taste or appearance. But in case of chemical preservatives, the regulation is important because all the chemicals are not allowed to be used in the food. According to the organization, national or international, that is available, the regulation is controlled. The components which are having the grass status are only allowed to be used in the food product.

The second thing that we can do with the chemical preservative is hurdle technology. It is the concept of combination of chemical preservatives or different treatments to preserve the food. With the use of the hurdle technology, we can make the optimum use of the chemical preservatives and the other available technique to meet the consumer demands for healthy and safe foods. Which are being continuously developed to satisfy the increasing current demands of consumer satisfaction and economic preservation in various aspects like nutritional aspects, sensory aspects, absence of preservatives, convenience, low demand of energy, and environmental safety.

To develop high-quality and safety products, we need a better understanding and safety products by better control of processes and selection of ingredients.

The greed of food adulterers to maximize higher profits in short time so that they are making fool to consumers by mal practices like food adulteration. Although adulteration is of three types, intentional, metallic, or incidental, major properties of intentional adulteration are seen everywhere because many of our citizens are not aware what is adulteration, how they are been cheated by the vendors and also they are not aware about side effects of adulterated food. Consumers are usually impressed by physical appearance of food, if the food is looking good, then it is good food and free from adulterants that is what consumers thought before buying. Although several steps have been taken by the government to stop food adulteration various laws are enacted as several seminars have been conducted, articles are published to spread awareness among people. Fraudsters are suing new technologies because in present era adulteration contributes to industries to enormous economic gain. So, it is a duty of consumers to take care of their health itself by analyzing common food items at home, by taking food which is prepared under hygienic condition. Never buy low quality food items which are available in cheap price at local grocery store, always buy food products with a proper barcode billing and having FSSAI certification. Check food labels. If you catch any vendor red handed who is selling adulterated food items, then do not stay quiet file complaint against them because it is a major crime. There are various rules for consumer safety so follow them keep healthy by eating healthy and spread awareness about food adulteration.

References

1. Restuccia, D., Gianfranco Spizzirri, U., Parisi, O.I., Cirillo, G., Curcio, M., Iemma, F., Puoci, F., Vinci, G., Picci, N., New EU regulation aspects and global market of active and intelligent packaging for food industry applications. *Food Control*, 21, 11, 1425–1435, 2010.
2. Dainelli, D., Gontard, N., Spyropoulos, D., Zondervan-van den Beuken, E., Tobback, P., Active and intelligent food packaging: legal aspects and safety concerns. *Trends Food Sci. Technol.*, 19, S103–S1125, 2008.
3. Appendini, P. and Hotchkiss, J.H., Review of antimicrobial food packaging. *Innovative Food Sci. Emerg. Technol.*, 3, 2, 113–126, 2002.
4. Brody, A.L., Bugusu, B., Han, J.H., Koelsch Sand, C., McHugh, T.H., Innovative food packaging solutions. *J. Food Sci.*, 73, 8, 107–116, 2008.

5. Vanderroost, M., Ragaert, P., Devlieghere, F., De Meulenaer, B., Intelligent food packaging: The next generation. *Trends Food Sci. Technol.*, 39, 1, 47–62, 2014.
6. Shin, J. and Selke, S.E.M., 11 Food Packaging, in: *Food Processing: Principles and Applications*, Second Edition, S. Clark, S. Jung, B. Lamsal, (Eds.), pp. 249–273, John Wiley & Sons, The Atrium, Southern Gate, Chichester, West Sussex, PO19 8SQ, UK, 2014.
7. Robertson, G.L., Packaging materials for biscuits and their influence on shelf life, in: *Manley's Technology of Biscuits, Crackers and Cookies*, pp. 247–267, Woodhead Publishing, Sawston, United Kingdom, 2011.
8. Robertson, G.L., Packaging and food and beverage shelf life, in: *The Stability and Shelf Life of Food*, pp. 77–106, Woodhead Publishing, Sawston, United Kingdom, 2016.
9. Newsome, R., Balestrini, C.G., Baum, M.D., Corby, J., Fisher, W., Goodburn, K., Labuza, T.P., Prince, G., Thesmar, H.S., Yiannas, F., Applications and perceptions of date labeling of food. *Compr. Rev. Food Sci. Food Saf.*, 13, 4, 745–769, 2014.
10. Dainelli, D., Gontard, N., Spyropoulos, D., Zondervan-van den Beuken, E., Tobback, P., Active and intelligent food packaging: legal aspects and safety concerns. *Trends Food Sci. Technol.*, 19, S103–S1125, 2008.
11. Kerry, J. and Butler, P. (Eds.), *Smart packaging technologies for fast moving consumer goods*, John Wiley & Sons, Sawston, United Kingdom, 2008.
12. Hellström, D. and Saghir, M., Packaging and logistics interactions in retail supply chains. *Packag. Technol. Sci.: An International Journal*, 20, 3, 197–216, 2007.
13. Sohrabpour, V., Oghazi, P., Olsson, A., An improved supplier driven packaging design and development method for supply chain efficiency. *Packag. Technol. Sci.*, 29, 3, 161–173, 2016.
14. Chauhan, O.P., Lakshmi, S., Pandey, A.K., Ravi, N., Gopalan, N., Sharma, R.K., Non-destructive quality monitoring of fresh fruits and vegetables. *Def. Life Sci. J.*, 20, 2, 103, 2017.
15. Sarig, Y., Potential applications of artificial olfactory sensing for quality evaluation of fresh produce. *J. Agric. Eng. Res.*, 77, 3, 239–258, 2000.
16. Aboonajmi, M. and Faridi, H., Nondestructive quality assessment of Agro-food products, in: *Proceedings of the 3rd Iranian international NDT conference*, 2016.
17. Abasi, S., Minaei, S., Jamshidi, B., Fathi, D., Dedicated non-destructive devices for food quality measurement: A review. *Trends Food Sci. Technol.*, 78, 197–205, 2018.
18. Kawakami, M., Sarma, S., Himizu, K. et al., Aroma characteristics of Darjeeling tea, in: *Proceedings of International Conference O-CHA (Tea) Culture Science*, Shizuoka, Japan, pp. 110–116, 2004.

19. Bhattacharyya, N., Seth, S., Tudu, B. *et al.*, Detection of optimum fermentation time for black tea manufacturing using electronic nose. *Sens. Actuators B Chem.*, 122, 627–634, 2007.
20. Lou, X., Ye, Y., Wang, Y., Sun, Y., Pan, D., Cao, J., Effect of high-pressure treatment on taste and metabolite profiles of ducks with two different vinasse-curing processes. *Food Res. Int.*, 105, 703–712, 2018.
21. Marcone, M.F., Wang, S., Albabish, W., Nie, S., Somnarain, D., Hill, A., Diverse food-based applications of nuclear magnetic resonance (NMR) technology. *Food Res. Int.*, 51, 2, 729–747, 2013.
22. Williamson, K. and Hatzakis, E., NMR spectroscopy as a robust tool for the rapid evaluation of the lipid profile of fish oil supplements. *JoVE (J. Visualized Exp.)*, 123, 123, e55547, 2017.
23. Hussain, A., Pu, H., Sun, D.W., Innovative nondestructive imaging techniques for ripening and maturity of fruits–a review of recent applications. *Trends Food Sci. Technol.*, 72, 144–152, 2018.
24. Ezeanaka, M.C., Nsor-Atindana, J., Zhang, M., Online Low-field Nuclear Magnetic Resonance (LF-NMR) and Magnetic Resonance Imaging (MRI) for Food Quality Optimization in Food Processing. *Food Bioprocess Tech.*, 12, 9, 1435–1451, 2019.
25. Caporaso, N., Whitworth, M.B., Fisk, I.D., Near-Infrared spectroscopy and hyperspectral imaging for non-destructive quality assessment of cereal grains. *Appl. Spectrosc. Rev.*, 53, 8, 667–687, 2018.
26. Dachoupakan Sirisomboon, C., Putthang, R., Sirisomboon, P., Application of near infrared spectroscopy to detect aflatoxigenic fungal contamination in rice. *Food Control*, 33, 1, 207–214, 2013.
27. Bindhu, M.R. and Umadevi, M., Antibacterial activities of green synthesized gold nanoparticles. *Mater. Lett.*, 120, 122–125, 2014.
28. Ethiraj, A.S., Jayanthi, S., Ramalingam, C., Banerjee, C., Control of size and antimicrobial activity of green synthesized silver nanoparticles. *Mater. Lett.*, 185, 526–529, 2016.
29. Gopinath, K., Shanmugam, V.K., Gowri, S., Senthilkumar, V., Kumaresan., S., Arumugam, A., Antibacterial activity of ruthenium nanoparticles synthesized using *Gloriosa superba* L. leaf extract. *J. Nanostruct. Chem.*, 4, 83, 2014.
30. Kujur, A., Kiran, S., Dubey, N.K., Prakash, B., Microencapsulation of *Gaultheria procumbens* essential oil using chitosan-cinnamic acid microgel: improvement of antimicrobial activity, stability and mode of action. *LWT-Food Sci. Technol.*, 86, 132–138, 2017.
31. Martins, F.C., Sentanin, M.A., De Souza, D., Analytical methods in food additives determination: compounds with functional applications. *Food Chem.*, 272, 732–750, 2019.
32. Maryam, I., Huzaifa, U., Hindatu, H., Zubaida, S., Nanoencapsulation of essential oils with enhanced antimicrobial activity: A new way of combating antimicrobial Resistance. *Int. J. Pharmacogn. Phytochem.*, 4, 3, 165, 2015.

33. Prakash, B., Kujur, A., Yadav, A., Kumar, A., Singh, P.P., Dubey, N.K., Nanoencapsulation: An efficient technology to boost the antimicrobial potential of plant essential oils in food system. *Food Control*, 89, 1–11, 2018.
34. Sun, B. and Wang, J., Food additives, in: *Food Safety in China: Science, Technology, Management and Regulation*, pp. 186–200, 2017.
35. Lei, T. and Sun, D.W., Developments of nondestructive techniques for evaluating quality attributes of cheeses: A review. *Trends Food Sci. Technol.*, 88, 527–542, 2019.
36. Alamprese, C., Casale, M., Sinelli, N., Lanteri, S., Casiraghi, E., Detection of minced beef adulteration with turkey meat by UV–vis, NIR and MIR spectroscopy. *LWT-Food Sci. Technol.*, 53, 1, 225–232, 2013.
37. Salguero-Chaparro, L., Gaitán-Jurado, A.J., Ortiz-Somovilla, V., Peña-Rodríguez, F., Feasibility of using NIR spectroscopy to detect herbicide residues in intact olives. *Food Control*, 30, 2, 504–509, 2013.
38. Xue, L., Cai, J., Li, J., Liu, M., Application of particle swarm optimization (PSO) algorithm to determine dichlorvos residue on the surface of navel range with Vis-NIR spectroscopy. *Proc. Eng.*, 29, 4124–4128, 2012.
39. Luna, A.S., da Silva, A.P., Pinho, J.S., Ferre, J., Boque, R., Rapid characterization of transgenic and non-transgenic soybean oils by chemometric methods using NIR spectroscopy. *Spectrochim. Acta A*, 100, 115–119, 2013.
40. Sarkar, M., Gupta, N., Assaad, M., Nondestructive Food Quality Monitoring Using Phase Information in Time-Resolved Reflectance Spectroscopy. *IEEE Trans. Instrum. Meas.*, 69, 10, 7787–7795, 2020.
41. Ebrahimi-Najafabadi, H., Leardi, R., Oliveri, P., Chiara Casolino, M., JalaliHeravi, M., Lanteri, S., Detection of addition of barley to coffee using near infrared spectroscopy and chemometric techniques. *Talanta*, 99, 175–179, 2012.
42. Leiva-Valenzuela, G.A., Lu, R., Aguilera, J.M., Prediction of firmness and soluble solids content of blueberries using hyperspectral reflectance imaging. *J. Food Eng.*, 115, 1, 91–98, 2013.
43. Suktanarak, S. and Teerachaichayut, S., Non-destructive quality assessment of hens' eggs using hyperspectral images. *J. Food Eng.*, 215, 97–103, 2017.
44. Sanchez, P.D.C., Hashim, N., Shamsudin, R., Nor, M.Z.M., Applications of imaging and spectroscopy techniques for non-destructive quality evaluation of potatoes and sweet potatoes: A review. *Trends Food Sci. Technol.*, 96, 208–221, 2020.
45. Vanoli, M., Rizzolo, A., Grassi, M., Spinelli, L., Verlinden, B.E., Torricelli, A., Studies on classification models to discriminate "Braeburn" apples affected by internal browning using the optical properties measured by time-resolved reflectance spectroscopy. *Postharvest Biol. Technol.*, 91, 112–121, 2014.
46. Vanoli, M., Grassi, M., Spinelli, L., Torricelli, A., Rizzolo, A., Quality and nutraceutical properties of mango fruit: influence of cultivar and biological age assessed by Time-resolved Reflectance Spectroscopy. *Adv. Hortic. Sci.*, 32, 3, 407–420, 2018.

47. Ibrahim, A., Grassi, M., Lovati, F., Parisi, B., Spinelli, L., Torricelli, A., Vanoli, M., Non-destructive detection of potato tubers internal defects: critical insight on the use of time-resolved spectroscopy. *Adv. Hortic. Sci.*, *34*, 1S, 43–51, 2020.
48. Djenane, D. and Roncalés, P., Carbon monoxide in meat and fish packaging: advantages and limits. *Foods*, 7, 2, 12, 2018.
49. Gaikwad, P.S., Yadav, B.K., Sugumar, A., Fabrication of natural colorimetric indicators for monitoring freshness of ready-to-cook idli batter. *Packag. Technol. Sci.*, 34, 1–8, 2020.
50. Fellows, P.J., *Food Processing Technology: Principles and Practice*, Elsevier, Woodhead Publishing, Sawston, United Kingdom, 2009.
51. Ghosh, T. and Dash, K.K., Modeling on respiration kinetics and modified atmospheric packaging of fig fruit. *J. Food Meas. Charact.*, 14, 1092–1104, 2020.
52. Minh, N.P., Influence of modified atmospheric packaging and storage temperature on the physico-chemical, microbial and organoleptic properties of cantaloupe (Cucumis melo) fruit. *Res. Crops*, 21, 3, 506–511, 2020.
53. Baswal, A.K., Dhaliwal, H.S., Singh, Z., Mahajan, B.V.C., Influence of Types of Modified Atmospheric Packaging (MAP) Films on Cold-Storage Life and Fruit Quality of 'Kinnow'Mandarin (Citrus nobilis Lour X C. deliciosa Tenora). *Int. J. Fruit Sci.*, 20, 1–18, 2020.
54. Junior, M.M., Castanha, N., Dos Anjos, C.B.P., Augusto, P.E.D., Sarmento, S.B.S., Ozone technology as an alternative to fermentative processes to improve the oven-expansion properties of cassava starch. *Food Res. Int.*, 123, 56–63, 2019.
55. Pandiselvam, R., Subhashini, S., Banuu Priya, E.P., Kothakota, A., Ramesh, S.V., Shahir, S., Ozone based food preservation: a promising green technology for enhanced food safety. *Ozone Sci. Eng.*, 41, 1, 17–34, 2019.
56. Porto, E., Alves Filho, E.G., Silva, L.M.A., Fonteles, T.V., do Nascimento, R.B.R. *et al.*, Ozone and plasma processing effect on green coconut water. *Food Res. Int.*, 131, 109000, 2020.
57. Brodowska, A.J., Nowak, A., Śmigielski, K., Ozone in the food industry: Principles of ozone treatment, mechanisms of action, and applications: An overview. *Crit. Rev. Food Sci. Nutr.*, 58, 13, 2176–2201, 2018.
58. Gallego-Juárez, J.A., Basic principles of ultrasound, in: *Ultrasound Food Process*, pp. 1–26, John Wiley & Sons, Woodhead Publishing, Sawston, United Kingdom, 2017.
59. Misra, N.N., Schlüter, O., Cullen, P.J. (Eds.), *Cold Plasma in Food and Agriculture: Fundamentals and Applications*, Academic Press, Cambridge, Massachusetts, 2016.
60. Knirsch, M.C., Dos Santos, C.A., de Oliveira Soares, A.A.M., Penna, T.C.V., Ohmic heating–a review. *Trends Food Sci. Technol.*, 21, 9, 436–441, 2010.

61. Kaur, N. and Singh, A.K., Ohmic heating: concept and applications—a review. *Crit. Rev. Food Sci. Nutr.*, 56, 14, 2338–2351, 2016.
62. Liu, Y., Tang, T., Duan, S., Qin, Z., Zhao, H. *et al.*, Applicability of Rice Doughs as Promising Food Materials in Extrusion-Based 3D Printing. *Food Bioprocess. Tech.*, 13, 3, 548–563, 2020.
63. Kalogeropoulos, N., Salta, F.N., Chiou, A., Andrikopoulos, N.K., Formation and distribution of oxidized fatty acids during deep-and pan-frying of potatoes. *Eur. J. Lipid Sci.*, 109, 11, 1111–1123, 2007.
64. Arvanitoyannis, I.S. and Dionisopoulou, N., Acrylamide: formation, occurrence in food products, detection methods, and legislation. *Crit. Rev. Food Sci. Nutr.*, 54, 6, 708–733, 2014.
65. Odueke, O.B., Farag, K.W., Baines, R.N., Chadd, S.A., Irradiation applications in dairy products: a review. *Food Bioprocess. Tech.*, 9, 5, 751–767, 2016.
66. Pati, S., Chatterji, A., Dash, B.P., Raveen Nelson, B., Sarkar, T. *et al.*, Structural Characterization and Antioxidant Potential of Chitosan by γ-Irradiation from the Carapace of Horseshoe Crab. *Polymers*, 12, 10, 2361, 2020.
67. Cserháti, T., Chromatography in authenticity and traceability tests of vegetable oils and dairy products: a review. *Biomed. Chromatogr.*, 19, 3, 183–190, 2005.
68. [15] Wang, M., Li, R., Zou, S., Determination of carbofuran residue in aquatic products by gas chromatography. *Chin. J. Chromatogr.*, 26, 6, 775–777, 2008.
69. McDowell, I., Taylor, S., Gay, C., The Phenolic Pigment Composition of Black Tea Liquors Part I: Predicting Quality. *J. Agric. Food Chem.*, 69, 467–474, 1995.
70. [22] Calabrese, M., Stancher, B., Riccobon, P., High-Performance Liquid Chromatography Determination of Proline Isomers in Italian Wines. *J. Agric. Food Chem.*, 69, 361–366, 1995.
71. Haughey, S.A., Graham, S.F., Cancouet, E., Elliott, C.T., The application of Near- Infrared Reflectance Spectroscopy (NIRS) to detect melamine adulteration of soya bean meal. *Food Chem.*, 136, 3–4, 1557–1561, 2012.
72. [31] Ozen, B.F. and Mauer, L.J., Detection of hazelnut oil adulteration using FTIR spectroscopy. *J. Agric. Food Chem.*, 50, 3898–3901, 2002.
73. Hohmann, M., Differentiation of Organically and Conventionally Grown Tomatoes by Chemometric Analysis of Combined Data from Proton Nuclear Magnetic Resonance and Mid-Infrared Spectroscopy and Stable Isotope Analysis. *J. Agric. Food Chem.*, 63, 43, 9666–9675, 2015.
74. [36] Drivelos, S.A. and Georgiou, C.A., Multi-element and multi-isotope-ratio analysis to determine the geographical origin of foods in the European Union. *TrAC - Trends Analyt. Chem.*, 40, 38–51, 2012.
75. Casale, M., Oliveri, P., Armanino, C., NIR and UV Vis spectroscopy, artificial nose and tongue: comparison of four fingerprinting techniques for the characterization of Italian red wines. *Anal. Chim. Acta*, 668, 143–148, 2010.

76. [49] Singh, V.P., Pathak, V., Nayak, N.K., Verma, A.K., Umaraw, P., Recent developments in meat species speciation – a review. *J. Livest. Sci.*, 5, 49–64, 2014.
77. Khan, S.K., Mirza, J., Anwar, F., Abdin, M.Z., Development of RAPD marker for authentication of Piper nigrum (L). *Environ. We Int. J. Sci. Tech.*, 5, 47–56, 2010.
78. [61] Babaei, S., Talebi, M., Bahar, M., Developing an SCAR and ITS reliable multiplex PCR-based assay for safflower adulterant detection in saffron samples. *Food Control*, 35, 1, 323–328, 2013.
79. Dhanya, K., Syamkumar, S., Jaleel, K., Sasikumar, B., Random amplified polymorphic DNA technique for the detection of plant based adulterants in chilli powder (Capsicum annuum). *J. Spices Aromat. Crops*, 17, 75–81, 2008.
80. [69] Cao, H., But, P.P., Shaw, P.C., Authentication of the Chinese drug "Ku-di-dan" (herba Elephantopi) and its substitutes using random-primed polymerase chain reaction (PCR). *Acta Pharm. Sin.*, 31, 543–553, 1996.
81. Martins-Lopes, P., Gome, S., Santos, E., Guedes-Pinto, H., DNA markers for Portuguese olive oil fingerprinting. *J. Agric. Food Chem.*, 56, 24, 11786–11179, 2008.
82. [75] Pereira, L., Martins-Lopes, P., Batista, C., Zanol, G.C., Clímaco, P., Brazão, J., Molecular markers for assessing must varietal origin. *Food Anal. Methods*, 5, 6, 1252–1259, 2012.
83. Dhanya, K., Syamkumar, S., Siju, S., Sasikumar, B., Sequence characterized amplified region markers: A reliable tool for adulterant detection in turmeric powder. *Food Res. Int.*, 44, 9, 2889–2895, 2011.
84. Schiefenhovel, K. and Rehbein, H., Differentiation of Sparidae species by DNA sequence analysis, PCR-SSCP and IEF of sarcoplasmic proteins. *Food Chem.*, 138, 1, 154–160, 2013.
85. Cheng, C.Y., Shi, Y.C., Lin, S.R., Use of real-time PCR to detect surimi adulteration in vegetarian foods. *J. Mar. Sci. Technol.*, 20, 5, 570–574, 2012.
86. Kesmen, Z., Yetiman, A.E., Sahin, F., Yetim, H., Detection of Chicken and Turkey Meat in Meat Mixtures by Using Real-Time PCR Assays. *J. Food Sci.*, 77, 2, C167–173, 2012.
87. Zhang, W.J., Qin, C.X., Guan, Q.C., Analytical Methods, Detection of peanut (*Arachis hypogaea*) allergen by Real-time PCR method with internal amplification control. *Food Chem.*, 174, 547–552, 2015.
88. Wu, Y., Chen, Y., Wang, Y.G.J., Xu, B., Huang, W., Yuan, F., Detection of olive oil using the Evagreen real-time PCR method. *Eur. Food Res. Technol.*, 227, 1117–1124, 2008.
89. Drummond, M.G., Brasil, B.S.A.F., Dalsecco, L.S., Brasil, R.S.A.F., Teixeira, L.V., Oliveira, D.A.A., A versatile real-time PCR method to quantify bovine contamination in buffalo products. *Food Control*, 29, 131–137, 2013.
90. Deng, H. and Gao, Z., Bioanalytical applications of isothermal nucleic acid amplification techniques. *Anal. Chim. Acta*, 853, 30–45, 2015.

91. [129] Carles, M., Cheung, M.K., Moganti, S., Dong, T.T., Tsim, K.W., Ip, N.Y., Sucher, N.J., A DNA microarray for the authentication of toxic traditional Chinese medicinal plants. *Planta Med.*, 71, 580–584, 2005.
92. Zammatteo, N., Lockman, L., Brasseur, F., De, P.E., Lurquin, C., Lobert, P.E., Hamels, S., Boon, T., Remacle, J., DNA microarray to monitor the expression of MAGE-A genes. *Clin. Chem.*, 48, 25–34, 2002.
93. [137] Burns, M., Wiseman, G., Knight, A., Bramley, P., Foster, L., Rollinson, S., Measurement issues associated with quantitative molecular biology analysis of complex food matrices for the detection of food fraud. *Analyst*, 141, 1, 45–61, 2016.
94. Zhang, R., Huo, W., Zhu, W., Mao, S., Characterization of bacterial community of raw milk from dairy cows during subacute ruminal acidosis challenge by highthroughput sequencing. *J. Sci. Food Agric.*, 95, 5, 1072–1079, 2015.
95. Garofalo, C., Osimani, A., Milanovic, V., Aquilanti, L., De, F.F., Stellato, G., Bacteria and yeast microbiota in milk kefir grains from different Italian regions. *Food Microbiol.*, 49, 123–133, 2015.
96. Willems, S., Fraiture, M.A., Deforce, D., De, K.S.C., De, L.M., Ruttink, T., Statistical framework for detection of genetically modified organisms based on Next Generation Sequencing. *Food Chem.*, 192, 788–798, 2016.
97. Colmenero, M.M., Martinez, J.L., Roca, A., Garcia, V.E., NGS tools for traceability in candies as high processed food products: Ion Torrent PGM versus conventional PCR-cloning. *Food Chem.*, 214, 631–636, 2017.
98. Mohamad, I., K.S.S., Shakeel, W., Rapid Detection of Adulteration in Indigenous Saffron of Kashmir Valley, India. *Res. J. Forensic Sci.*, 3, 7–11, 2015.
99. Dar, M.M., Detection of Sudan Dyes in Red Chilli Powder by Thin Layer Chromatography. *J. Allergy Ther.*, 2012. https://doi.org/10.4172/scientificreports. 2, 1–3, 586.
100. Tateo, F. and Bononi, M., Fast determination of Sudan I by HPLC/APCI-MS in hot chilli spices, and oven-baked foods. *J. Agric. Food Chem.*, 52, 655–658, 2004.

2
Additives and Preservatives Used in Food Processing and Preservation, and Their Health Implication

Sunita Adhikari (Nee Pramanik)

Department of Food Technology & Biochemical Engineering, Jadavpur University, Kolkata, India

Abstract

Food additives are the materials that are added in to the food product intentionally while processing or storage or transportation of food. They are added directly or indirectly in the food product. They may be either natural, nature identical or artificial. As per their purpose of addition in to food products they can be classified in to following four categories:

   i. Preservatives- They are added into the food product for particularly food preservation purpose.
   ii. Processing agents- They are added with the intention to maintain desired product property or aid the food processing.
   iii. Sensory agents- They improve the sensory properties of food products like food colour, flavour sweetener etc.
   iv. Nutritional additives- These additives are added during fortification or enrichment of food products to correct the dietary deficiency.

The use of additives and preservatives increased with the growing need of processed food. Excessive use of many of these additives have health effects. So, there was a need for the toxicological study and regulation of these additives. The safety of food additives and its usage levels both are evaluated and recommended by some international body.

Email: sunitapramanik@gmail.com; sunita.adhikari@jadavpuruniversity.in

Mousumi Sen (ed.) Food Chemistry: The Role of Additives, Preservatives and Adulteration, (43–72)
© 2022 Scrivener Publishing LLC

Keywords: Additives, preservatives, processed food, natural, nature identical, artificial, health hazard

Abbreviations

FDA: Food and Drug Administration
JECFA: Joint FAO/WHO Expert Committee on Food Additives
FAO: Food and Agriculture Organization
WHO: World Health Organization
FSSAI: Food Safety and Standards Authority of India

2.1 Introduction

Throughout history, human beings are cultivating, harvesting food, and processing it with sophisticated methods. In ancient Greece, bread, olive oil, and wine were three products. These were the products in which the hardly edible raw materials were transformed to safe, nutritious, and stable food. Historically, there is a tradition to add ingredients or substance to foods for a particular function like ancient Egyptian used spices in food for flavoring. They used yeast in bread as leavening agent. They also used coriander and castor oil for medicinal application, cosmetics, and as preservative. Color was also used in candy and wine. In Japan, during 8th century, color was used in soybean and cakes.

Foods are the substances that are consumed in solid and/or liquid form to get nutritional benefits. Food additives are used in food from ancient times like salt is used in preservation of meat, herbs and spices are used for the improvement of food flavor, sugar is used as a preservative of fruits, and pickles were prepared with spices and vinegar. Use of processed food is increasing day by day. As a result, there is an increased use of food additives and preservatives. Additives are any substances that are used at any stage of food processing with various objective like the following:

i. To improve the keeping quality of the final product.
ii. To improve the appearance of the product.
iii. To improve the nutritional value of the product.

The additives might be natural or synthetic. Natural additives are extracted from any natural source like plant, animal, or mineral. Synthetic additives are synthesized by chemical or enzymatic reaction. These synthetic additives might be either identical to any natural compound or a pure synthesized compound. Table 2.1 gives an idea about the source of different food additives.

The additives might be added directly or indirectly in to food product. When the additives are added with particular intention and mentioned in the food label, they are called direct additives. Sometimes, food materials are exposed to additives by chance during processing, storage, or packaging. They are called indirect additives.

Preservatives are substances which is able to retard the growth of microorganism or able to mask the deterioration caused by microorganism. Preservatives like salt, sugar, herbs, spices, and vinegar are used from centuries [1].

There are hundreds of food additives currently used in the food industry. FDA maintains a list of over 3,000 additives in its database. Product label contains list of additives along with other ingredients. As per convention, all the additives are mentioned in the list with their category name

Table 2.1 Food additives and its sources.

Source of food additives	Example of food additives
Plant origin	i. Acidulant as tartaric acid from fruit ii. Color from fruits and vegetables iii. Thickening agent from seaweed, fruit and seeds
Nature identical product	i. Antioxidant Ascorbic acid from fruit ii. Tocopherol from vegetable oil iii. Food color carotenoid
Additives obtained by modification of natural substances.	i. Emulsifier from edible oils ii. Thickening agent as modified starch, modified cellulose iii. Sweetener as sorbitol
Man-made product	i. Antioxidant like BHA ii. Color like Quinoline yellow iii. Sweetener like saccharin

along with their specific name or with a number preceded by E number. Letter E indicates that the additive is included in the European list of food additives. Countries outside Europe use either number or name in the food labels. The E number does not indicate whether the additive is a natural or synthetic compound but category of the compound can be identified with the E number. For example, coloring agents have E number between 100 and 180, 500, to 1,520, which includes acids, alkalis, flavor enhancer, and additives with various other functions.

Risk assessment of any new food additives is carried out by autonomous international committee JECFA (Joint FAO/WHO Expert Committee on Food Additives). The committee assesses the health risk of additives on human being since 1956. The committee gives approval to only those additives that are found having no appreciable health risk to consumer. National authorities on the basis of JECFA assessment or national assessment approve the level of additives for a particular food. To use a new food additive in the processed food, the manufacturer needs to get its approval from national or international authorities. For that purpose, they need to file a petition to FDA for approval. The petition includes proper evidence that the substance is safe for they ways it is used. To approve the new additives under European Union, the procedure has four parts as follows:

a. Chemistry and specification of the additive: The additive is identified as per their origin, as a single, complex mixture, derivative of plant source, or synthesized material.
b. Existing authorization and approval: It is found that whether the additive already exists and what are the previous data of the additive.
c. Proposed use and exposure assessment: Estimation of dietary exposure based on proposed use and proposed age level of the person.
d. Toxicological study: Toxicological study regarding the effect of additive on the basis of in vitro and in vivo.

After approval of the new additive, the FDA issues regulation for the additive that includes the type of food in which it can be used, the maximum amount it can be used, and how it should be mentioned on food labels.

2.2 Merits and Demerits of Food Additives and Preservatives

2.2.1 Merits of Food Additives and Preservatives

i. Some additives are added to improve or maintain the nutritive value of food. For example, iodine is added in salt to improve its nutritional value. Enrichment of cereal products with B vitamin is another example of it [2].

ii. Additives are added to preserve food and without which the food would be spoiled at accelerated rate. Addition of food additives and preservatives reduces the amount of food waste and incidence of food poisoning e.g., additives are added in bread so it does not arrive mouldy when it arrives at grocery store.

iii. Addition of natural and synthetic color and flavor improves the appearance of the product.

iv. Additives can change the texture; mouth feel and sensory properties of food products.

v. Food can make the food processing easier, e.g., chemical defoamer can minimize the foaming in foods having high fat content.

2.2.2 Demerits of Food Additives and Preservatives

i. Some additives and preservatives have allergic reaction. For example, artificial color Red 40, Yellow 5, and preservative like benzoate can cause allergic reaction.

ii. Some of the food additives and preservatives have potential link to cancer. For example, artificial sweetener aspartame has potential to cancer development when applied on rat. Butylated hydroxy anisole (BHA) is used as antioxidant but it was found to cause cancer in animal study.

iii. They can replace the real ingredient present in the original food. For example, a chicken soup may contain little amount of chicken and the flavor can be managed with the addition of monosodium glutamate.

iv. Presence of food additives like corn syrup in the processed food may cause obesity and diabetes [3–5].

2.3 Types of Food Additives and Preservatives

Food additives and preservatives can be classified in various ways. Sometimes, it is possible that there might be an overlap between these groups [6]. They can be broadly divided into groups as follows (see also Figure 2.1):

1. Preservatives
2. Nutritional Additives
3. Flavoring Agent
4. Coloring Agent
5. Texturizing Agent
6. Miscellaneous Additives [7]

2.3.1 Preservatives

The food additives under this category are used for the preservation of the food. Food may be easily spoiled by the microorganism or by chemical or

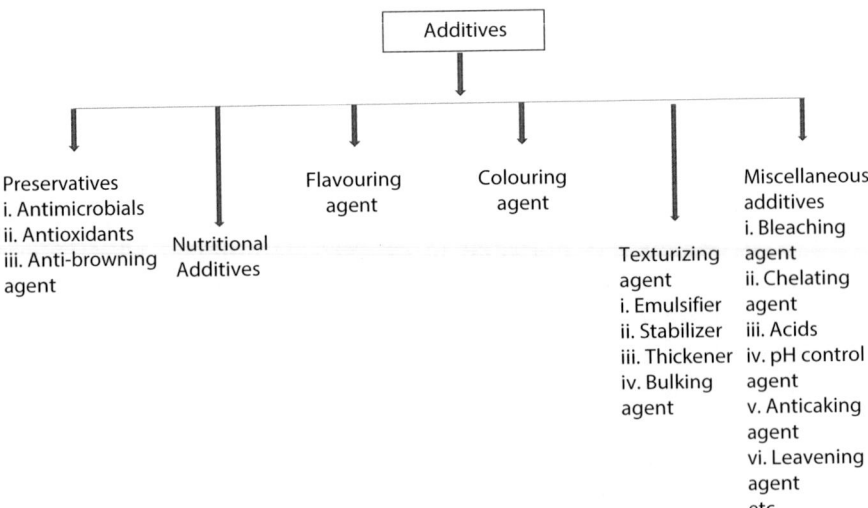

Figure 2.1 Types of food additives.

enzymatic reaction. The addition of preservative may inhibit the growth of spoilage microorganism or they may inhibit chemical and enzymatic reactions. The food preservatives may be of following three types.

   i. Antimicrobials: These additives are added for their ability to inhibit the growth of microorganism. Common additives include benzoates, sorbate, acetic acid, nitrate and nitrites, sulfur dioxide, antibiotics, and phosphates. The overall mode of action of these antimicrobial compounds includes reduced water availability and increased acidity. Natural antimicrobials have the dual benefit that they can inhibit the microorganism as well as make the food safe for eat. They may be derived from microorganism or animal or plant source. Some examples of natural alternative of food preservative include Natamycin, which is an antibiotic selectively that inhibits mold but the growth of bacteria required for ripening of cheese. Nisin and Pediocin are bacteriocin produced by microorganism that is effective on other microorganism. Lysozyme, enzyme lactoperoxidase, and lactoferrin are examples of antimicrobial derived from animal source. Antimicrobial extracted from plant source includes essential oils.
   ii. Antioxidants: Antioxidants are additives that are added in to the food to prevent the oxidation reaction. In food material, two types of oxidation reaction may occur: a. enzymatic oxidation of fruits and vegetables; and b. oxidation and rancidity of fats and oils. Antioxidants used in the food material are of following five types:

   a. radical scavenger or chain breaking antioxidant that is primary antioxidant and it binds with the free radical formed during oxidation reaction, thus preventing the chain reaction;
   b. chelator binds with metal and prevents them from radical formation;
   c. quencher that inactivate high energy oxidant;
   d. oxygen scavenger that removes oxygen from the system;
   e. antioxidant regenerator that regenerates the antioxidants.

   Polyphenols are natural antioxidants like flavonoids, tannins, lignans, and coumarins. Natural antioxidant also includes Ferulic acid, Catechin, Ascorbic acid, Tocopherol, Carotenoids like Lycopene, and β-carotene. They are generally used as plant extracts. Synthetic antioxidants include Butylated Hydoxy Anisole (BHA), Butylated Hydroxy Toluene (BHT), tbutylhydroquinone (TBHQ), gallate, and stannous chloride [8–10].

In contrast to synthetic antioxidant scientists are finding alternative source of antioxidant in nature. They found that algae contain natural antioxidative compounds that affect various pathogenic organisms with low-cost isolation process. The main antioxidant includes vitamin C and E, carotenoids, polyphenols, and chlorophyll. The red and brown algae contain high levels of folic acid and folate. Two species of cyanobacteria *Spirulina platensis* and *Spirulina maxima* are considered as supplements. They have the capacity of lowering lipid content in human body, antioxidant and anti-inflammatory effect. They have high nutritional value. They are considered as important food during long term space mission by NASA. Other important algae include *Chlorella vulgaris, Phaeodactylum tricornutum, Gayralia oxysperma, Chaetomorpha antennina, Sargassum vulgare,* and *Chondrus crispus*. They have high antioxidant capacity and have a potential to act as a source of natural antioxidant.

iii. Anti-browning agent: The development of brown or gray or black color in food product is due to browning reaction that may be of two types.

a. Enzymatic browning that makes the formation of quinones by the oxidation of polyphenols in presence of polyphenol oxidase. It can be prevented by blanching, acidification, and addition of sulfites.

b. Non-enzymatic browning occurs due to the reaction between reducing sugar and free amino acids resulting melanoidin pigment. It can be prevented by avoiding excessive heat, controlling moisture content, and application of sulfites. Sulfites are used in the form of sodium (or potassium) sulfite, bisulfite, or metabisulfite. Other anti-browning agent includes ascorbic acid, cysteine, and 4-Hexylresorcinol.

2.3.2 Nutritional Additives

Nutritional additives mean addition of pure compounds like vitamin, mineral, amino acid, and fatty acid to maintain or improve the nutritional quality of food. It is also added for other purposes, like vitamin C and E are added as antioxidant, and carotenoids are added to improve color.

Nutritional additives include fortification of salt with iodine, enrichment of cereal and dairy products with vitamin A, B, and D, enrichment of fruit beverages with vitamin C, essential fatty acid, mineral, and fiber [11].

2.3.3 Flavoring Agent

Flavoring agent includes flavorings and flavor enhancers. Flavorings added in the food produce flavor in the food, while flavor enhancer intensifies the existing flavor. Food flavoring may be of three types.

   a. Natural flavoring: These are single or mixed substances obtained from vegetable and animal source. This type of flavoring includes spice oleoresin and essential oils from citrus oils.
   b. Nature identical flavoring: These are the chemical substances but the identical substances are present in nature.
   c. Artificial flavoring: These types of compounds are completely synthetic and they are not identified in natural products.

Natural flavor includes spices, herbs, essence, and essential oil. The synthetic flavor includes esters, aldehydes, ketones, alcohols, and ethers. Monosodium glutamate is an example of widely used flavor enhancer [12, 13].

2.3.4 Coloring Agent

Coloring agents are added into the food to make it attractive and to influence the flavor. The other reasons for the addition of food color are as follows:

   a. Color may be lost during processing and storage, so to compensate it, food color is used.

b. Food material containing natural color may show variation of color food color is added to correct that variation.
c. Food color is added to improve the color imparted by the natural color.
d. Food color is used for the food items having no color.

As per FDA regulation, the colors added may be with or without certification. The certified colors are man-made color. They are more effective than the natural color. They do not introduce any off flavor in the food. The natural colors are exempted from certification but they are more expensive and can produce off flavor. Color additives used may be of three types.

a. Pure pigment or straight color.
b. Colors obtained by mixing the straight colors with other substances.
c. Mixtures produced by mixing two or more pure colors.

Natural food color includes annatto, betanin, saffron, lycopene, and anthocyanin. The certified food color includes Erythrosine, Tartrazine, Sunset Yellow, and Brillant Blue [14]. Natural color additives have some health benefits. For example, curcumin has antimicrobial and antioxidant activity. Riboflavin has antioxidant activity.

The natural and synthetic food colors permitted by FSSAI (Food Safety and Standards Authority of India), 2011, are mentioned in Tables 2.2 and 2.3.

Table 2.2 List of natural food colors permitted by FSSAI.

Name of the color	Type	Color
Carotene and carotenoid	Natural	Orange
Chlorophyl		Green
Riboflavin		Yellow-orange
Caramel		Dark brown
Annatto		Orange-red
Saffron		Yellow or orange
Curcumin or turmeric		Yellow

Table 2.3 List of synthetic food colors permitted by FSSAI.

Name of the color	Color	Type of the color	ADI (mg/kg of body weight)
Ponceau 4R	Red	Synthetic	4.0
Carmoisine	Red		4.0
Erythrosine	Red		0.1
Tartrazine	Yellow		7.5
Sunset Yellow FCF	Yellow		2.5
Indigo Carmine	Blue		5.0
Brilliant Blue FCF	Blue		12.5
Fast Green FCF	Green		25.0

As per FSSAI, 2011, the non-permitted synthetic food color includes Rhodamine, Amaranth, Metanil Yellow, Malachite Green, Acid Magenta, Orange G, and Butter Yellow.

2.3.5 Texturizing Agent

The role of texturizing agent as food additive is to modify the overall texture and mouthfeel of food. It includes emulsifier, thickener, stabilizer, and bulking agent.

a. Emulsifier: It allows water and oil to mix together to form an emulsion. Naturally occurring emulsifying agent is lecithin, milk protein, and legume peotein. Emulsifiers are molecules having both hydrophobic and hydrophilic end.
b. Thickener: It increases the viscosity of food without affecting other properties. Starch, albumen are used as thickener.
c. Stabilizer: Its function is to stabilize the foodstuff structure. Gelatin, pectin, and alginate are used as stabilizer to stabilize the emulsion formed. These additives prevent separation of emulsion, suspension, and foam. They are used in frozen dessert to improve mouth feel. Some of the stabilizers are antioxidants, some are emulsifiers, some may be ultraviolet stabilizer, and some may be sequestrants.

d. Bulking agent: It increases the bulk of food without upsetting the nutritional value. Starch is used as bulking agent [15, 16].

Other than these, phosphates are used as the meat binder in sausage and in processed cheese to get smooth consistency.

2.3.6 Miscellaneous Additives

2.3.6.1 *Acidity Regulator*

Acidity regulators are food additives that are used to change or keep the pH. They can be acid, bases, and buffering agents. Commonly used acidity regulators are citric acid, lactic acid, and acetic acid. They are used as they are organic acids [17].

2.3.6.2 *Anti-Caking Agent*

Anti-caking agents are anhydrous compounds that are added in dry food in small quantity. The purpose for the addition of this additive is to prevent the particles caking together and keep the product dry, free flowing. Without anti-caking agents, the dry products like cake and biscuit mix and soup powder will be clumped. The anti-caking agent either adsorbs excess moisture or coats particles and makes them water repellent. Some anti-caking agents are water soluble and some are soluble in organic solvent. Different anti-caking agents are starch, magnesium carbonate, and silica. They are added to food products with table salt, flour, coffee, and sugar. Calcium silicate is an anti-caking agent that is added to table salt. It is able to adsorb water and oil both.

2.3.6.3 *Antifoaming Agent*

They are also known as defoamer. They reduce and hinder the formation of foam in industrial processed liquid. They are classified as oil based, powder based, water based etc. When used, they reduce the effervescence in a preparation. Silicone oil is added in commercial cooking oil to prevent the froth formation in deep frying food.

2.3.6.4 *Flour Treatment Agents*

They are also called improver or flour improving agent as they improve the baking functionality of flour. These agents are of following four types:

a. oxidizing and reducing agent
b. enzymes
c. bleaching agents
d. emulsifier

These agents are used to speed up the process of dough rising and also to expand the strength of dough.

a. Oxidizing and reducing agents: Aging of flour occurs naturally when it is exposed to atmosphere. The oxidizing agents affect the sulfur-containing amino acids and aid to form disulfide bridges between the gluten molecules. The dough becomes stronger with the addition of these compounds. Examples of oxidizing agent include dehydroascorbic acid and potassium bromate, and they particularly oxidize the glutathione molecule. Potassium bromate acts directly and number of conversion steps are also less.
Reducing agents make the flour weak by breaking the protein network. Benefits of the addition of reducing agents are reduction in mixing time, dough elasticity, proofing time, and improved machinability. Examples of reducing agent include cysteine and bisulfite.
b. Enzymes: Different enzymes are used to improve the processing characteristics of dough. Examples of enzymes that are added for this purpose includes the following
- Amylases: It breaks down the starch into simple sugars. Yeast uses it up and can ferment quickly. Malt is used as it is the natural amylase source.
- Proteases: It improves the extensibility of dough by breaking the gluten.
- Lipoxygenase: It is used to oxidize flour.
c. Bleaching agents: Objective of addition of this additive is to make the flour whiter. Nitrogen dioxide is used for that purpose. In presence of high voltage, it is converted to nitrogen tetra oxide which is effective bleaching agent. Benzoyl peroxide is also used and it is better than nitrogen dioxide.
d. Emulsifier: Lecithin, monoglyceride, and diglyceride are used as emulsifier to disperse fat evenly through the dough that helps to trap carbon dioxide more. Addition of emulsifier produces softer crumb and fine grain with increased baked volume.

2.3.6.5 Fat Replacers

There is a growing consciousness developing in world to the ratio of saturated fatty acid to poly- and mono-saturated fatty acids in the diet. It is concluded that calories from fat in the diet should not be more than 30%. There is growing market demand for low fat food that encourage research on low fat content of food. Whenever fat content is reduced, it will automatically affect organoleptic property and physical property of food. To get desirable character, there is a need to use fat replacers. Fat replacers are carbohydrate, protein, and fat-based compound that replace fat to reduce calories in food. They can be synthetic fat substitute, emulsifier, starch derivative, hemicellulose, bulking agent, etc. They can be classified as follows:

a. Fat substitute: It is a synthetic compound that can replace fat on weight by weight basis. These compounds are having similar chemical structure to fat but they are resistant to digestive juices.
b. Fat mimetics: These are the substances that imitate the physical or organoleptic properties. They are also known as "texturizing agent". They require high amount of water for their action.
c. Fat analogs: These compounds have similar characteristics of fat but digestibility and nutritional value are different.
d. Fat extenders: They allow decrease in the amount of fat in the product [18].

Carbohydrate-based fat replacers are not used for frying purpose. Protein-based fat replacers are made from milk protein or egg protein. They can be heated and be converted to microparticle that give the same mouth feel as fat. They are used in butter, cheese, mayonnaise, and salad dressings. They cannot be used for frying and baking. Fat-based fat replacers are fat molecules that are modified so that they can be partially absorbed.

2.3.6.6 Sweeteners

Sweeteners are added to improve the taste of food. They also contribute to energy value. Some sweeteners like polyols are the modified sweetener that are popular nowadays. They are natural compounds and often added to boost flavor. They can provide some nutrition but they do not cause tooth decay or affect insulin response. On the other hand, artificial sweeteners have very little or no nutritional value. Sweeteners can be classified as in Figure 2.2.

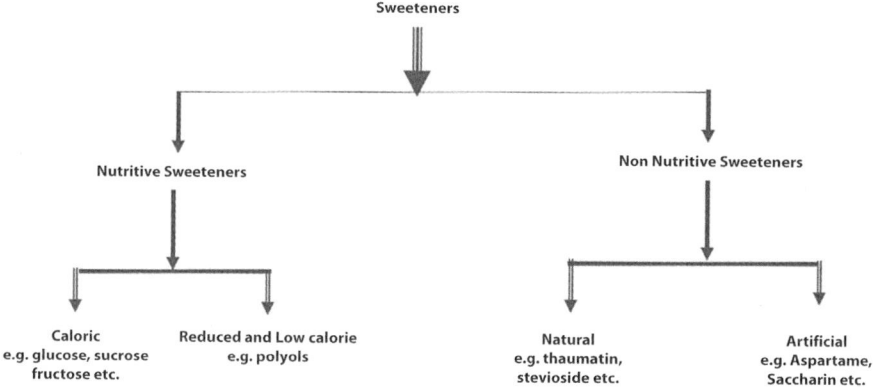

Figure 2.2 Types of sweetener.

2.3.6.6.1 Nutritive Sweeteners

Nutritive sweeteners are sweet tasting and have nutritional value. The natural sweetener includes mono- and disaccharides like sucrose, high fructose corn syrup, honey, molasses, and sugar alcohols. Properties of natural sweetener are follows:

- They provide calories and energy to diet.
- They are natural sweetener because they are extracted from natural products without any chemical modifications during the production or extraction process.
- The major ingredient of natural sweeteners is either mono- or disaccharides.

Demerits of natural sweetener are follows:

- Tooth decay
- Diabetes/hypoglycemia
- Hyperactivity or misbehavior in children
- Heart disease
- Obesity

Reduced-calorie and low-calorie sweeteners are Polyols or "sugar alcohols" or sugar replacers are naturally occurring in many fruits and beverages, but for commercial uses, they are made from other carbohydrates, such as starch, sucrose, and glucose. They are classified as follows:

- Monosaccharide-derived (e.g., sorbitol, erythritol, xylitol, and mannitol)
- Disaccharide-derived (e.g., maltitol, isomalt, and lactitol)
- Polysaccharide-derived (e.g., hydrogenated starch hydrolysates).

They are added to sugar free cookies, candies, chewing gum, baked goods, ice cream, toothpastes, mouthwashes, breath mints, and pharmaceuticals.

Merits of polyols are follows:

- Provides fewer calories than regular sugars because they are not completely absorbed.
- Add bulk and texture to foods.
- Provide a cooling effect or "cool" taste.
- Help to retain moisture in foods.
- Do not lose sweetness, cause browning when heated.
- Molds do not grow well on polyols; they may contribute to longer shelf life of foods.
- Bacteria in the mouth do not metabolize and convert the sweetener into plaque or harmful acids that cause tooth decay.

Demerits are as follows:

Excess amounts of polyols may cause gastrointestinal symptoms. So, food products containing sugar alcohols carry a label "Excess consumption may have a laxative effect."

2.3.6.6.2 Non-Nutritive Sweetener

Non-nutritive sweeteners have very little or no nutritional value. They are either natural or synthesized compounds. Other than providing sweet taste, they can act as humectant, can give the mouth feel in soft drinks, and can reduce the freezing point of ice cream. Artificial sweetener can limit the calorie intake and prevent dental cavities.

Merits of non-nutritive sweetener are follows:

- Can contribute pleasurable sweet sensations without increasing energy intake.
- Do not increase the incidence of dental caries and may even prevent cavities.
- Do not affect blood sugars.
- Decrease calorie content of food.

- Can be used in non-food items to make them taste more pleasurable.
- Only require small amount to sweeten foods and beverages.
- May aid in weight control.

Natural non-nutritive sweeteners are present in nature. Some of the examples of natural non-nutritive sweeteners are listed in Table 2.4.

Table 2.4 List of natural non-nutritive sweeteners.

Sweetener	Source	Chemical nature	Potency	FDA approval
Brazzein, Pentadin	West African Fruit Oubli	Protein	500–2,000	Not Approved
Neohesperidin dihydrochalcone	A derivative of the bitter constituent of grapefruit	Glycoside	300–1,800	Not Approved
Glycyrrhizin	Root of Liquorice	Glycoside	30–50	Not Approved
Thaumatin	Katemfe fruit	Protein	2,000	Not Approved
Stevia	Leaves of *Stevia rebaudiana*	Glycoside	300	Approved
Curculin	Malaysian fruit *Curculigo latifolia*	Protein	430–2,070	Not Approved
Inulin	Various Plants	Polysaccharide	10% of Sucrose	Not Approved
Mabinlin	Seed of a Chinese plant Mabinlang	Protein	100–400	Not Approved
Miraculin	Fruit of *Synsepalum dulcificum*	Glycoprotein	modifies taste receptors	Not Approved
Monatin	Plant *Sclerochiton ilicifolius*	Indole derivative	3,000	Not Approved
Monellin	Fruit of West african shrub *Dioscoreophyllum cumminsii*	Protein	800–2,000	Not Approved
Osladin	Rhizome of *Polypodium vulgare*	Saponin	500	Not Approved
Luo han guo	Fruits of **Siraitia grosvenorii**	Glycosides	300	Approved

2.3.6.6.3 Artificial Non-Nutritive Sweeteners

They are derived from the chemical synthesis of organic compounds. These compounds may or may not be found in nature. A list of artificial non nutritive sweeteners along with their potency and use are listed in Table 2.5. New artificial sweeteners are researched. These compounds are low-cost product and can be synthesized easily. So, in future, they will be the primary sweetening compound [19].

Table 2.5 List of artificial non-nutritive sweeteners.

Chemical name of sweetener	Trade name of the sweetener	Potency	Caloric value (cal/g)	Use
Acesulfame	Sunett, Sweet One	200 times sweeter than sugar	0	Used in different candies, desserts, baked goods.
Aspartame	Nutra Sweet	180–200 times sweeter than sugar	4	Used in soft drinks, chewing gum, frozen desserts, pharmaceuticals.
Neotame	-	8,000–13,000 times sweeter than sugar	0	Used in beverage, dairy products, baked goods, gums.
Saccharin	Sweet and Low, Sweet Twin, Sugar Twin,	300–700 times sweeter than sugar	0	Used in beverages and sometimes used together with aspartame.
Sucralose	Splenda	600 times sweeter than sugar	0	Used in frozen desserts, cookies, candies, baked products.

2.3.6.7 Leavening Agent

These types of additives are added to increase the surface area of a dough or batter by creation of gas bubbles inside the dough or batter. They also make the product light in weight. During the baking process, the gas bubbles get expanded so volume of the product increases and a desirable porous structure is obtained. The functions of leavening agents are as follows:

1. It makes the final product light and spongy.
2. It makes the product digestible.
3. It makes it soft.
4. It makes the product tastier and appetizing.

The leavening agents used as additives are mainly of two types.

a. Biological agent: Biological leavening agents are harmless microorganism like yeast *Saccharomyces cerevisiae*. The yeast cell added ferments the sugar and produces carbon dioxide and ethanol. The yeast added may be active dry yeast or compressed yeast.
b. Chemical agents: Three types of chemical leavening agents are used. They are baking soda, baking powder, and ammonium bicarbonate. Baking soda contains sodium bicarbonate ($NaHCO_3$) that releases carbon di oxide when heated. Baking soda is always used with acid like buttermilk, vinegar, lemon juice, molasses, honey, vinegar, buttermilk, cocoa fruits, fruit juices, and cream of tartar. Baking powder contains baking soda, dry acids, and starch. Baking powder may be either single acting powder or double acting powder. In single acting baking powder carbon dioxide is released as soon as it is added in the liquid. Double acting baking powder releases carbon dioxide in two stage: first is when it is mixed in cold liquid and second is during baking. Ammonium bicarbonate is another leavening agent that produces carbon dioxide when it decomposes. They are used in the products which are finally dry and porous so that all the ammonia liberated during heating gets released otherwise it will affect the taste.

2.3.6.8 Firming Agent

These are the additives that give firming action to food. Fruits and vegetables contain pectin that is relatively insoluble and form firm gel around the tissues of fruit. Firming agents are additives that help to precipitate residual pectin. It strengthens the supporting tissue and prevent its collapse during processing. Calcium salts are used as firming agent. It forms calcium pectate gel that protects the tissues against softening during processing. Some of the other examples of firming agents include aluminium sulfate, ammonium aluminium sulfate, calcium chloride, calcium sulfate, and sodium/potassium aluminium sulfate.

2.3.6.9 Glazing Agent

Food glazing agents are additives also known as polishing agent. When they are applied on food material, it gives a protective and shiny coating on fruits, bakery and confectionary items. These additives are prepared from either natural or synthetic sources. Common glazing agents are beeswax, lac, mineral oil, etc.

2.3.6.10 Humectant

These are hygroscopic substances that promote moisture retention. It prevents the food from drying out. Polyols including propylene glycol, glycerol, sorbitol, and mannitol are some examples of humectants. They contain several hydroxyl groups that makes them hydrophilic that enables them to bind with water. They are also known as moisture retention agent or wetting agent. Examples of products where humectants are used include shredded coconut, cookies, glazed and dried fruit, and cake.

2.3.6.11 Sequestering Agent

Sequestering agents are also known as chelating agent. They protect food products from chemical, oxidative, and enzymatic reactions that promote deterioration of food through chelation during processing and storage. They form chelate with polyvalent metal ions like copper, iron, and nickel. Even at very low concentration of these metal ions as low as 0.05 ppm, they can cause rancidity, discoloration, spoilage in texture, odor, and flavor. Copper can affect ascorbic acid, folic acid, vitamin E, and thiamine.

Vitamin A will be destroyed by both copper and iron. Fats and oils are also oxidized in presence of metal ions that act as catalyst. Examples of chelating agents are EDTA, polyphosphate, citric acid, and tartaric acid.

2.3.6.12 Gelling Agent

These are the food additives that are used with the purpose to thicken and stabilise various foods like jelly, desert, and candies. The change the texture of a food by forming gel. Typical gelling agents are natural gum, starches, pectin, agar, and gelatin. The provide the food a texture by the formation of gel. Some stabilizers and thickening agent are also gelling agent. These agents are polysaccharide or protein as per their nature.

2.3.6.13 Propellants

Propellants are the food additives that are used in foods like carbonated drinks to expel it out of their container. A propellant creates more pressure to liquid food so that it will come out from its container. It is also used in whipped creams. Common propellant includes carbon dioxide, nitrogen, nitrous oxide, propane, and butane. They are also used to reduce the amount of oxygen in contact with the food in the container. Nitrous oxide and carbon dioxide are better propellants as they dissolve in the liquid food and expand during its release from its container.

2.3.6.14 Foaming Agent

These additives are also known as aerating agent. It facilitates the formation of foam and helps in the formation of uniform dispersion of gaseous phase over liquid or solid food. They are used in toppings, cake mixes, whipped creams, beverages, and soft drinks. Examples are propylene glycol ester of fatty acid, glycyrrhizin, alginic acid, and polysorbate.

2.3.6.15 Seasoning

They are the additives from plant origin. They are added in food and drinks to improve the taste and aromatic fragrance. They are obtained from different parts of the plant like root, leaves, flower, bud, fruits, seeds, and whole plants. A list of some common spices along with their bioactive compounds are presented in Table 2.6. They also add phytonutrients, essential oil, antioxidants, vitamin, and minerals to the food where it is applied [20].

Table 2.6 List of some common spices and their bioactive compounds [21].

Spice	Bioactive compound	Benefits
Black pepper	Piperine, piperidines, and pyrrolidines	When applied in meat products, it decreases lipid peroxidation.
Paprika	Capsaicin	It reduces inflammation and stimulate digestion.
Turmeric	Curcumin, curcuminoid	It reduces join pain and stomach pain.
Ginger	Gingerol, shogaols, paradols	It has anti-blood clotting activity, stimulates digestion, and reduces bacterial and fungal infection.
Cardamom	Eucalyptol, limonene, cineole, and α-terpinyl acetates	It has anti-microbial properties and can improve cardiovascular health.
Vanilla	Vanillin	It has antioxidant and anti-microbial property.
Black cumin	Thymoquinone, dithymoquinone, and thymohydroquinone	It has anti-microbial properties.
Clove	Eugenol	It has antioxidant and anti-microbial property.
Cinnamon	Trans-cinnamaldehyde, eugenol, and linalool	It has antioxidant and anti-microbial property.

2.3.6.16 Curing Agents

These are the food additives that are used to preserve meat. They give desirable color and flavor to the meat. They prevent the growth of microorganism and prevent toxin formation.

2.3.6.17 Probiotics

Probiotics are live microbial additives that are used in many food products like fermented milk. It can balance intestinal microbial composition, reduce the serum cholesterol, improve the immunity and metabolic process, and reduce the food allergy symptom of infants. Probiotic food comprises 60%–70% of total available functional food in the market. A wide variety of microorganism is considered as potential probiotics but the predominant bacteria are *Lactobacillus* and *Bifidobacterium* species. Other probiotic species include *Lactococcus, Enterococcus, Saccharomyces, Propionibacterium*, and *Aspergillus* species. Lactic acid bacteria can contribute to safety of food, offer organoleptic, nutritional, and health advantages. Viability of probiotic microorganism is dependent on factors like food ingredients, processing operation, packaging and storage, fermentation conditions, and presence of protective agents. Lactic acid bacteria are able to produce exopolysaccharides that play a major role in the production of fermented dairy products like yoghurt, cheese, and milk-based dessert. The exopolysaccharide is used as a texturizing and stabilizing agent. Nowadays, consumers prefer safe and healthy food without additives. The exopolysaccharide has the potential for development of improved products like low fat/low milk solid yoghurt. These exopolysaccharides have anti-tumor and immune modulating capacity so with these the lactic acid bacteria have the potential for development of functional food [22].

2.3.6.18 Other Food Additives

Other food additives include clarifying agent like gelatine and synthetic resin that remove the sediment and haziness of fruit juice. Enzymes are added to get desirable changes like pectinase used for clarification of juice, papain for meat tenderization, and renin for cheese making. Freezing agents like liquid nitrogen that evaporates very quickly and used to quick chilling of food. Solvents like alcohol and glycerine are used to dissolve flavor, color, and inert gases for packaging to prevent oxidative changes.

2.3.6.19 Indirect Food Additives

These are the food additives that become the part of food due to its packaging, storage, or other handling process. They are generally present in trace amount. Its regulation guideline by FDA includes that all food packaging material manufacturers must ensure that all materials coming in contact with food are safe before getting permitted for its use. These additives are also termed as incidental additives. For example, these groups of additives may include chemical substances used in packaging material often in detectable level but they can migrate into the food. FDA found that a chemical known as acrylonitrile used for the fabrication of plastic beverage bottles may migrate into food and may cause cancer. So, then its approval was withdrawn.

2.4 Health Effect of Food Additives and Preservatives

Additives and preservatives are essential for food the storage of food. With the increase of the use of processed food, there is an increasing use of food additives and preservatives. As some of these additives were found to have adverse health effect, there was a need to approval of these food additives and preservatives after the details study of the effect of them on human being. Some of the food additives were found to cause allergies, hyperactivity, asthma, and reactions like rashes, vomiting, and headache. They can be the cause of different health hazards like the following:

a. Artificial food colors are found to be carcinogenic agent. They can cause hyperactivity and allergies.
b. Nitrites and nitrates can develop carcinogen nitrosamine.
c. Sulfites can cause asthmatic reaction.
d. Sweeteners can cause obesity, dental cavity, diabetes, and increase blood fat. Artificial sweetener can cause hyperactivity and some of them are found carcinogen. Government cautions children and pregnant women against the use of it.
e. Artificial flavor and monosodium glutamate can cause allergic reactions.
f. BHT can be toxic to nervous system.
g. Salt can increase the blood pressure [23].

Table 2.7 shows some specific additives and their health effect.

Exposure to hazardous chemicals through food can create adverse health effects that depends on various factors like the following

a. Type of chemical
b. Dose of chemical
c. Duration of exposure
d. Frequency of exposure [25]

The safety of food additives is generally evaluated by international organization as well as local government. The goals for the assessment of the intake of food additive are as follows:

a. Monitoring the actual intake of food additives with respect to its acceptable daily intake (ADI).

Table 2.7 List of additives and preservatives that can affect human health [24].

Additive	Role of the additive	Food material where it is added	Possible health hazard
Erythrosin	Food color	Canned food, confectionary, and dairy products	Possible risk of thyroid tumor in animal study
Blue 1, Red 40, Yellow 5 and Yellow 6	Food color	Different candy and confectionary product	Allergic reaction to some people.
Tartrazine	Food color	Ice cream, carbonated drinks	Skin rashes, asthma, and headache
Carmosine	Food color	Dairy products	Cause tumor to animals and DNA damage
Amaranth	Food color	Alcoholic beverages	Can cause asthma and allergies
Indigo Carmine	Food color	Bakery, confectionery, and dairy products	Can cause skin rashes, DNA damage and tumor to animals
Brilliant Blue	Food color	Dairy products and sweets	Can cause skin rashes, DNA damage and tumor to animals
Sodium benzoate	Preservative	Carbonated beverages, sauce, pickles	Can cause asthma and probable carcinogen
Sulfur dioxide	Preservative	Dried food juice	Nausea, diarrhoea, DNA damage to animals

(Continued)

Table 2.7 List of additives and preservatives that can affect human health [24]. (*Continued*)

Additive	Role of the additive	Food material where it is added	Possible health hazard
Sodium metabisulfite	Antioxidant	Processed food	Can cause asthma
Potassium nitrate	Curing agent	Cured meat and meat products	Can form nitrosamine that is carcinogenic agent
BHA/BHT	Antioxidant	Fats/oils	Considered as carcinogenic agent
Monosodium glutamate	Flavor enhancer	Processed food	Can cause cancer, DNA damage
Aspartame	Sweetener	Processed food	Can cause neurological damage
Acesulfame K	Sweetener	Processed food	Can cause cancer to animals
Saccharine	Sweetener	Processed food	Can interfere DNA coagulation process
Bromate	Preservative	Processed food	Destroy nutrients in food
Caffeine	Flavorant	Processed food	Can cause heart defects
Chelating agent	Sequestering agent	Processed food	They affect mineral balance in our body

(*Continued*)

Table 2.7 List of additives and preservatives that can affect human health [24]. (*Continued*)

Additive	Role of the additive	Food material where it is added	Possible health hazard
Auramine	Food color	Used in beverages	Retard growth and damage kidney, liver
Metanil yellow	Food color	Used in sweets	Responsible for weakness and food poisoning
Lead chromate	Food color	Added to turmeric for bright color	Anemia, abdominal pain

b. Identify the consumer group who are at risk for the intake of additives close or more than the ADI value.
c. To offer information to the international organization for the reassessment of the regulations of food additives in case of high intake.

The safety assessment of the additives is based on the toxicity of the additives. On the basis of these toxicological data, food additives are classified as follows:

Group A: Additives having established ADI value
Group B: Additives generally considered as safe.
Group C: Additives with insufficient data.
Group D: Flavoring components.
Group E: Natural components with very few or no scientific safety data.

ADI value is determined by JECFA. ADI value is expressed on the basis of body weight, and it is the amount of food additive consumed over a lifetime without considerable health hazard. Only the food additives having JECFA safety assessment and having no appreciable health risk to consumers can be used. These additives might be natural or synthetic. Either on the basis of JECFA assessment or assessment of national authority authorized the use of particular additives at specified level for specific food. JECFA evaluation is based on review of biochemical, toxicological, and other relevant data that includes mandatory animal study and observation on human being. The toxicological test includes acute, short-term and long-term studies. The test determines how the additives are absorbed, distributed, and excreted. It also determines the harmful effect of the additives and its by-products. The Codex Alimentarius Commission establishes the guidelines on food labelling. The standards are implemented in most countries. Food manufacturers are obliged to indicate the additives used in their products. In European Union, there is legislation for labelling of food additives as per E numbers. People sensitive to certain food additives can check labels.

2.5 Conclusion

In the last 30 years, the use of food additives has been increased in huge amount. In present situation, over 200,000 tonnes of additives are used per year. It is estimated that, today, about 75% of western diet is made up of processed food and each person is consuming 3.6–4.5 kg of food additive per year. With the increase in the consumption of food additives, there is a concern with the various physical and mental disorders. More concern is with the effect of additives on children as consumption of additives is related to hyperactivity and hypersensitivity. Synthetic additives should be avoided more due to their potential toxic effect. Consuming excessive additives causes various adverse reactions like gastrointestinal problem, respiratory disorder, and neurological adverse reaction. Worldwide, there is a search for new and safe additive from natural sources to prevent spoilage of food with no adverse effect on human being. The demand for probiotic functional food is increasing day by day due to awareness of consumer about the impact of these foods on human health. There is a continuous development and reduction of the need of chemical additives. There are huge number of natural compounds present in nature that are still unknown. These natural compounds have a tremendous potential to act as new safe food additive. Synthetic food additives can react with the cellular components that lead to various effect on human health. If we have to use additives in food, then we should try to use the natural one which has minimum effect. Otherwise, we should use those additives that are generally recognized as safe (GRAS), and if they are not of GRAS category, then the ADI should not be exceeded. Before purchase of food the ingredient, it should be checked to avoid the foods containing artificial additives.

References

1. Abdulmumeen, H.A., Risikat, A.N., Sururah, A.R., Foods: Its preservatives, additives and applications. *Int. J. Chem. Biochem. Sci.*, 1, 36–47, 2012.
2. Vaclavik, V.A. and Christian, E.W., Food Additives, in: *Essentials of Food Science. Food Science Texts Series*, Springer, New York, 2008.
3. Pasca, C., Coroian, A., Socaci, S., Risks and Benefits of Food Additives – Review. *Bull. UASVM Anim. Sci. Biotechnol.*, 75, 2, 71–79, 2018.
4. Carocho, M., Barreiro, M.F., Morales, P., Ferreira, I.C.F.R., Adding Molecules to Food, Pros and Cons: A Review on Synthetic and Natural Food Additives. *Compr. Rev. Food Sci. Food Saf.*, 13, 377–399, 2014.
5. Sharma, S., Food Preservatives and their harmful effects. *Int. J. Sci. Res. Publ.*, 5, 4, 1–2, 2015.
6. Awuchi, C.G., Twinomuhwezi, H., Igwe, V.S., Amagwula, O., Food Additives and Food Preservatives for Domestic and Industrial Food Applications. *J. Anim. Health*, 2, 1, 1–16, 2020.
7. Nagasinduja, V. and Shahitha, S., Food preservatives-types, uses and side effects. *Int. J. Basic Appl. Res.*, 9, 4, 632–637, 2019.
8. Franco, R., Navarro, G., Martinez-Pinilla, E., Antioxidant versus Food Antioxidant Additives and Food Preservatives. *Antioxidants*, 8, 542, 2019.
9. Atta, E.M., Mohamed, N.H., Abdelgawad, A.A.M., Antioxidants: An Overview of the Natural and Synthetic Types. *Eur. Chem. Bull.*, 6, 8, 365–375, 2017.
10. Silva, M.M. and Lidon, F.C., An overview on applications and side effects of antioxidant food additives. *Emir. J. Food Agric.*, 28, 12, 823–832, 2016.
11. Branen, A.L., Davidson, P.M., Salminen, S., Throngate III, J.H., *Food Additives*, Marcel Dekker, Inc., New York, 2002.
12. Shukla, P., Sharma, A., Sharma, A., Food additives from an organic chemistry perspective. *MOJ Biorg. Org. Chem.*, 1, 3, 70–79, 2017.
13. *The Food Safety and Standards (Food Product Standards and Food Additives) Regulations*, pp. 503–507, 512, 513, 529, 530, Food Safety and Standards Authority of India, New Delhi, https://www.fssai.gov.in/dam/jcr:99067191-c774-4c81-b9c8-708b9e72b770/Food_Additives_Regulations.pdf, 2011.
14. Sezgin, A.C. and Ayyıldız, S., Food additives: Colorants. In: *Food Analysis: From Structure, Chemistry and Flavour to Foodomics*, Vilela, A., T. Pinto, B. Gonçalves, E.A. Bacelar, A.C. Correia, A.M. Jordão and F. Cosme (Eds.), pp. 87–94, Formatex Research Center, Spain, 2017.
15. Faustino, M., Veiga, M., Sousa, P., Costa, E.M., Silva, S., Pintado, M., Agro-Food Byproducts as a New Source of Natural Food Additives. *Molecules*, 24, 1056, 2019.
16. Khodjaeva, U., Bojnanska, T., Vietoris, V., Sytar, O., About Food Additives as Important Part of Functional Food. *J. Microbiol. Biotechnol. Food Sci.*, 2, 1, 2125–2135, 2013.

17. Kesava, R.C., Sivapriya, T.V.S., Arun, K.U., Ramalingam, C., Optimization of Food Acidulant to Enhance the Organoleptic Property in Fruit Jellies. *J. Food Process. Technol.*, 7, 11, 2016.
18. Chavan, R.S., Khedkar, C.D., Bhatt, S., Fat Replacer, in: *The Encyclopedia of Food and Health*, vol. 2, Caballero, B., Finglas, P., Toldrá, F. (Eds.), pp. 589–595, 2016.
19. Periyasamy, A., Artificial Sweeteners. *Int. J. Res. Rev.*, 6, 1, 120–128, 2019.
20. Elizabeth, D.L.T., Gassara, F., Kouassi, A.P., Brar, S.K., Belkacemi, K., Spices used in food: Properties and benefits. *Crit. Rev. Food Sci. Nutr.*, 57, 1078–1088, 2017.
21. Asfaw, D., Review on Use of Bioactive Compounds in Some Spices in Food Preservation. *Food Sci. Qual. Manage.*, 87, 6–21, 2019.
22. Abdel, G.T.M., Safe Food Additives. *J. Biol. Chem. Res.*, 32, 1, 402–437, 2015.
23. Baig, S.K.M.S. and Kasim, S.S., Study of Harmful Effects of Consuming Food Additives and Public Awareness. *IJSRST*, 2, 4, 1071–1074, 2018.
24. Pandey, R.M. and Upadhyay, S.K., *Food Additive*, Y. El-Samragy (Ed.), pp. 1–29, Egypt, www.intechopen.com/books/food-additive/food-additive, 2012.
25. Sharma, M., Rajput, A., Rathod, C., Sahu, S., Food Chemicals Induces Toxic Effect on Health: Overview. *UK J. Pharm. Biosci.*, 6, 4, 33–37, 2018.

3

Role of Packaging in Food Processing

Bhasha Sharma[1], Susmita dey Sadhu[2], Rajni Chopra[3] and Meenakshi Garg[4]*

[1]*Department of Chemistry, Netaji Subhas University of Technology, Dwarka, Delhi, India*
[2]*Department of Polymer Science, Bhaskaracharya College of Applied Sciences, Dwarka, Delhi, India*
[3]*Department of Food Science and Technology, National Institute of Food Technology Enterpreneurship and Management, Sonipat, India*
[4]*Department of Food Technology, Bhaskaracharya College of Applied Sciences, Dwarka, Delhi, India*

Abstract

The employment of food packaging is a socioeconomic indicator of the escalated spending ability of the populace as well as regional food availability. The zealous objective to elevate sustainability or recyclability of plastic packaging via enhanced recovery and selection implicit redesigning packaging through a viewpoint of the end of life thinking. Food packaging is a pre-eminent unit of operations amidst the process of food production as it imparts food products with expropriate conditions able to obviate mechanical destruction and decelerate biochemical retrogression. Progression in the efficacy of packaging materials can succor to obviate food pathogens and alleviate environmental waste by maintaining food quality. This chapter explains the aseptic packaging modified atmospheric packaging and active packaging technology which have gained increasing importance in the food industry in recent years. Sustainability of food packaging can be trailed through operations emphasizing the augmentation of efficiency and other triggering effectiveness. Intelligent and active packaging is of utmost importance for the storage of food which facilitates improved food safety and quality. Furthermore, the meticulous compendium of smart packaging along with its impediment has been also outlined as packaging systems have soared to be smarter with the amalgamation of emerging wireless communications and electronics.

Keywords: Food packaging, processing, sustainability, life cycle assessment

---

*Corresponding author: meenakshi.garg@bcas.du.ac.in

Mousumi Sen (ed.) Food Chemistry: The Role of Additives, Preservatives and Adulteration, (73–96)
© 2022 Scrivener Publishing LLC

3.1 Introduction

The progression in the environmental impacts of packaging is frequently synthesized for the packages alone, which shuns the fundamental functions of packaging. The myriad of packaging in the food industry has expanded colossal fostering since its origination in the 18th century with the most ingenious inventions prevailing during the past century. The conception of packaging is under vehement scrutiny globally which is regarded as the origin of pollution and detrimental waste. The packaging role is to enable distribution, inform, protect, and assist in displaying products on the shop itself. The packaging which dwindles to fulfill these functions legitimately will lead to enormous wastage owing to the disfigured products and consequently give rise to colossal and completely unavoidable fatalistic environmental repercussions. It has been affirmed that approximately 15% to 25% of the climate impacts of exhaustion are produced by nutrition and food [1]. Packaging also decreases the post contamination and imparts assurance of food safety during marketing. Furthermore, packaging makes an imperative contribution to the acceptability and marketability of products [2]. In food processing, the role of packaging is just one of several ways of preservation of food. The food technologists must have cognizance of food, labor equipment, materials, research, capital, and management. Food safety is defined as cooperation with requirements during production, processing, storage, transport, and distribution of food and taking essential measurements to ensure healthy food production. It is ensuring the safety of materials in contact with food from field to table as the final link of the food safety chain, determination of its effect on food quality and safety, and identification of health risks resulting from these materials are compulsory for the protection of consumer rights. Food safety is an important aspect to maintain our health and well-being. One of the common ways we can be affected is through our consumption of food. The carelessness toward food handling and preparation safety procedures can expose hundreds, thousands, or millions of people to health issues. In case of an outbreak of disease because a particular food has been consumed, correctly packaged and labeling food products will make it easier for the food to be recalled. Figure 3.1 depicts the cycle of food packaging from cradle to grave. Exploration of the impact of food packaging has led to numerous stipulations to sustainability but has substantially focused on the direct influence of materials. The burgeoning of degradable, edible, and virulent-free packaging materials that are safe for the environment and mankind should be a key challenge in terms of regulations and governing the assessment. Biopolymers have recently been instigated as an alternative to replace conventional plastics which pose a critical ecological menace due to the diminution of fossil fuel resources and non-biodegradability.

The objective of food packaging is to maintain safety, to contain food in a cost-efficient way, minimizes the environmental impact that satisfies consumer desires and industrial requirements [3]. The properties of food packaging materials and systems to comprehend cognizance of quality of the essential packaging materials including paper, plastic, metal, and glass which can affect the quality of food and its shelf life has been contemplated. Food packaging can impede deterioration of products, extend shelf life, and preserve propitious impacts of processing, and augment or maintain the safety and quality of food. The construction and design of packaging play an eminent role in estimating the shelf life of food products. The selection of the right packaging technology materials that maintain the freshness and quality of the product during storage and distribution. Recently, food packages frequently juxtapose numerous materials to utilize materials aesthetic and functional properties [4]. Processing of food is the transformation of raw ingredients by chemical or physical means into food. The amalgamation of

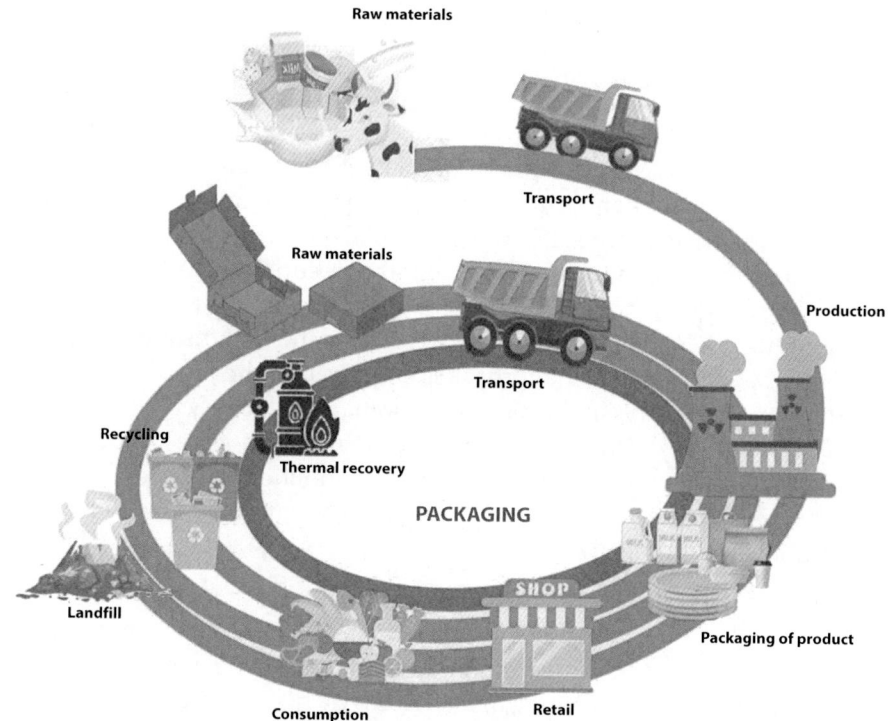

Figure 3.1 Life cycle of food packaging.

raw food materials to develop marketable products that can be effortlessly synthesized and served by consumers. The activities involved in food processing typically are macerating and mincing, emulsification, liquefaction, cooking, pasteurization, pickling followed by preservation, and canning (a type of packaging). Packaging of food has become an essential part of the current civilization; mercenary processed food could not be distributed, handled safely, and efficiently without packaging [5]. This stalwart indagation will shed light on the effectiveness of nanotechnology in food packaging to truncate the putrefaction of food products. Nanotechnology has the inherent potential to persuade the food packaging sector expeditiously. The integration of nanomaterials in biobased polymers to improve the numerous packaging properties such as antimicrobial properties and prevention toward foodborne pathogens by providing a pronounced enhancement in biobased materials. The utmost objective of this robust contrivance is to decipher the sustainable packaging comprised of recycling, biodegradation, stable materials, and designing with the framework of life cycle assessment which has negligible environmental repercussions. Additionally, the amalgamation of sustainability would bestow innovative and new-fangled dimensions in the framework of food security.

3.2 State-of-the Art

The functions of current trends of changed lifestyle and retail practices are the perks for the innovation of novel packaging techniques without compromising food quality and safety [6]. The imperative innovation in food packaging is the rising implications of microbial outbreaks which are foodborne behest the employment of packaging with antimicrobial properties by retaining the quality of food [7]. In the 20th century, the progression in packaging technology emerged as smart, intelligent, or active packaging [8]. Figure 3.2 represents the approaches employed in the food chain. The transpired modifications in the packaging domain will fortify the economy by improvising food quality and safety and by reducing losses of product [9]. Therefore, it is mandatory to analyze the functions of food packaging which can be categorized into four such as containment, preservation, communication, and utility.

> a. *Containment*: It contains product and enable them to store or move; it is a key factor amidst functions of packaging. The products must be contained carefully to avoid contamination from the environment before reaching their end destination. Previously, leaves, baskets, and animal skins were

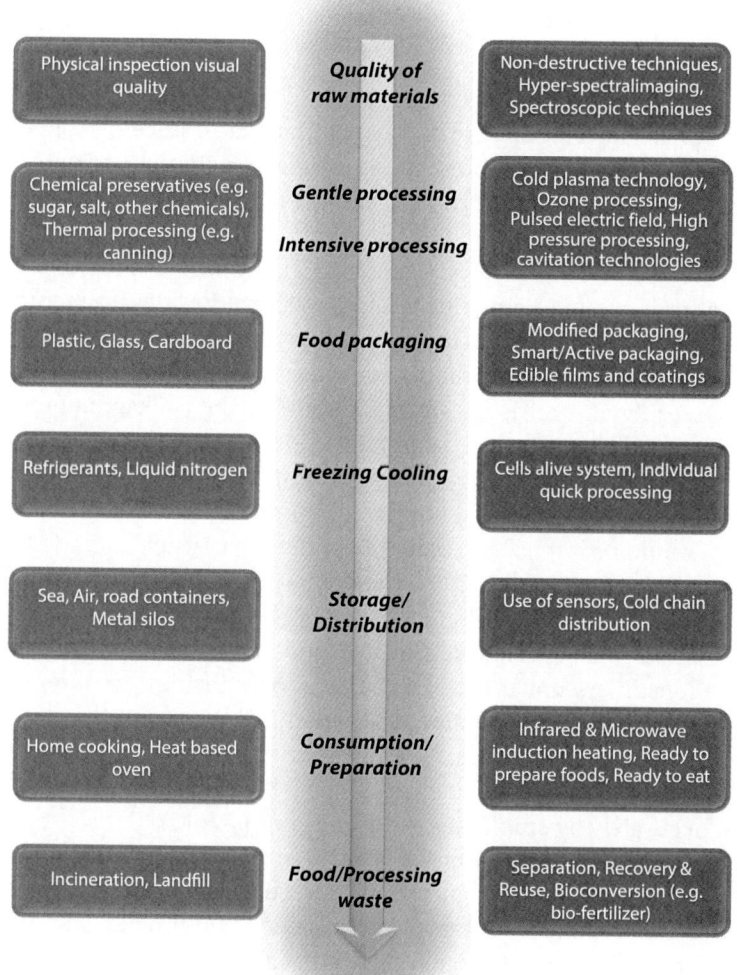

Figure 3.2 Conventional and emerging technologies strategies employed for the food chain.

employed to contain powder, liquid, and grains. The process of containment remarkably contributes to preserving and protecting products while distribution [10].

b. *Preservation*: The damage that processed and fresh food sustain during transportation and storage is of two types—the one is environmental damage which occurs due to exposure of light, microorganisms, odor, gases, and water, and another is physical damage like compressive forces, shock, and vibration. The fine packaging system will bestow protection or minimize the damage to the contents of the package. For instance, an essential flavor or aroma in juice or coffee may easily get evaporated or oxidized without optimum packaging barrier properties. When packaging provides protection, stable or shelf food in a pouch or can sustain stability. Contrarily, for fresh food, the hunch of protection is arduous to attain with packaging alone. Besides, the temperature also influences the degradation in food which can be controlled by modifications in the supply chain such as freezing and refrigeration. Furthermore, the packaging here can add a specific level of protection to regulate the temperature change. Packaging can also influence the mitigation of microbial contaminates at consumption time.

c. *Communication*: The recent techniques of consumer marketing would dwindle if the information communicated by the package through different labeling, distinct branding, and facilitating supermarkets to function is not accurately done. Consumers make procuring decisions utilizing the numerous suggestions provided by different shapes and graphics of the packaging. Other functions of communication include Universal Product Code which can be read rapidly and more accurately by employing scanning equipment.

d. *Utility*: This function bestows packaging attributes that impart value-added and benefits to consumers; for instance, steam in pouch vegetables, pump-action condiments, microwave entrees, and meat pouches that are oven safe, etc.

3.3　Raw Materials Used in Food Packaging

A broad variety of packaging materials has been employed in fresh and processed food to store, distribute, and handle these products from the field to the consumers. The materials such as plastics, glass, metals, and cardboard have been employed to produce containers that depend on the nature of food products as diverse packaging materials have variation in performance characteristics that has a crucial impact on the shelf life of

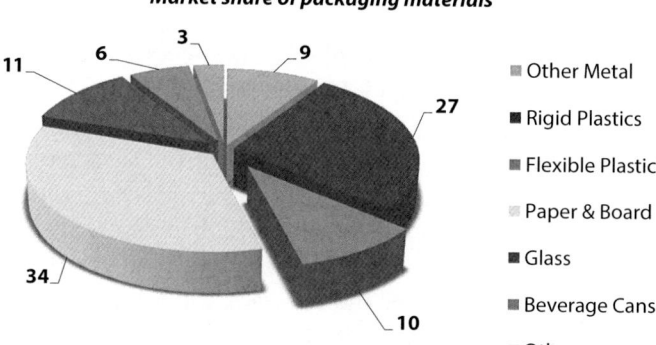

Figure 3.3 The global market share of packaging materials [14].

food [11]. Processed vegetables and fruits are generally packed in metal containers (airtight) to impede oxidation which results in spoilage of food through microbial growth [12]. Glass jars and bottles are usually employed for liquid food packaging, while plastics and cardboard have been used for solid foodstuff [13]. Typically, the packaging materials are of two types: flexible and rigid. Using appropriate packaging will ensure durability and protection during the shelf life of food products. Figure 3.3 illustrates the market share of packaging materials.

3.3.1 Metals

The recyclability and physical protection of metals are broadly preferred in a myriad of food packaging applications such as tuna cans and rack systems. The metals which are commonly employed are aluminum, steel, chromium, and tin. Aluminum is employed in the form of alloys that contains small amounts of manganese and magnesium [15]. Copper is utilized as a wire electrode in containers of three-piece tin plate. The materials like tinplates have been globally sued for the manufacturing of food, beverage closures, and containers owing to significant characteristics like low toxicity, superior gas barrier properties, mechanical strength, light, moisture, and ability to withstand high temperatures. The disadvantage of using aluminum is its high cost in the food packaging system. Contrarily, steel can be reused and recycled several times without losing quality and its cost is comparatively lower, and therefore, it is highly employed in food packaging applications.

3.3.2 Glass

Glass has been employed in packaging where oxygen and moisture barrier properties are required. The glass which is used for packaging systems is generally made up of silica (68%–73%), alumina (1.5%–2%), soda ash (12%–15%), and limestone (10%–13%). The inactive characteristics of glass to edible and non-edible products, such as transparency, rigidity, strong barrier properties, and ability to withstand high pressure, make it useful for packaging applications. The major stumbling block in the glass is its delicacy and heavy weight. Over the 50 years, the weight of jars and bottles has been reduced by 25% to 50%. The chemical property and odorless properties of glass ensure flavor and unimpaired taste. The recyclability and reusability of glass derived packaging materials have fewer environmental implications but their weight increases the cost of transportation of food.

3.3.3 Plastics

Plastics are the most employed material in the food packaging system which integrates the biophysical properties by maintaining nutrition and quality of food products. It can easily be molded into a variety of shapes on regulating pressure and heat at low temperatures. It is easy to maintain transparency in plastic food packaging for visibility of products by which visual quality can be assessed before purchase. However, the permeability toward gases and light is the major limitation of using plastics [16]. The polymer used in plastic packaging is polyethylene, low-density polyethylene, laminated aluminum foil, polypropylene, high-density polyethylene, etc. The interaction and performance of these plastics are different for different foods [17].

3.3.4 Paper and Cardboard

Cardboard and paper are synthesized from cellulosic fibers derived from plant and wood fibers by utilizing sulfite and sulfate [18]. Besides, several fibers have been used in the manufacturing of paper like bamboo, leaves, grasses, flax, and cottonseed hairs. Approximately, 97% of board and paper are synthesized from wood pulp (~85% of wood pulp is used from pines and firs and spruces). The plain paper has poor barrier properties which makes it unsuitable for long-term storage applications. The protective properties of paper can be improvised by filling or laminating and coating with resins or waxes. The high recyclability and low cost are major advantages of employing paper in food packaging applications. These are widely used in paper plates, milk cartons, corrugated boxes, etc. The paper board packaging is enormously used in the horticulture domain. Here, the employment of ventilated packaging is imperative to deliver the cold air inside the package during pre-cooling and refrigerated storage to make fruits and vegetables remain alive after harvesting.

3.4 Packaging Footprints on Quality, Shelf Life, and Safety of Food

The quality of beverages and foods decreases withholding or storage time. In some foods, there is a finite time before it becomes deplorable. The time of production to deplorable is referred to as the "*shelf life*" of the product. Modifications in the flavor of food due to the exchange of undesirable flavors and aroma sorption in polymer-derived packaging materials are important deterioration mechanisms. Vigilant considerations must be accorded to the factors influencing such interactions with the host material to ensure food quality, safety, and shelf-life and to avoid undesirable changes. Besides, storage considerations should also be contemplated like processing methods, temperature, and time [19]. The three prime factors which can affect the shelf life are product characteristics, storage, and packaging material. The food can be decomposed on exposure to high oxygen levels. The loss in the quality of food may be due to the processing at a low temperature. Insufficient shelf life leads to consumer complaints and dissatisfaction which, as a result, affects the sale and acceptance of these products and can produce malnutrition and illness in human beings. Table 3.1 displays the development to enhance the shelf life and quality of food with suitable examples.

3.5 Prolegomenon on Active and Smart Packaging Systems

The ubiquitous globalization among the members of the European Union emanated in a dire need to augment the packaging sector to protect the consumers as well as to preserve our environment and ecosystem selection of a suitable packaging system while maintaining the economical aspect is imperative. Besides, environmental consequences also need to be contemplated while selecting packaging. The innovations in communication technologies and sensors that can impart desirable attributes to packaging facilitate greater interactions between package and product, and it also

Table 3.1 The innovations to increase the quality and shelf life of food.

Product	Primal packaging	Current packaging	Outcome
Grapes	Sold loose	Perforated bags	Leads to a 20% reduction in store waste
Cucumbers	Sold loose	Polyethylene shrinkwrap	3 to 14 days of shelf life extended
Banana	Sold loose	Perforated polyethylene bags	36 days of shelf life extended
Beef	Polystyrene foam wrap	Vacuum packaging in an oxygen barrier film	4 to 30 days of shelf life extended
Cheese	Paper	Polyester tray with polyethylene lid	Waste reduced from 5% to 0.14%
Bread	Paper	Biaxially oriented polypropylene film	Waste reduced from 11% to 0.8%
Bell peppers	Sold loose	Perforated polypropylene film with a modified packaging atmosphere	Can last 20 days in packing

enables the end-user to select the product based on quality and safety in the package. In recent years, smart and active packaging systems have created a quest amidst the research fraternity in the myriad packaging sector [20]. This packaging may extend shelf life and maintenance and monitor the quality of food products. This kind of packaging includes designs that adopt chemical, mechanical, electronic, and electric solutions to impart quality and to bestow information about goods [21].

3.5.1 Active Packaging

Active packaging is the class of materials that emphasize consumer expectations. They are designed to sustain or meliorate the condition of food

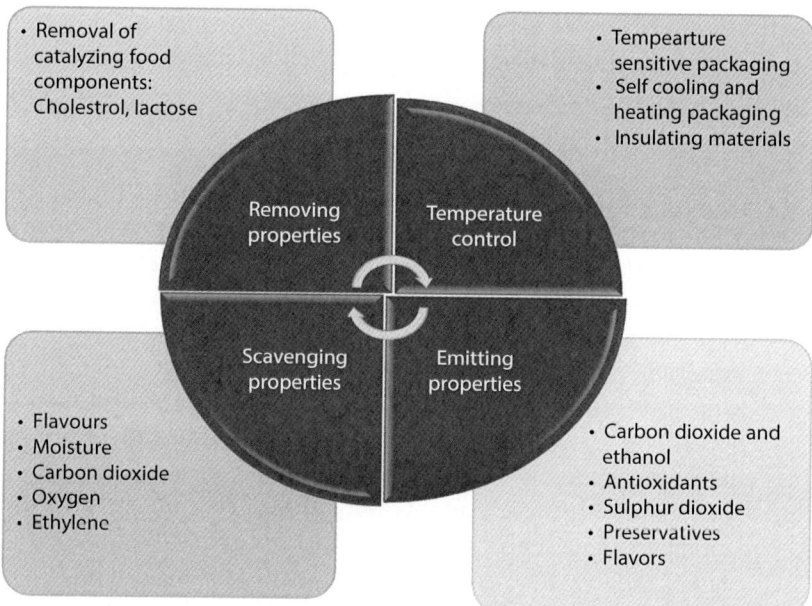

Figure 3.4 Characteristics of active packaging in food applications.

by eradicating unwanted materials from the food or the package and by releasing active components into food. It results in improved safety, an extension of shelf life, maintaining quality, and food attributes [22]. It plays a dynamic role in the protection/preservation of food. The main characteristics of active packaging are to mitigate oxidation, regulating moisture migration, respiration rates, microbial growth, aromas, and volatile flavors. This can also regulate permselectivity which is the selective permeation of packaging material toward gases. Permselectivity, co-extrusion, coating, polymer blending, and lamination can be regulated to meliorate the gases atmospheric concentration inside the package, relative to respiration kinetics, and oxidation of food. Figure 3.4 elucidates the characteristic features of the active packaging system.

3.5.2 Intelligent Packaging System

It refers to the employment of packaging as an intelligent messenger to provide and regulate the quality information and condition of packed foods to consumers. Microbial growth, temperature, pack integrity, and product authenticity are examples of indicators employed in intelligent packaging. Leakage and freshness indicators are commercially available for quality monitoring. This packaging includes gas emission indicators, time-temperature indicators, sensors, physical shock indicators, as well as theft and counterfeit prevention methods [23]. Typically, there are three prime factors employed in intelligent packaging technologies:

   a. *Quality indicator:* It directly regulates the quality attributes in the food itself; for instance, freshness indicators and biosensors. These devices are incorporated into the packaging.
   b. *Environmental conditions*: It regulates the conditions which lead to a modification in food characteristics quality. For example, relative humidity sensors, gas leakage indicators, and time-temperature indicators, but these indicators depend upon monitoring by which it can allocate outside or inside in packaging.
   c. *Data carriers:* These systems are utilized to transfer and store data while sensors and indicators are employed to regulate the external environment and convey the information afterward [24, 25]. The frequently employed data carriers are Radio Frequency Identification Technology (RFIT) tags and barcodes.

The intelligent packaging system generally increases the safety of the product and diminishes the unnecessary wastage of food [26, 27]. Furthermore, consistent quality also decreases the cost of material and time in the analysis of packaging food. The intelligent systems must be compatible with the food to be regulated or monitored. Hence, it must be clarified which indicator or sensor is appropriate for the specific food.

3.6 Aseptic Packaging in Food Processing

This technique thermally sterilizes liquid food products at elevated temperature and then cooling followed by packing them into sterilized containers to develop shelf-stable products under sterile conditions that do not require refrigeration. The sterilization of food products ensures no microorganisms and eradicates the decaying or spoiling of food. There are three essential steps in aseptic processing: sterilization of packaging material, thermal sterilization of product, and during packaging, conservation of sterility. Aseptic packaging has low gas transmission rates and low water vapor transmission rate to oxygen which is required to preserve color, nutritional constituents, and the flavor in the products. There are three major advantages of employing aseptic packaging as follows:

a. To increase the shelf life at normal temperature of food products by aseptic packaging.
b. To employ high-temperature short-time processes for sterilization which usually produces superior quality food products and are thermally efficient in contrary to the products processed for a longer time at lower temperatures.
c. To facilitate containers to be employed which are unsuitable for in-package sterilization.

The two specific functions of aseptic packaging are one is non-sterile packaging of the product to circumvent infection of microorganisms, and another is pre-sterilized and sterile packaging of the product. For instance, these applications include fruits, vegetable sauces, desserts, pudding, dairy products, soups, and products with particulates (fresh products such as dairy products like yogurt which are fermented).

3.7 The Paradigm in Strategies for Improvement of Food Packaging

Owing to the commercialization of trends and escalating consumer demands, packaging has an eminent role in preserving consumer goods. The inquisitiveness in methods to meliorate the safety and quality of food, as well as the food supply management, is gigantic. There is a progression in exigency for the information provided on food and packaging products.

3.7.1 Bequest of Packaging Into the Cycle of Food Chain Sustainability

To produce innovative value-added services, the supply chain imparts provision for enhancing efficiency by automating valuable and simple data flows. The progression in attention toward sustainable packaging results in the employment of relationships of the packaging value chain for competitive advantage and to augment the evolving role of the packaging of foodservice. The frequently employed approach in estimating sustainability in the packaging of food is the life cycle assessment. It is a technique to evaluate environmental burdens and resource consumption associated with the process, product, or activity [28]. Figure 3.5 shows the development

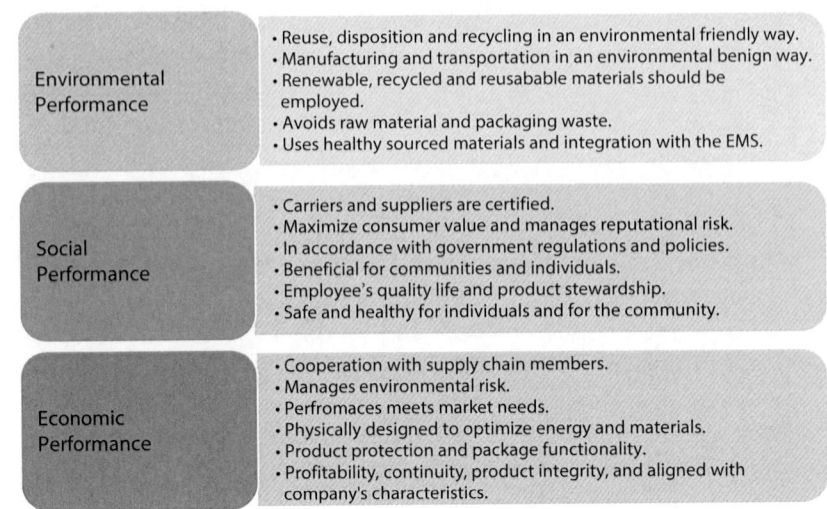

Figure 3.5 The framework of dimensions of the food packaging industry of supply chain.

of supply chain sustainability which is associated with mainly three factors such as environmental, social, and economic performance. Therefore, the appropriate design of the packaging should contribute to the three dimensions of sustainability in the supply chain [29, 31]. The best packaging selection from an economic point of view is connected with an ameliorated reduction in cost and sales. The packaging should be fabricated and designed to enhance product efficiency at the productive and logistic level. The efficiency in the design of packaging could be contemplated in terms of the influence of processes such as packing, handling, supplying, transporting, and storing [32, 33].

3.7.2 Selection of Materials With the Objective of Recyclability

The sustainability of materials can be attained by three factors: (a) production level via energy-efficient processes, (b) raw materials, and (c) waste management level which considers biodegradation, recycling, and reuse. The recycling transfer packaging is by the waste stream to recovery through processing, sorting, collection, and fabrication into recycled products. Plastic packaging recyclability depends on the efficiency of sorting and collection systems. To eradicate hazardous implications on the environment, the emphasis is to augment the recyclability of plastic packaging which implies redesigning packaging via end-of-life approaches. The adoption of materials such as high-performance composites has contributed to eradicate the amount of packaging. Composite films are fabricated to amalgamate different materials performances into a single component. Specifically, multi-layered structures like composites are designed to enhance the moisture and gas barrier properties of films in applications to modify vacuum or atmosphere packaged food and also employ in moisture sensitive foods which facilitate shelf life extension by reducing the transfer of moisture and keeping the modified moisture for a long time respectively. The plastic packaging domain has the responsibility to regress materials that assure good recyclability, e.g., mono polyethylene terephthalate (PET).

3.7.3 Escalating Protective Role of Packaging

The packaging system designs implicit modification of packaging functions for elective requirements in food. The protective function in packaging can be enhanced by utilizing the novel concepts of active packaging which includes devices and materials capable to proffer auxiliary protection to the packaged food by tailoring interactions among package and food itself. The packaging potential has been broadened by active packaging to maintain the quality of food and to extend the shelf life of perishable and fresh food. Protection in packaging can be provided by external factors like (i) physical, (ii) chemical, and (iii) biological.

(i) The physical protection preserve food from mechanical destruction which includes cushioning against vibration and shock encountered during distribution. Corrugated and paperboard materials are usually employed in this type of packaging and can be used as the packaging of delicate foods such as fresh fruits and eggs and also in shipping containers.

(ii) The chemical protection reduces the change in compositions triggered by environmental factors such as light, moisture, and exposure to gases like oxygen. Metals and glass provide chemical barrier properties and resistance toward environmental agents. For instance, plastic caps have permeability toward vapors and gases.

(iii) Biological protection facilitates barrier properties toward microorganisms such as insects, pathogens, animals, rodents, and spoiling agents by preventing spoilage and disease. Furthermore, biological protection controls conditions of aging and ripening, also called senescence.

3.7.4 How Biodegradable Polymers can Mitigate the Plight of Packaging in Food Processing

Traditionally, food packaging materials are dependent on non-renewable materials. Bio-based materials are referred to as materials derived from renewable resources which is the prime target to achieve sustainability. The employment of synthetic plastics has posed a critical threat to the ecological plight owing to the depletion of fossil resources and nondegradability. Currently, biobased plastics are fabricated from biopolymers like cellulose, starch, and monomers produced by fermented organic materials. The carbohydrates such as polysaccharides consist of monosaccharides joined by glycosidic bonds together and are the most abundant materials in nature. They have a potential role in plant energy storage like starch. These polysaccharides have also been employed in edible packaging and are economically viable, easy to handle, abundant in nature with excellent film forming properties. Starch, a mixture of amylose composed of monomer α-(1→4)-D-glucopyranosyl monomers and amylopectin which is highly branched containing amylose backbone. They are used as protective edible coatings as encapsulating agents and food to impart oxygen and lipid barrier properties to enhance handling, texture, and appearance. The edible starch coatings and films are frequently employed in batters, confectionaries, bakeries, and products made up of meat [34]. On other hand, cellulose is composed of linear chains of β-(1→4)-D-glucopyranosyl monomers with no branching or coiling. The several hydroxyl moieties in celluloses make it water insoluble in an aqueous solution and strong hydrogen bonding results in the strong physical network which turns into non-fusible material. Methyl cellulose has high solubility, outstanding film making properties, and good lipid barriers properties. They are commonly used to impart barriers toward moisture oil and oxygen. Hemicellulose and its derivatives have been used as additives to fabricate edible coating composite films in wheat gluten [35]. Figure 3.6 represents the classification of biobased polymers which can be sued in food packaging applications.

The most abundant polysaccharide, chitosan, is found in the cell wall of insect cuticles and the exoskeleton of arthropods. It is composed of N-acetyl-D-glucosamine groups, an acetylated polysaccharide attached by β-(1→4)-linkages. It is non-antigenic and non-toxic polysaccharide derived from the deacteylation process with alkalis from chitin. Due to broad antimicrobial properties by which *gram positive* and *gram negative* bacteria are highly susceptible. The influence of coatings of chitosan on the fresh products shelf life like carrots, strawberries, pineapple, mushroom, and mango have been illustrated and results elucidated that shelf life was increased

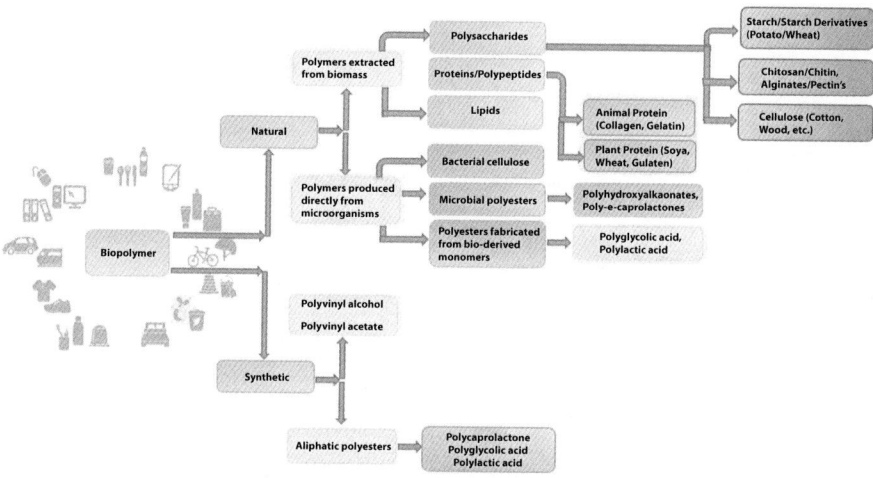

Figure 3.6 The classification of biopolymers [36].

and antimicrobial growth was inhibited [37]. Alginates are obtained from brown seafeeds (*Phaeophyceae* family) and alginic acid salts composed of linear copolymer of L-guluronic and D-mannuronic acids. Alginate coatings are helpful in reducing desiccation in enrobed meats which act as sacrificing agents and have good oxygen barrier properties and also protect food products against oxidation.

3.8 Integration of Nanotechnology to Ameliorate Food Packaging

Nanotechnology has emerged as the most propitious ingenious technique by facilitating oxygen barrier, mechanical and antimicrobial properties, and by detecting pathogens. Nanotechnology has improved physiochemical properties due to high aspect ratio and small size of nanoparticles. The employment of nanomaterials in packaging enhances thermal stability, electrical properties, processability, durability, and optical properties by imparting active packaging functions [38, 39]. Presently, in the market, aluminum nanolayer is being utilized to coat the interior surface of food packaging. The nanodimensions of cellulose and chitosan have been employed as food additives. The obtained nanoparticles may have diverse properties, morphologies, and dimensions as per the cellulose broke down, they are referred as cellulose nanocrystals. These materials have been

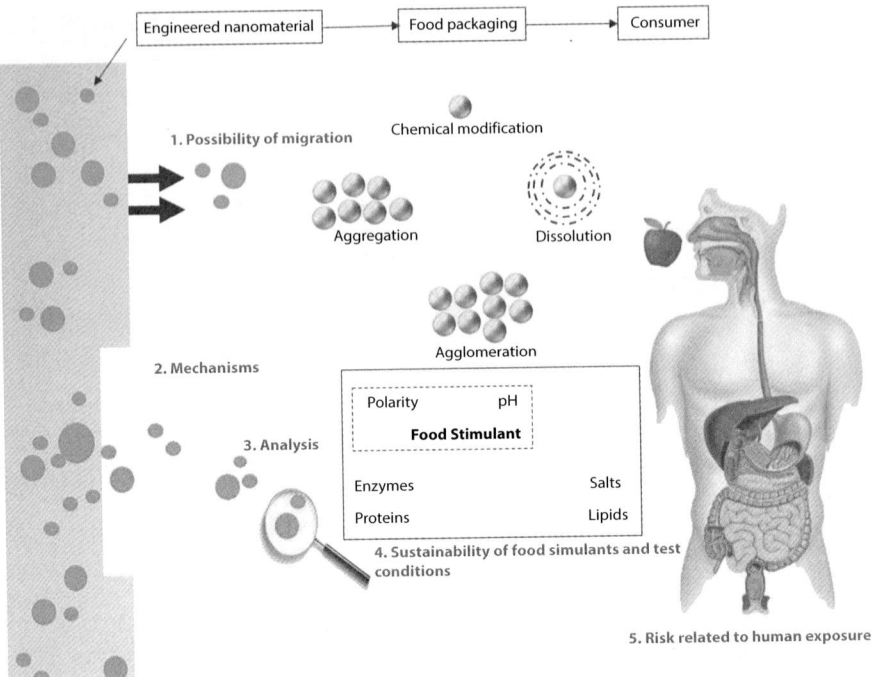

Figure 3.7 Probable harmful effects of exposure of nanoparticles on human health.

employed to enhance the charge density in packaging applications [40]. Nanotechnology-derived packaging has inherent potential to develop effectiveness and efficiency in food packaging. The most imperative enhancement is water and gas barrier properties of materials employed to package food products. The utilization of nanomaterials has been employed to enhance barrier properties and mechanical performance. Nanoparticles impart a tortuous pathway for water and gas molecules which are enforced to traverse long path to diffuse through the material. The employment of which is applied in the form of thin coating on several substrates and results in reducing the gas permeability and enhancing transparency, anti-fog properties, and coefficient of friction [41]. Figure 3.7 displays the influence of nanotechnology and exposure of nanoparticles on human health.

3.9 Life Cycle Assessment (LCA)

The process to examine environmental burdens that are associated with the product by utilizing materials, wastes, or emissions exposed throughout the cycle by quantifying the energy. The general framework and minimum standards were outlined by ISO 14044 and 14040 for the execution of an assessment of the life cycle of packaging [42, 43]. In packaging design, life cycle assessment has become a decision-supporting tool. The Life Cycle Initiative to harmonize LCA on international level was conducted to bestow science based global forum, building consensus process [44]. The European Commissions stated the recommendations of employment of life cycle assessment of materials in 2013. The methods of end of life allocation and impact assessment are included in these recommendations [45, 46]. There are two parameters of packaging of food waste and loss in life cycle analysis:

a. Analysis of food packaging ratio.
b. Inclusion of wasted and lost food in LCA of packaging.

3.10 Deciphering the Challenges for Sustainable Food Packaging

The inclusion of sustainability should be a supplementary and futuristic dimension of food security. A multidisciplinary strategy is requisite to perceive social norms, consumption patterns, lifestyles, and behavior. This strategy will assist in developing acceptable transitions to minimize waste, lessen ecological footprints, and emissions of carbon to evolve food ecosystems for sustainable urbanization.

- The major challenge is to acquire agreement on sustainability parameters for food processors. This comprises the exigency to address the life cycle analysis of processed foods and food processes as elucidated by Smetana *et al.* [47]. The matter of transportation of raw materials and food also needs enlarged attention.
- Another challenge is the escalating urban population and metamorphosis in urban-rural linkages. This needs the inclusion of better connectivity in cities and rural sectors while producing opportunities for rural food security, employment, and economic development.
- The corroboration of health-related contributions and nutrition of processed food is in dire need [48]. The extensive communication and interaction amidst the field of nutrition and food science are requisite.

➢ Futuristic resource-efficient processing of food will also require to emphasize and take dominancy of the existing biosystems from food-microbiome human interactions as well as from microorganisms-host (animal, plant, and human). It requires to include edible microbial biomass as well as techniques and enzymes or the employment of the fermentation process to eradicate toxins [49–51].

3.11 Conclusion and the Way Forward

The packaging is an imperative segment of the food system and plays an analytical in preservation, conservation, and containing food from the field to consumers. The primary objective of food packaging is to maintain the quality, safety, and wholesomeness of food. The implications of packaging on environment can be minimized by selecting potential materials and following guidelines of EPA (Environmental Protection Agency). Nanotechnology will play eminent role in future by imparting enhancement in properties. The intelligent and active packaging system can address the safety as well as quality issues to bestow eco-friendly packaging materials. With the advent of active food packaging, innovation in stimuli-responsive polymers provide functional features in the food packaging applications. The emerging trends in biodegradable packaging results in improved quality and safety of food by reducing environmental repercussions. The design of packaging should employ life cycle thinking approach and recyclability with the objective of reducing environmental impacts.

Acknowledgement

The authors would like to pay gratitude toward Bhaskaracharya College of Applied Sciences, University of Delhi, India.

References

1. Seppälä, J., Mäenpää, I., Koskela, S., Melanen, M., Katajajuuri, J.-M., Nissinen, A., Virtanen, Y., Usva, K., Myllymaa, T., Härmä, T., Environmental impacts of material flows caused by the Finnish economy Market Opportunities in Life Cycle Thinking, in: *First symposium of the Nordic Life Cycle Association*, October 9-10, 2006, Lund, Sweden/Jørgensen, Andreas, Molin, SE, 2006. https://jukuri.luke.fi/handle/10024/463910

2. Yusof, N.A.A., Zain, N.M., Pauzi, N., Synthesis of ZnO nanoparticles with chitosan as stabilizing agent and their antibacterial properties against Gram-positive and Gram-negative bacteria. *Int. J. Biol. Macromol.*, 124, 1132–11365, 2019.
3. Verghese, K., Lewis, H., Lockrey, S., Williams, H., *The role of packaging in minimising food waste in the supply chain of the future*, RMIT University, Melbourne, Australia, 2013.
4. Shin, J. and Selke, S.E.M., 11 Food Packaging, in: *Food Processing: Principles and Applications*, Second Edition, S. Clark, S. Jung, B. Lamsal, (Eds.), pp. 249–273, John Wiley & Sons, Menomonie, Wisconsin, USA, 2014.
5. Restuccia, D., Gianfranco Spizzirri, U., Parisi, O.I., Cirillo, G., Curcio, M., Iemma, F., Puoci, F., Vinci, G., Picci, N., New EU regulation aspects and global market of active and intelligent packaging for food industry applications. *Food Control*, 21, 11, 1425–1435, 2010.
6. Dainelli, D., Gontard, N., Spyropoulos, D., Zondervan-van den Beuken, E., Tobback, P., Active and intelligent food packaging: legal aspects and safety concerns. *Trends Food Sci. Technol.*, 19, S103–S1125, 2008.
7. Appendini, P. and Hotchkiss, J.H., Review of antimicrobial food packaging. *Innovative Food Sci. Emerg. Technol.*, 3, 2, 113–126, 2002.
8. Brody, A.L., Bugusu, B., Han, J.H., Koelsch Sand, C., McHugh, T.H., Innovative food packaging solutions. *J. Food Sci.*, 73, 8, 107–116, 2008.
9. Vanderroost, M., Ragaert, P., Devlieghere, F., De Meulenaer, B., Intelligent food packaging: The next generation. *Trends Food Sci. Technol.*, 39, 1, 47–62, 2014.
10. Shin, J. and Selke, S.E.M., 11 Food Packaging, in: *Food Processing: Principles and Applications*, Second Edition, S. Clark, S. Jung, B. Lamsal, (Eds.), pp. 249–273, John Wiley & Sons, Michigan, USA, 2014.
11. Robertson, G.L., Packaging materials for biscuits and their influence on shelf life, in: *Manley's Technology of Biscuits, Crackers and Cookies*, pp. 247–267, Woodhead Publishing, University of Queensland, Australia, 2011.
12. Robertson, G.L., Packaging and food and beverage shelf life, in: *The Stability and Shelf Life of Food*, pp. 77–106, Woodhead Publishing, Brisbane, Australia, 2016.
13. Newsome, R., Balestrini, C.G., Baum, M.D., Corby, J., Fisher, W., Goodburn, K., Labuza, T.P., Prince, G., Thesmar, H.S., Yiannas, F., Applications and perceptions of date labeling of food. *Compr. Rev. Food Sci. Food Saf.*, 13, 4, 745–769, 2014.
14. https://www.foodpackagingforum.org/food-packaging-health/food-packaging-materials
15. Marsh, K. and Bugusu, B., Food packaging—roles, materials, and environmental issues. *J. Food Sci.*, 72, 3, R39–R55, 2007.
16. Pathare, P.B., Opara, U.L., Al-Julanda Al-Said, F., Colour measurement and analysis in fresh and processed foods: a review. *Food Bioprocess Technol.*, 6, 1, 36–60, 2013.
17. Siracusa, V., Food packaging permeability behaviour: A report. *Int. J. Polym. Sci.*, 1–11, 2012, 2012.

18. Robertson, G.L., *Food packaging: principles and practice*, CRC Press, Australia, 2016.
19. Behe, B.K., Campbell, B.L., Khachatryan, H., Hall, C.R., Dennis, J.H., Huddleston, P.T., Thomas Fernandez, R., Incorporating eye tracking technology and conjoint analysis to better understand the green industry consumer. *HortScience*, 49, 12, 1550–1557, 2014.
20. Dainelli, D., Gontard, N., Spyropoulos, D., Zondervan-van den Beuken, E., Tobback, P., Active and intelligent food packaging: legal aspects and safety concerns. *Trends Food Sci. Technol.*, 19, S103–S1125, 2008.
21. Kerry, J. and Butler, P. (Eds.), *Smart packaging technologies for fast moving consumer goods*, John Wiley & Sons, USA, 2008.
22. Yam, K.L., Takhistov, P.T., Miltz, J., Intelligent packaging: concepts and applications. *J. Food Sci.*, 70, 1, R1–R10, 2005.
23. Dobrucka, R. and Cierpiszewski, R., Active and intelligent packaging food–research and development–a review. *Polish J. Food Nutr. Sci.*, 64, 1, 7–15, 2014.
24. Heising, J.K., Dekker, M., Bartels, P.V., Van Boekel, M.A.J.S., Monitoring the quality of perishable foods: opportunities for intelligent packaging. *Crit. Rev. Food Sci. Nutr.*, 54, 5, 645–654, 2014.
25. Han, J.H., Ho, C.H.L., Rodrigues, E.T., Innovations in Food Packaging-PDF Free Download, Elsevier Book, 2005.
26. Verghese, K., Lewis, H., Lockrey, S., Williams, H., Packaging's role in minimizing food loss and waste across the supply chain. *Packag. Technol. Sci.*, 28, 7, 603–620, 2015.
27. Müller, P. and Schmid, M., Intelligent packaging in the food sector: A brief overview. *Foods*, 8, 1, 16, 2019.
28. Heller, M.C. and Keoleian, G.A., Assessing the sustainability of the US food system: a life cycle perspective. *Agric. Syst.*, 76, 3, 1007–1041, 2003.
29. Azzi, A., Battini, D., Persona, A., Sgarbossa, F., Packaging design: general framework and research agenda. *Packag. Technol. Sci.*, 25, 8, 435–456, 2012.
30. Grönman, K., Soukka, R., Järvi-Kääriäinen, T., Katajajuuri, J.-M., Kuisma, M., Koivupuro, H.-K., Ollila, M. *et al.*, Framework for sustainable food packaging design. *Packag. Technol. Sci.*, 26, 4, 187–200, 2013.
31. Molina-Besch, K. and Pålsson, H., Packaging for eco-efficient supply chains: why logistics should get involved in the packaging development process, in: *Sustainable Logistics*, Emerald Group Publishing Limited, 2014.
32. Hellström, D. and Saghir, M., Packaging and logistics interactions in retail supply chains. *Packag. Technol. Sci.: An International Journal*, 20, 3, 197–216, 2007.
33. Sohrabpour, V., Oghazi, P., Olsson, A., An improved supplier driven packaging design and development method for supply chain efficiency. *Packag. Technol. Sci.*, 29, 3, 161–173, 2016.
34. Janjarasskul, T. and Krochta, J.M., Edible packaging materials. *Annu. Rev. Food Sci. Technol.*, 1, 415–4485, 2010.
35. Hansen, N.M.L. and Plackett, D., Sustainable films and coatings from hemicelluloses: a review. *Biomacromolecules*, 9, 6, 1493–1505, 2008.

36. Sharma, S., Sharma, B., Manral, A., Bajpai, P.K., Jain, P., Biopolymers in the automotive and adhesive industries, in: *Biopolymers and their Industrial Applications*, pp. 261–280, Elsevier, India, 2021.
37. Tamer, C.E., Incedayi, B., Yönel, S.P., Yonak, S., Copur, O.U., Evaluation of several quality criteria of low calorie pumpkin dessert. *Not. Bot. Horti Agrobot. Cluj-Napoca*, 38, 1, 76–80, 2010.
38. Han, J.W., Ruiz-Garcia, L., Qian, J.P., Yang, X.T., Food packaging: a comprehensive review and future trends. *Compr. Rev. Food Sci. Food Saf.*, 17, 860–877, 2018.
39. Sharma, C., Dhiman, R., Rokana, N., Panwar, H., Nanotechnology: an untapped resource for food packaging. *Front. Microbiol.*, 8, 1735, 2017.
40. Zheng, J., Aziz, T., Fan, H., Haq, F., Khan, F.U., Ullah, R., Ullah, B., Khattak, N.S., Wei, J., Synergistic impact of cellulose nanocrystals with multiple resins on thermal and mechanical behavior. *Z. Phys. Chem.*, 1, no. ahead-of-print 1–16, 2020.
41. Li, F., Mascheroni, E., Piergiovanni, L., The potential of nanocellulose in the packaging field: a review. *Packag. Technol. Sci.*, 28, 6, 475–508, 2015.
42. ISO, *Environmental Management—Life Cycle Assessment—Requirements and Guidelines*, International Organization for Standardization, Geneva, Switzerland, 2006.
43. ISO, *Life Cycle Assessment–Principles and Framework: 14040:2006*, International Organization for Standardization, Geneva, Switzerland, 2006.
44. Ren, J. and Toniolo, S. (Eds.), *Life Cycle Sustainability Assessment for Decision-making: Methodologies and Case Studies*, Elsevier, Hong Kong, 2019.
45. EC-European Commission, Commission Recommendation of 9 April 2013 on the use of common methods to measure and communicate the life cycle environmental performance of products and organisations, in: *ANNEX II: Product Environmental Footprint (PEF) Guide*, European Commission, Europe, 2013.
46. Wolf, M. and Chomkhamsri, K., *The "Integrated formula" for modelling recycling, energy recovery and reuse in LCA-White paper*, Berlin, Germany, 2014.
47. Metcalfe, R., *Food Routes: Growing Bananas in Iceland and Other Tales From the Logistics of Eating*, MIT Press, Cambridge, MA, 2019.
48. Sybesma, W., Blank, I., Lee, Y.-K., Sustainable food processing inspired by nature. *Trends Biotechnol.*, 35, 4, 279–281, 2017.
49. Raman, Z., Arjun S., J.L., Venkatesh, S., Chang, H.-W., Hibberd, M.C., Subramanian, S., Kang, G. *et al.*, A sparse covarying unit that describes healthy and impaired human gut microbiota development. *Science*, 365, 6449, eaau4735, 2019.
50. Linder, T., Edible Microorganisms—An Overlooked Technology Option to Counteract Agricultural Expansion. *Front. Sustain. Food Syst.*, 3, (325, 2019.
51. Everett, J.A.C., The 12 item social and economic conservatism scale (SECS). *PLoS One*, 8, 12, e82131, 2013.

4
Laws Impacting Chemicals Added to Food

Preeti Khanna[1], Rajni Chopra[2*] and Meenakshi Garg[3]

[1]Department of Food and Nutrition, Institute of Home Economics, University of Delhi, New Delhi, India
[2]National Institute of Food Technology Entrepreneurship and Management, Sonepat, Haryana, India
[3]Department of Food Technology, Bhaskaracharya College of Applied Sciences, University of Delhi, New Delhi, India

Abstract

Food additives are defined as chemicals added to the food products. The main classes of food additives include antioxidants, colors, flavor enhancers, sweeteners, emulsifiers, stabilizers, and preservatives. These additives help in sustaining and enhancing shelf life, nutritional quality and sensory attributes of a food product. All additives are thoroughly assessed for safety before they are allowed to be used for consumption by the general public. The joint FAO/WHO expert committee on Food Additives (JECFA) is the apex body for assessing the health risks associated with them. Food additives which have undergone a JECFA safety assessment and do not pose any potential health risks are certified by JECFA for technological needs. Irrespective of the origin JECFA safety assessment applies to all the food additives. After the safety assessments are completed by JECFA, Codex Alimentarius Commission uses them to establish the maximum levels of usage in food and beverage items. Safety assessment of all additives is performed before they are considered to be used in specified quantities. These quantities are based on an acceptable daily intake (ADI) from the results of safety tests. The current chapter highlights the national and international laws associated with food additives used in the food industry.

Keywords: Additives, food regulations, JECFA, codex, USFDA, FSSAI

Corresponding author: rajnichopra.niftem@gmail.com

Mousumi Sen (ed.) Food Chemistry: The Role of Additives, Preservatives and Adulteration, (97–116)
© 2022 Scrivener Publishing LLC

4.1 Introduction

Chemicals added into the edible products are categorized as food additives. "World Health Organization [1] defines food additives as substances that are added to food to maintain or improve the safety, freshness, taste, texture, or appearance of food are known as food additives". They are either derived from natural sources like plant and animal cells and minerals or can be synthetically processed. They are purposely added to the foods because they serve the need for some technological purposes, like making a food item more appealing. Traditionally, food additives have been used for the purpose of preservation such as salt (in pickles, chutneys, and dehydrated and canned fish), sugar (in jam, jellies, and fruit preserves), or sulfur dioxide (alcoholic beverages like wine).

Over the period of the time, as the large-scale production of the processed food increased, the need for food additives also increased. Food additives have an important role to play in large-scale production and manufacturing of the processed food items. In order to ensure the safety and maintenance of shelf life and sensory attributes of a food product until it reaches its desired consumers, food additives are added to processed food items.

Chemicals added to the food or food additives are characterized by the following attributes:

a. Its usability is a technological necessity.
b. It does not misguide the consumers.
c. It performs technological functions such as to preserve the sensory attributes of a product (for example, coloring agent), to enhance the taste and flavor profile of a product (flavoring agent), to enhance the nutrient quality of a product (fortificants), and to expand the shelf life of a product (preservatives)

4.2 Functions of Food Additives

Additives perform a variety of useful technological functions in the food industry. They also improve the aesthetic appeal of a product and also provide convenience benefits to the consumers. The functions of the food additives are listed below.

4.2.1 Sustain or Enhance the Shelf Life and Freshness of a Product

Some additives like preservatives can slow down the product spoilage caused by air and microorganisms. For example, "antioxidants are a class of preservatives which prevent rancidity and production of off flavor and odor in fats and oils and products made with them as the base ingredients". Antioxidants also prevent enzymatic browning in cut apple slices and prevent them from turning brown.

4.2.2 Sustain or Enhance the Nutritional Quality of a Product

Fortificants are a group of additives which are added to a food product for enrichment and restoration of nutrients (vitamins and minerals) lost during processing or which were not naturally present in the food product. For example, addition of wheat bran to wheat flour restores the fiber content lost during processing of wheat and addition of vitamins and minerals to breakfast cereal enhances it nutritional quality.

4.2.3 Improve the Aesthetic Appeal and Sensory Attributes of a Product

Natural, nature-identical, and artificial coloring and flavoring agents are used as food additives to enhance the aesthetic appeal and sensory characteristics of a product. Also, food additives like emulsifiers, stabilizers, and thickeners provide the food with the desired texture and consistency. On the other hand, additives and chemicals like leavening agents allow the baked products to rise during baking, releasing carbon dioxide gas and making the product soft, spongy, and fluffy.

4.3 Classification of Food Additives

4.3.1 Classification Based on Functionality

Based on their functionality WHO-FAO classifies food additives into three broad categories, namely,

    a. Flavoring agents
    b. Enzyme preparations
    c. Other additives

4.3.1.1 Flavoring Agents

"Flavoring agents are food additives which improve aroma, taste and flavor of a food product". They are of prime importance in bakery and confectionary industries for enhancing the taste profile of the products. Flavoring agents are further divided into three classes, based on their source of origin:

    a. Natural flavor
    b. Nature-identical flavors
    c. Artificial flavors

4.3.1.2 Enzyme Preparations

"Enzyme preparations are a group of additives that boost the biochemical reactions by breaking down the larger molecules into their smaller precursors". They are naturally occurring proteins or they can be obtained by extraction from plants, animals, or microorganisms such as bacteria. Enzyme preparations find a number of uses in many important processes in the food industry for manufacturing of various food products. The main functions are, as dough conditioners in bakery products, to increase the yield of fruit based beverages, to enhance the process of fermentation in wine making and brewing, and to improve the curd formation process in the cheese manufacturing industries.

4.3.1.3 Other Additives

In the food industry, "additives are used to perform other technological agents to perform functions like preservation, coloring and balance out sweetness". They can be added to a food when it is prepared, packed, transported, stored, and ends up becoming a component of the particular food product.

    Preservatives can enhance the shelf life of a product by slowing the rate of decomposition caused by air, molds, bacteria, and yeast. They also help in controlling contamination which can lead to foodborne illness like botulism (food poisoning) caused by Clostridium botulinum.

    Coloring agents are added to enhance the sensory attributes and appearance of a food product by replacing the natural coloring pigments lost during preparation and processing of a food product.

No-calorie (or sugar) sweeteners are considered as an alternative to sugar as they can maintain the sweetness of a food product and contribute fewer or no calories when added to food.

4.4 Classification Based on Primary and Secondary Technological Roles—Direct and Indirect Additives

Direct food additives are added intentionally to a food product to perform a specific technological need and purpose. They are identified on the ingredient label of the food item to which they are added. For example, Aspartame is a low-calorie sweetener, which is added intentionally to sugar sweetened beverages, puddings, soft drinks, yogurt, and breakfast cereals in order to reduce the calorie content of these products [2].

Indirect additives are not added but form constituent of the food product in minute amounts during cycles of manufacturing, packing, and storing. They may or may not perform and fulfill any technological needs. They are usually by-products of agriculture and packaging materials. Harmful chemicals used in food products and minute particles that are adsorbed on food contact points and packaging material into the food products are some of the examples of indirect food additives [2].

4.5 Evaluating the Health Risk of Food Additives

The joint FAO/WHO expert committee on Food Additives (JECFA), an international and independent expert scientific group, is responsible for assessing the risks of food additives on human health. Food additives which have been assessed by JECFA safety assessment and do not pose any potential health risks to the consumers are certified by JECFA for technological needs. JECFA safety assessment applies to all the food additives irrespective of their origin.

JECFA safety assessment certifications are validated by scientific reports and data and on the biochemical, toxicological, and other relevant information available for usage of a particular additive like animal studies and clinical and research trials in humans. As per JECFA's requirements the toxicological data must include specific brief and extended time span studies that evaluate and study the absorption, distribution, excretion, and other possible health risks of additives or their by-products at exposure (specific) limits.

The first step in safety assessment procedure is to establish the acceptable daily intakes (ADIs) for that particular food additive. "The ADI is an estimate of the amount of a food additive present in a food product (including beverages and drinking water) that is safe for daily consumption over lifetime without any adverse effects on the health of the consumers" [1].

In addition to JECFA assessment, other international and national regulatory bodies work in conjunction to authorize and establish the efficacy of food additives at specified levels for usage in particular food products based on the health safety risk assessment.

4.6 International Regulations for the Efficacy of Food Additives

Once the safety approvals are completed by JECFA for the use and efficacy of particular food additive, Codex Alimentarius Commission (joint intergovernmental food standard setting body of FAO and WHO) uses them to further establishes the maximum levels of additives added in food and beverage items.

Codex norms and standards are the benchmark standard limits for protecting the consumer and safety regulations for international trade in food and beverage industry. Codex standards further ensure that food product meets international standards for safety and quality, irrespective of the place of manufacture.

Once the safety and efficacy of a food additive has been established by JECFA and the maximum amounts for addition to any product as prescribed by Codex General Standard for Food Additives, national food regulatory bodies have to further implement national laws permitting the use of that particular additive.

4.7 International Laws

4.7.1 US Food and Drug Administration

The Food Additive Amendment (1958) to the Food, Drug, and Cosmetic Act (FDCA) authorizes Food and Drug Administration (FDA) to enquire and gather data from producers exemplifying that the additive does not have any harmful effects before being added into the food products [2].

US food regulations determine food ingredients by the intentional use in particular food categories, people who are expected to consume those foods and anticipated health claims made, if any. Ingredients can be added legally either as direct or indirect additives once they are approved safe by the manufacturer. As per these regulations, if a new ingredient becomes a component of the food or in future it might affect the sensory attributes of any food product, then it can be assessed as a direct food additive or a GRAS food ingredient. Furthermore, color additives are defined as additives whose designated use is to impart color when added or applied to a food product.

As per the 21 CFR regulations nos. 175, 176, 177, and 178, indirect food additives are classified as a separate category. These regulations further define a novel food ingredient as an "ingredient whose purposeful use is to adjunct the diet by increasing the total dietary intake of the substance".

The Food Additive Amendment of 1958 further makes it compulsory for the manufacturers and petitioners to adhere to and qualify the FDA's safety norms before its addition to the edible products and its marketing as a food additive. FDA further elaborates on the safety criteria. Data and documents regarding the purity, chemistry, intake, and exposure as a result of suggested usage within the various sources of diet, and specific mechanisms of toxicity of the ingredient are required by the FDA before granting approvals. It also confirms that a requisite allowance of safe intake between the FDA further confirms an ample allowance of safe intake between the anticipated level of the substance that might produce unfavorable repercussions in animals and the anticipated vulnerability to the human beings which also includes vulnerable population groups of babies, pregnant and lactating mothers, and elderly.

As per the section 409 of the act, any additive whose intentional usage might have a technological impact in the products is considered dubious until it adheres to the safety standards of its accepted application or is excused for its use. As per the 402(a)(2)(C) of the 1938 Federal FDCA food products which might contain a high-risk food additive is scrutinized and contaminated.

4.8 Indian Regulations—Food Safety and Standards Authority of India (FSSAI), Additives Regulations (Regulation 3.1)

Under the Food Safety and Standards Act, 2006 (34 of 2006), the Food Safety and Standards Authority of India (FSSAI) has laid down the

Food Safety and Standards Regulations (Food Products Standards and Food Additives) for food additives.

According to the FSSR (2011) regulations [3], a food additive is defined as the mentioned below:

"the quantity of the additive added to food shall be limited to the lowest possible level necessary to accomplish its desired effect; the quantity of the additive becomes a component of food as a result of its uses in the manufacturing, processing or packaging of a food and which is not intended to accomplish any physical or other technical effect in the food itself; is reduced to the extent reasonably possible; and the additive is prepared and handled in the same way as a food ingredient."

"A carry over additive is the one present in a food, other than by direct addition, as a result of carry-over from a raw material or ingredient used to produce that food."

Within the provision of these regulations, the authority allows some carry-over additives:

a. the additive is verified for use in the uncooked food material or other components (including additives);
b. amount of the additive is below its specified maximum use level;
c. the food itself must not include the carry over ingredients in a large amounts than the quantity which is intended to be established by the use of raw material or components
d. technological conditions or processing practice, compatible with the preview of these regulations.

Exception:
Some food categories for infants and young children are exempted from regulations. For example, infant follow-up feeding formulas, complementary foods, and formulas for specialized therapeutic uses for babies and children are exempted unless a food additive provision is specifically mentioned.

The regulations further specify 22 categories of food additives. They are mentioned as follows:

a. Anticaking agent
b. Foaming agent
c. Antifoaming agent

d. Gelling agent
e. Antioxidant
f. Glazing agent
g. Bleaching agent
h. Humectant
i. Bulking agent
j. Packaging gas
k. Carbonating agent
l. Preservative
m. Carrier propellant
n. Color raising agent
o. Color retention agent
p. Sequestrant
q. Emulsifier stabilizer
r. Emulsifying salt
s. Sweetener
t. Firming agent
u. Thickener
v. Flavor enhancer

4.9 Safety Assessment: Redbook's Principles of Safety Evaluation

"Toxicological Principles for the Safety Assessment of Direct Food Additives and Color Additives used in Food", also known as "the Redbook", which was originally published in 1982, is a compilation of safety assessment approaches of food additives. It is foundation document for the safety regulations of food additives [4]. In the revised and updated versions of the red book, the approaches for assessment safety standards of food additives is formulated on four broad principles as follows:

a. Presumption by the agency that every food additive has some toxicological information.
b. Level of concern (LOC), i.e., the safety limits needed for a particular food is calculated.
c. LOC of additives is assessed on the extent of potential human intake and its molecular structure: exposure data, if available.

d. As per the data available, if unexpected adverse effects are reported with the ingestion of a particular additive, the initial evaluation of the testing requirements can be adjusted.

This is followed by the calculation of ADI from the results of the toxological researches, which is further compared to the estimated daily intake (EDI). If the value of EDI is less than the ADI, then the proposed additive is determined to be safe under proposed conditions of use. As per the USFDA (2014) regulation, "food additive petitions (FAPs)" deposited to the FDA must contain the following data about the additive [4]:

a. Specifications and composition
b. Expected role
c. Usage limit
d. Data pertaining to its purposeful effects
e. Quantitative techniques of detection in the anticipated food
f. Anticipated vulnerability from the usage
g. Complete documentation of all safety assessments
h. Suggested tolerance limits
i. Environment safety data – "requirement by National Environmental Policy Act (NEPA), as revised (62 FR 40570; July 29, 1997"
j. Reliable data must be provided for the chemistry, toxicology, microbiology and environmental sciences.

4.10 Levels of Concern for Direct Food Additives

As defined in the FDA's Redbook, LOC is a probabilistic assessment of the possibility that a hazard presented by a distinct additive may cause harm. As mentioned in the Redbook II, expected toxicity levels are designated to a compound on the basis of its molecular composition into three classes: low toxicity—class A, moderate toxicity—class B, and high toxicity—class C [5, 6]. These allotments are assigned on a basis of a decision tree. It is based on the chemical structure, number, and amount of unknown particles present in the additive and its anticipated metabolites.

Category: "Category A includes the following compounds such as acyclic, simple aliphatic, and monocyclic hydrocarbons, simple aliphatic and

Table 4.1 Concern levels and study requirements for compound categories as per USFDA.

Concern level	Study requirements
CL I compounds	28 days of short-term feeding study in a rodents and short assessment tests for carcinogenic effect/property
CL II compounds	90 days of feeding study in rodents and other species. "In addition, multiple generation reproduction study with a toxicity period, and short-term assessment tests for carcinogenic effect/property"
CL III compounds	This involves more extensive testing. Apart from the studies mentioned for a CL II substance, "carcinogenicity studies in two rodent species, and a persistent feeding phase of 1 year in duration in a nonrodent species are mandatory for approvals"

noncyclic (saturated), carbohydrates and lipid metabolites, fats and fatty acids, monofunctional alcohols, ketones, aldehydes, acids, esters, and ethers. Category B includes compounds such as inorganic salts of iron, zinc, copper and tin, amino acids, proteins and polypeptides, and non-conjugated olefins (excluding unsaturated fatty acids and fats). Category C includes structurally different compounds such as N-nitroso, azide, and purine groups, organic halides, amides and imines, polycyclic aromatic hydrocarbons, and conjugated alkenes."

LOC for the direct food additives is based on anticipated human intake. This foundation document also enlists types of researches that are considered as a least requirement to ensure safety approvals for the compounds enlisted in the concern levels I, II, and III as discussed in Table 4.1.

4.11 Threshold Regulation Exemption for Indirect Food Additives

"As per USFDA regulations an indirect food additive is any ingredient whose intentional use is as a component of products used in processing phases like manufacturing, packing, etc. [7]."

According to the 21 CFR §170.39 regulation, if it is exemplified that any component which was used in a the proximity of the food and is anticipated to transfer into the product, and if its amount is at or below 0.5 ppb, then this component will be regarded as safe with no health or safety concerns by the regulatory agency. Since these substances are present at a level below the threshold of regulation (TOR), they are exempted from the regulations [7].

More importantly, carcinogens do not classify themselves under the category of TOR exemptions. The data findings and documents, on the basis of which TOR is calculated, are also required by the FDA for approvals. Post the approval of the FDA, the ingredient might be considered for addition to the approved list TOR exemptions [7].

4.12 Estimated Daily Intakes

In case of direct food additives estimation of the daily intake (EDI) of the ingredient is calculated. Manufacturers need to provide sufficient safety assessment documents to establish a reliable EDI. It is calculated by multiplying the dietary amount of the ingredient by the total quantity of the food consumed by a person in a day. Recommended amounts of direct additives is the concentration recommended for each ingredient's technological purpose. As per USFDA regulation, each food product that contains the additive must be approved by the authority. The safety assessment documents on the maximum concentration of the additive and on human consumption amounts for the food ingredient, along with the 90th percentile consumption, must be deposited by the manufacturer for approval. Thus, if the EDI for the particular additive does not exceeds the ADI, then the additive is accepted as safe for consumption and should be approved. For calculating intake of ingredients in the diet for the various population groups, FDA uses the National Health and Nutrition Examination Survey (NHANES) data.

In case of indirect food additives, the EDI is calculated by following modus operandi documented in the "Recommendations for Chemistry Data for Indirect Food Additive Petitions and the Guidance for Industry: Preparation of Premarket Notifications for FCSs: Chemistry Recommendations [8]". It is calculated by assessing the amount of the ingredient that could transfer from the food contact surfaces into various products, further calculating the concentration of foods that might be consumed by an individual daily. Further cumulative EDI (CEDI) is calculated by adding up the other uses of the additive. The FDA further uses CEDI to substantiate a "LOC" of the product which determines the magnitude of harmful assessments as documented in the "FDA document entitled Preparation of Premarket Notifications for FCS: Toxicology Recommendations [9]".

4.13 Human Data and Clinical Studies

As per the FDCA, clinical safety data is not required for food additives. Findings from the animal studies can be used for safety analysis process of the food additives. Wherever possible, human documents should be included into the safety description of the food ingredient. In case of macro-ingredient food additives, for example, non-caloric fat substitutes, clinical researches may be required because high level of consumption of such components in rodents has been shown to produce modifications in the organ systems leading to harmful consequences [10]. Also, human studies may be required to further understand the micronutrient homeostasis.

4.14 GRAS Substances

As per the FDCA (1958) amendments, any ingredient is deliberately added to food and is subjected to premarket approval by the FDA, unless it is designated as GRAS. Prior to January 1958, food ingredients like vinegar, salt, baking soda, and pepper are exempted from regulation considering their traditional use in the food products. Scientific evaluation procedure is carried out for ingredients to be labeled as GRAS. "The principal criterion for GRAS status is documentation that a substance is generally recognized, among experts qualified by scientific training and experience to evaluate its safety, as having been adequately shown through scientific procedures or experience based on common use in food) to be safe under the conditions of its intended use [11]."

In October 2016, FDA had issued a guidance document titled, "Guidance for Industry: Frequently Asked Questions about GRAS for Substances Intended for Use in Human or Animal Food". It answers common questions about the regulative procedures and GRAS designated human and animal ingredients. This document has been updated and has replaced a document, entitled "Frequently Asked Questions about GRAS," that was issued by Center for Food Safety and Applied Nutrition in December 2004. This updated (81 Fed. Reg. 54960) enlists substances used in human and animal food."

4.15 European Union Legislation

The European Union legislation defines food additives as "any substance not normally consumed as a food in itself and not normally used as a characteristic ingredient of food, whether or not it has nutritive value, the intentional addition of which to food for a technological purpose in the manufacture, processing, preparation, treatment, packaging, transport or storage of such food results, or may be reasonably expected to result, in it or its by-products becoming directly or indirectly a component of such foods."

4.16 Categorization of Food Additives

4.16.1 Additives Can Be Used for the Following Purposes

a. Colors: they enhance or restore color of any food product.
b. Preservatives: they provide protection from microorganisms and increase the shelf life of the foods.
c. Antioxidants: they provide protection against oxidation reactions substances (rancidity of fat and color changes).
d. Flour treatment agents: they improve baking quality of the flour or dough.

Apart from the above-mentioned categories, as per EU food additives must also prove to be advantageous and beneficial for its end consumers. Therefore, they must serve following purposes and functions [12]:

a. Protect the nutrients of a food product to which they are added.
b. Supply requisite components for special products made for special need consumers.
c. Protect the shelf life and stability of a food product.
d. Enhance its organoleptic and sensory properties.

e. Aid in the manufacturing, processing, packaging, transportation, and storage.

The Annex II of Regulation (EC) No. 1333/2008 provides a list of food additives along with their conditions of use as authorized by EU. This list of food additives was approved by the European Parliament and of the Council of 16 December 2008 on food additives [13]. In this list, the additives are grouped on the "basis of the categories of food to which they might be added, e.g., fish and fish products, fruit and vegetables, dairy products, confectionery, etc. This list allows for easy identification of the additives authorized for use in a certain foodstuff, further offering greater transparency. The new list is more accessible for everyone involved in any component of the food chain, be it as a consumer, the control authorities or the food industry. The improved transparency allows correct and therefore safer use of food additives [13]. This guidance document describes the different categories to enhance uniform application and enforcement [13]."

4.17 Safety Assessment of Food Additives

European Food Safety Authority (EFSA) and Scientific Committee on Food (SC) currently authorize the safety of all the food additives. Only food ingredients whose intentional uses are concerned safe are classified in the EU list. Based on the advices of EFSA revisions of the present condition of use of the food, additives may be proposed by the commission, and if required, an additive may be removed from the approved list.

Continuous revaluation of the food additives is a rigorous exercise undertaken by the authority for establishing the safety among its population groups. Due to various review programs, EFSA decreased the ADI of three food colors. In 2001, the maximum levels of usage of these colors (E 104 Quinoline yellow, E 110 Sunset Yellow, and E 124 Ponceau 4R) that can be used in the food were further lowered. Therefore, a food additive might be authorized if it fulfilling the requirements:

a. According to the current scientific documents, it does not cause any harmful effects to consumers at recommended usage.
b. A rational transformational demand which cannot be met by other sources.

c. Its usage is not misleading for the consumers.
d. It must be beneficial to the consumers.

Other relevant factors like ethics, environment, and traditional practices may also be considered while authorizing the food additives.

4.18 Safety Evaluation Process and Authorization

The safety evaluation procedure for food additives is assessed by the EFSA. The manufacturer or the anticipated user of the food additive provides EFSA with a dossier of documents. This document must contain the following information about the food additive:

a. Chemical identifications
b. Process of manufacturing
c. Assessment methods
d. Reactions with in the food product
e. Need and importance
f. Toxicological data
g. Proposed uses

Furthermore, the toxicology study document must provide details on persistent toxicity, carcinogens, metabolism, genotoxicity, and developmental toxicity. Based on all the available information, EFSA further defines the safe limit of intake of the additive which can be considered safe—that is known as the ADI. The European Authority also determines the various anticipated uses of the additive in different food products, to conclude if the ADI will be exceeded or not. In case the ADI is not above this value, then the food additive is considered safe among the population groups. In case the ADI of the food additive is exceeded, the authorities can control the amount of the additive or may not even reject the proposal. The ultimate goal of EFSA is to ensure that the consumption of food additives should be designated as safe even for the population groups which consume substantial quantities of food products with the additives used at maximum sanctioned levels.

"Food additives causing minimum toxicological concern can be added in almost all processed food products. For example: calcium carbonate (E 170), pectins (E 440), lactic acid (E 270), fatty acids (E 570), citric acid

(E 330), fatty acids (E 570), and nitrogen (E 941)". But some of these additives are to be used in more restrictive in nature. For example:

a. "Natamycin (E 235) - to be used as preservative for the surface treatment of cheese and dried sausages only
b. Erythorbic acid (E 315) - to be used as antioxidant in certain meat and fish products only
c. Sodium ferrocyanide (E 535) - to be used as anti-caking agent in salt and its substitutes only"

Authorization procedure for additives is document in Regulation EC No. 1331/2008 (https://eurlex.europa.eu/LexUriServ/LexUriServ.do?uri=OJ:L:2008:354:0001:0006:EN:PDF) [12].

EFSA is requested by the commission to assess the safety concerns of the new additive. From 9 months following the request for approval, EFSA usually gives it opinion. The Commission along with food additive experts of all member states will consider the possible authorization based on EFSA's recommendation. Safety concerns, transformational need, possibility of abuse, and the benefits to consumers are also considered. In case the additive is appraised suitable as per the reports, the Commission will draft a proposal for viable authorization of the food additive and then it will be presented for referendum at the Standing Committee on the Food Chain and Animal Health (SCoFCAH). Further, if the proposal is supported by the SCoFCAH, then it is referred to the council and parliament of Europe. They have the right to reject it in case they consider the "approved status" inappropriate as per the purview of the EU legislation.

4.19 Use of Food Additives in Food Products

4.19.1 Traditional Foods

"The traditional foods, namely, Snacks of Savory snacks and fried products like Chiwda, Bhujia, Dalmoth, Kadubale, Kharaboondi, Spiced and fried dals, banana chips and similar fried products sold by any name, Sweets, desserts, and milk product based, such as Halwa, Mysore Pak, Boondi Ladoo, Jalebi, Khoya Burfi, Peda, Gulab Jamun, Rasogolla, and similar milk product based sweets sold by any name, Instant Mixes Powders only of Idli mix, dosa mix, puliyogare mix, pongal mix, gulab jamoon mix, jalebi mix, vada mix, Rice and Pulses based Papads, Ready-to-Serve Beverages (tea/coffee based only)" may contain food additives permitted in these regulations and in Table 2 of Appendix A of FSSR document [3].

4.19.2 Restricted Provisions

As per the FSSR regulation 3.3.1 (4), the following flavors are prohibited from addition in edible product:

"Coumarin and dihydrocoumarin, Tonkabean (Dipteryl adorat), β-asarone and cinamyl anthracilate, Estragole, Ethyl methyl ketone, Ethyl-3-phenylglycidate, Eugenyl methyl ether, Methyl β napthyl ketone, p-Propylanisole, Saffrole and isosaffrole, Thujone and isothujone (α & β thujone)"

4.20 Labeling Regulations and Guidelines

The Codex Alimentarius Commission also provides with the labeling regulations and guidelines for food additives. Codex norms are executed globally and the food companies are obligated to declare and to provide details of the additives which have been added to a particular food item during the course of processing and packaging. In European countries, there is a legislation governing declaration of food additives as per pre-defined "E-numbers", whereas in India, we follow the International Numbering System (INS) for declaring the additives on the food labels.

4.21 Conclusion

Food additives thus form a very important part of the food industry. If used in the recommended amounts by the manufacturers, then they can aid in the development of many novel products benefitting the consumers. Any faulty practices may result in serious complications on the health of the consumers. Also, the food manufacturers breaching national and international rules and guidelines regarding food additives may have to face serious consequences, including imprisonment. Therefore, considering the laws regulating these food additives is extremely important for safe use and application.

References

1. World Health Organisation (WHO), https://www.who.int/news-room/fact-sheets/detail/food-additives, 2018.

2. Pressman, P., Clemens, R., Hayes, W., Reddy, C., Food additive safety: A review of toxicologic and regulatory issues. *Toxicol. Res. Appl.*, 1, 1–22, 2017.
3. Food Standards and Safety Authority of India (FSSAI), https://www.fssai.gov.in/cms/food-safety-and-standards-rules–2011.php, 2011.
4. US FDA, Redbook II. Overview of food ingredients, additives and colors, International Food Information Council (IFIC) and U.S. Food and Drug Administration (FDA), 2014. https://www.fda.gov/food/food-ingredients-packaging/overview-food-ingredients-additives-colors
5. US FDA, Redbook II. Overview of food ingredients, additives and colors, International Food Information Council (IFIC) and U.S. Food and Drug Administration (FDA), 2014. https://www.fda.gov/food/food-ingredients-packaging/overview-food-ingredients-additives-colors
6. US FDA, Toxicological principles for the safety assessment of direct food additives and color additives used in food "Redbook II." Center for Food Safety and Applied Nutrition, Washington, DC, 1993.
7. US FDA, Indirect food additives: paper and paperboard components, United States, (21CFRPart 176), vol. FedReg. Vol. 81, No. 1, pp. 5–6, Monday, 4 January 2016, Rules and Regulations, Federal register. The Daily Journal of United States Government (federalregister.gov/documents/2016/11/22/2016-28116/indirect-food-additives-paper-and-paperboard-components) 2016.
8. US FDA, *Preparation of premarket notifications for food contact substances: chemistry recommendations*, Center for Food Safety and Applied Nutrition, Washington, DC, 1999a.
9. US FDA, *Preparation of premarket notifications for food contact substances: toxicology recommendations*, Center for Food Safety and Applied Nutrition, Washington, DC, 1999b.
10. Munro, I.C., Issues to be considered in the safety evaluation of fat substitutes. *Food Chem. Toxicol.*, 28, 751–753, 1990.
11. SCOGS (Select Committee on GRAS Substances), Evaluation of health aspects of GRAS food ingredients: lessons learned and questions unanswered. *Fed. Proc.*, 36, 2519–2562, 1981.
12. EU, Regulation (EC) No 1333/2008 of the European Parliament and of the Council of 16 December 2008 on food additives (Text with EEA relevance). *OJ L*, 354, 31.12.2008. p. 16–33 (BG, ES, CS, DA, DE, ET, EL, EN, FR, GA, IT, LV, LT, HU, MT, NL, PL, PT, RO, SK, SL, FI, SV) Special edition in Croatian: Chapter 13 Volume 034 P. 67–84, 2008.
13. EU, Guidance document describing the food categories in Part E of Annex II to Regulation (EC) No 1333/2008 on Food Additives, June (https://ec.europa.eu/food/sites/food/files/safety/docs/fs_food-improvement-agents_guidance_1333-2008_annex-2.pdf), Official Journal of European Union https://eur-lex.europa.eu/legal-content/EN/TXT/?uri=CELEX: 32008R1333, 2017.

5

Detection of Food Adulterants in Different Foodstuff

Aditi Negi[1], P. Lakshmi Praba K.[2], R. Meenatchi[1]* and Akash Pare[2]†

[1]Department of Primary Processing Storage and Handling, Indian Institute of Food Processing Technology, Ministry of Food Processing Industries, Govt. of India, Thanjavur, Tamilnadu, India
[2]Department of Academics and Human Resource Development, Indian Institute of Food Processing Technology, Ministry of Food Processing Industries, Govt. of India, Thanjavur, Tamilnadu, India

Abstract

Globally, consumers are demanding more healthier food for consumption. Adulteration of food has become a major threat that imposes health issues and affects the quality of the food. The estimated economic loss of food due to adulteration is 30–40 billion dollars per year. Therefore, in order to provide adulterant free food stuff to consumers, various detection techniques such as physical, chemical, analytical, and immuno-based methods are employed to identify the adulterants in food. This book chapter covers various techniques used for the detection of adulteration with their advantages and disadvantages. Microscopic and macroscopic techniques are the basic physical detection methods, to visualize the structural variation among the food products due to adulteration. Spectroscopic techniques have shown great potential for the assessment of food quality and safety due to their non-destructive properties. Chromatographic-based techniques provide rapid and predictable separation of chemically similar compounds in complex food matrix. Molecular techniques provide accurate results regardless of physiological status of the sample with high accuracy, selectivity, and sensitivity in food control laboratories. Next-generation sequencing is generating massive data with minimal quantity of DNA. Nanoparticle-based approach ensures the fast and effective detection of adulterants, toxins, dyes, and pesticides in food materials.

Keywords: Food, adulteration, detection methods, physical, analytical, molecular

---

Corresponding author: meena@iifpt.edu.in
†*Corresponding author*: akashpare@iifpt.edu.in

5.1 Introduction

Adulteration and contamination are ongoing concerns in the food industries, as they can result in unsafe products and damage the brand integrity. Food adulteration is an incorporation or addition of low-grade, unsafe, poor quality, or extraneous substances to foods. This not only degrades the quality and originality of food items but also causes harmful effects to the consumers including chronic health problems like paralysis, cancer, and weakening of the immune system through the inclusion of toxic or lethal adulterants in the human body [1]. The estimated economic loss of food due to adulteration to the food industry is 30 to 40 billion dollar per year [2]. Adulteration results in significant economic losses like mistrust about brand; significant loss for industry; costs to government from public health, regulatory agency and law enforcement response activities, and consumer mistrust about the food system. Adulterant detection and authenticity testing are important for export market, religious conviction (halal/vegetarian), traceability of food products, and prevention of commercial frauds. Due to more attention on food safety and implementation of food regulation, various techniques for identification of adulterants, which includes physical, biochemical/immunological, and molecular methods have been developed. Molecular methods are suitable for the finding of biological adulterants in food [3].

Major adulteration cases reported by JRC food fraud summary (based on media coverage) by European commission during year 2017–2018 [4] are summarized in Table 5.1.

5.2 Types of Adulteration

Adulteration in food and food products represent an attractive category, because the products have a high value by weight and consumers have a limited capacity to detect adulteration. The adulteration in food is categorized based on inclusion of unapproved substances intentionally or unintentionally to imitate the extension. These are mainly as follows.

Intentional adulteration mimics in the properties of food in which it is added such as sand, marble, saw dust, dyes, mineral oil, and floral parts to enhance the organoleptic qualities of the food and food materials. Such adulterants pose threat to public health and are difficult to

Table 5.1 Cases of food fraud reported by media and summarized by European commission [4].

S. no.	Country	Adulterated food
1	India	Addition of toxicant in alcohol/spices (artificial enhancement), extraneous matter in tea, fish, and milk
2	European country	Horse meat (mislabeling)
3	Italy	Frozen food, vegetables, sea food, coffee, cheese, ice cream and dairy products, vinegar, bakery items, and wine (mislabeling)
4	Brazil	Olive oil (substitution)
5	Spain	Wine, fish, saffron, meat, and olive oil
6	UK	Fish and meat substitution
7	Canada	Fish mislabeling and sugar dilution
8	Pakistan	Spices, milk dilution, confectionary, and bakery
9	China	Chocolate and fruit
10	USA	Beef, lobster, olive oil, honey, and fish

detect when it is added as water in milk, extraneous material in powdered spices, etc.

Unintentional adulteration can occur due to lack of knowledge during food processing and handling. This adulterant can be acquired from the spoilage from bacteria, fungi, rodents dropping, harmful materials from the packaging, or inherent in nature such as presence of toxic compound or radical which is naturally occurring in some variety of mushrooms or sea food. The most common type of incidental adulteration is presence of pesticide residue in food.

Metallic contamination is contamination of food products with heavy metals or its compounds like lead, mercury, arsenic, cobalt, and cadmium. These elements are toxic and can accumulate in body, which can cause organ damage. Fish grown in mercury contaminated water, fruits sprayed with lead arsenate, and barium from the rodent baits (barium carbonate) are some examples of the metallic contamination.

Microbial contamination occurs when vegetables are grown under the unhygienic condition such as sewage water. This contamination also occurs in food if they are not stored under proper condition like stored under high moisture or humid environment. Generally, ready-to-eat and poultry products are contaminated with the microbes but these contaminations can be destroyed during food processing and adequate cooking.

Accidental adulteration includes insects, larvae, insect excreta, fragments, and secretion which promote the microbial activity and spoilage. Spices are more prone to get adulterated by insects and their products due to presence of moisture content. Past few years, 7% of spice lot have been rejected due to accidental adulteration (TIMES). Apart from these pesticide residues present in the plant products are the accidental adulterant present in spices.

5.3 Impact of Adulteration on Health

The main motivation behind the addition or substitution of the authentic product is to gain more economic profit; however, detrimental result of this criminal behavior is responsible for number of health risks pose to public. The effect of food adulteration is ranging from immediate mild to life-threatening symptoms. The consumption of adulterated food lead to various health risks such as stomach pain, ulcer, diarrhea, dysentery, vomiting, heart problem, kidney problem, and loss of vision [1].

Intentional adulteration such as addition of argemone seeds to mustard seeds or addition of argemone oil to groundnut oil will lead to loss of vision and heart diseases. Addition of washing powder to jaggery, ice creams, chalk to sugar, water to honey, chicory to coffee, colored leaves and iron fillings to tea, vanaspati to pure ghee or butter, white powered stone to Common salt, etc., causes appendicitis and small intestine problems in humans. Health problems such as anemia, epilepsy, and neurotoxicity arise due to presence of metanil yellow to turmeric powder and in pulses, juices, jams, and sweets. Calcium carbide used in ripening process for fruits, oxytocin hormone used for faster growth of fruits and vegetables, metallic lead (chemical containing green colors) applied to give fresh color to bitter gourd and leafy vegetables, pesticides and herbicides, etc., has been proven as disastrous for digestive system, eyes, and liver.

Several dyes such as Sudan1, Sudan 4 Para Red, orange 2, Methyl yellow, and Rhodamine used in spices such as pepper, turmeric, and chili are potentially genotoxic and possible carcinogenic in nature.

Metallic adulteration such as lead present in water affects the quality food leads to lead poisoning causing foot drop, anemia, constipation, mercury present in fish food cause brain damage, paralysis and even death. Soft drink and fruit juices prepared or stored in cadmium plated vessels cause gastritis, kidney, and liver damage and prostate cancer. Non-food grade or contaminated packing material used in food cause blood clot and angiosarcoma cancer.

The contaminated water acts as vehicle for several pathogens and causes health problems in vulnerable individuals. The milk adulterated with contaminated water is carrier of several etiologies that include bacteria, virus, protozoa, and helminths. Food infection arises by bacteria causes abdominal pain, nausea, vomiting, and diarrhea.

Aflatoxin present in food such as groundnuts and cotton seed contaminated by *Aspergillus flavus* causes liver damage and cancer. Machupo virus present in foods such as cereals which is contaminated with rodent's urine, and Bolivian hemorrhagic fever is one of the examples of health risks associated with the adulterated food.

5.4 Approaches for Adulterant Authentication in Food Materials

Numerous methods were developed mainly based on physical and chemical testing to authenticate food products to differentiate from original to non-original. In general, three basic strategies are used to demonstrate the adulteration:

    a. Identification of unusual profile of component
    b. Differentiation from the normal standard compound
    c. Identification of marker or foreign material in the product [3]

In above-mentioned strategies, implementing the last strategy for the identification of foreign material in food commodity is comparatively easy and more target oriented. Physical methods of adulterant identification performed by using shape, color, odor, and texture of food materials and microscopic examination of tissue structure and arrangement of food and food products. However, these methods are simple and cost effective; they do not guarantee the quality of the product. Several chemical and biochemical methods are also being used to detect food adulterants. Chemical or biochemical methods are more precise and receptive than physical methods such as chromatography, spectroscopy, electrophoresis, and immunology. Industrial application of these methods is lagging behind due to high cost and technical challenges and requires trained man power [5].

5.5 Physical Authentication Techniques

Physical methods mainly use microscopic techniques where morphology recognition of surface structure, color, shape, and arrangement of tissue and structure along with sensory method for aroma and flavor for food and food products. Figure 5.1 depicts the various physical authentication methods. These methods are simple and cost effective but do not guarantee qualitative adulterant detection [3]. Various imaging techniques for microscopic food analysis are now available. Light microscopy is one of the most commonly used technique to study the microstructure, and with the advancement of Transmission Electron Microscopy (TEM) as well as Scanning Electron Microscopy (SEM) with greater magnification (imaging) added capabilities to study the ultrastructure of food products. Food microscopy efficiently uses a variety of staining techniques. This microscopy works with sample prepared with specific or basic staining to separate

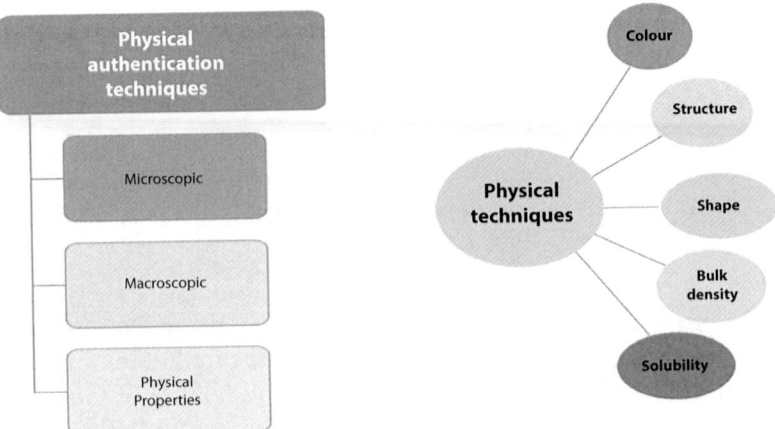

Figure 5.1 Physical authentication techniques.

the individual internal structures. Normally, the basic staining is used to intensify all the structures present in the final product, which are then recognized by their structures morphological features (shape, size, and configuration of the mutual cell, presence of the crystals, grains, or other constituents).

Adulteration detection in food materials of animal origin are meat, milk, eggs, and honey by microscopy has been well documented in literature [175]. Meat is comprised of different cells and tissues, and it is the most complex and nuanced raw material in general. A standard microscopic examination in meat products can detect different types of muscle tissue, the individual types of connective tissue, adipose tissue, bone fragments, and parts of organs, all of which can be used to assess a potential adulteration of food. These structures are clearly and apparently visible under a light microscope and basis of identification of meat correctly. During the assessment, technological processing (tearing of muscle fibers, swelling, etc.) should also be taken into account because vigorous mechanical processing considerably complicated the diagnosis under microscopic examination. Advanced electron microscopic techniques used to analyze anatomical features such as presence or absence of hairs (trichomes), starch grains, oil glands, canals, seed or pollen morphology, and vascular traces reveal the adulteration. Electronic microscopy helps in adulterant detection in honey such as Parenchyma cells, sclerous rings, and sugar constitutive cells, which shows the presence of cane sugar and its products in the product. Similarly, surface pattern of pollen grains also helps in identification of botanical origin of honey by electron microscopy.

Adulteration in vegan food can be easily identified by normal microscopic staining especially in their traditional form such as flours and starches. Their presence can be identified based on their certain structural elements such as aleurone cells of cereal seed coat or palisade cells of soya. Vegetable additives can be easily confirmed on the basis of typical appearance and structure in the microscopic image by basic staining techniques. In a microscopic examination, wheat starch grains show small and large grains shaped like lentils with tiny central layering, while corn starch shows multi-faceted, tile-shaped grains with a star-like core.

Microscopy techniques also be helpful to analyze adulteration in spices. Several anatomical features of spice are presence or absence of oil glands, epidermal cell, trichrome, seed and pollen structure, starch grains, and extraneous matter like insect's fragments and faces. SEM helps in the identification of *Illicium verum* (star anise from its adulterant *Illicium anisatum* by displaying the differences in their epicarp morphologies). Joshi et al. [12] differentiated star anise adulterant using fluorescence microscopy.

Microscopic analysis is not cost effective for routine analysis due to its meticulous sample preparation. Moreover, structural analysis of processed food sample like fine powders mixed with other plant material requires expertise. Similarly, adulterant differentiation is possible in meat products based on pigmentation or less intramuscular fat based on gross appearance. These methods are not reproducible and quantitative identification is difficult. Adulterant identification through morphological characteristics such as size, shape, and appearance is difficult in case of highly processed food sample like meat, fishes, powder spices, and saffron [12]. Physical techniques used for the authenticity of some food products are resumed in Table 5.2.

5.6 Application of Biochemical and Analytical Methods in Adulterant Authentication

Analytical methods are employed to separate different component from the mixture using suitable solvent. Analytical methods fall into various categories based on the principle involved. Chromatographic-based method is reliable and helps in separation of components. These techniques help to overcome several difficulties arousing from complex food matrices. They provide high resolution of results in identifying the authenticity of food stuffs. Various techniques include UV, FTIR, NMR, various spectroscopic techniques, HPLC, and GCMS and LCMS. Figure 5.2 shows various analytical methods involved for adulteration detection.

TLC is the simplest, most versatile, and economic way of obtaining the chemical fingerprints of multiple herbal samples. The technique has been used to detect adulteration of herbs [156] and was proven to be the best approach in testing saffron purity [157]. The presence of Sudan dye in chilli powder was detected using TLC by Dar *et al.* [158].

5.6.1 Adulterant Authentication Through HPLC

HPLC is the key tool helps in separating various components from mixtures. HPLC can separate and identify amino acids, proteins, phenolic acids, etc., present in a compound. HPLC is the common method used for detection of food adulterants [159]. Coffee adulteration has been studied by Wai Lok Cheah and Mingchih Fang using HPLC-based chemometric analysis technique which provides information on separated compounds [160]. Detection of adulterants in cereal flours using HPLC by Ehling *et al.*

Table 5.2 Physical methods used for authentication for various food products.

Food products	Physical method	Food adulterant	Parameter studied	References
Powdered spices mainly cumin, coriander, chilies, and cloves	Microscopic	Lower grade material	Extraneous starch	[6]
Black pepper	Floatation test followed by visual and microscopic examination	Papaya seed	Shape and color	[7]
Powdered chili	Microscopic examination	Tomato waste and added starch	Starch	[8]
Turmeric	Microscopic examination	Cheaper vegetable substance	Starch and oleoresin	[9]
Honey	Electronic or optic microscopy	Cane sugar/cane sugar product	Parenchyma cells, sclerous rings, sugar constitutive cells	[10]
Honey	Electron microscopy	Botanical origin of honey	Surface pattern of pollen grains	[11]

was conducted successfully for wheat flour and can be explored on corn flour, rice flour, and soy flour. This method helps in detection of melamine, ammeline, and ammelide [161]. UHPLC has been used to detect skimmed milk powder adulterants such as soy, pea, and hydrolyzed wheat protein [162]. Chen *et al.* [163] quantified cow milk adulteration in caprine milk

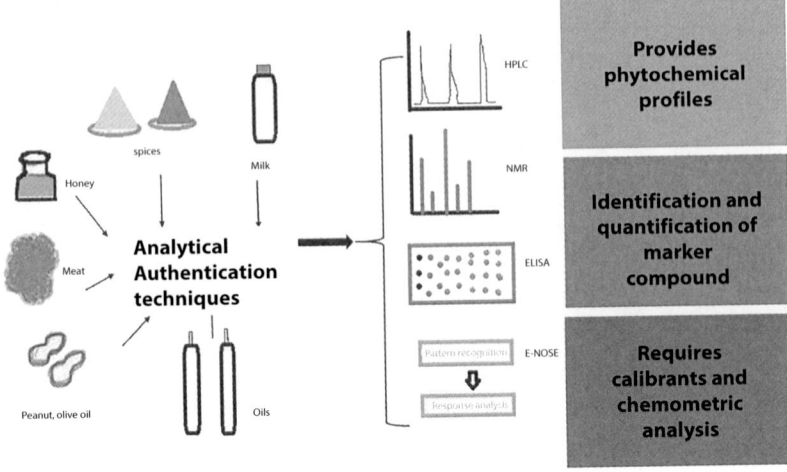

Figure 5.2 Analytical techniques for food authentication.

using High-Performance Liquid Chromatography/Electrospray Ionization Mass Spectroscopy (HPLC/ESI-MS). Oil adulteration such as olive oil with hazelnut oil, citric juices adulterated with flavones glycosides and polymethoxy flavones, tea liquors containing phenolic pigments, and amino acid in wines are detected using HPLC [17, 163]. This is used in the detection of olive oil adulteration with other oils such as soybean oil, corn oil, and sunflower oil [164]. Identification of black pepper powder adulterated with chilli [165], para red adulteration in red chilli powder, pickles, etc., could be done using HPLC. Detection of saffron components: crocins, picrocrocin, and safranal using HPLC has proved to be effective for determining the quality of the spice grown under different environmental and storage conditions [166]. Sproll *et al.* [167] developed a water HPLC system for analyzing the coumarin content of cinnamon, and the same was used for adulteration detection of true cinnamon with cassia cinnamon in Czech Republic retail markets. The high coumarin content in the market samples analyzed showed the adulteration of true cinnamon with cassia cinnamon [168]. The presence of Sudan dyes in hot chilli and spices could be detected using HPLC [169] with a detection limit of 3 mg/kg of the sample.

Recent advances in the field of liquid chromatography such as micellar electro kinetic capillary chromatography (MECC), high-speed countercurrent chromatography (HSCCC), low-pressure size-exclusion chromatography (SEC), reversed-phase ion pairing HPLC (RP-IPC-HPLC), and strong anion-exchange HPLC (SAX-HPLC) are reported to provide better separation of specific compounds found in herbal extracts [170]. Using these techniques, levels of different components present in the spices/herbs can be detected, which denote the quality of the spices and herbs.

Cow milk adulteration in caprine milk has been quantified by HPLC/ESI-MS. This method identifies molecular masses to differentiate between proteins in the milk of cow and goat. Detection of addition of cow milk to goat and ewe milk has been described by Abrantes and his co-workers [171, 172]. Quantitative detection of adulteration of Notoginseng root extract with other Panax species was done by HPLC coupled with principal component analysis (PCA) [173].

5.6.2 Adulterant Authentication Through GCMS

Gas chromatography (GC) is used for separating volatile organic compounds. GC along with mass spectroscopy (MS) and Fourier transform infrared spectroscopy (FTIR) has been widely used for adulterant detection as these are non-destructive techniques with respect to the sample. Gas chromatography is generally used to discriminate among different varieties of the same product, adulteration detection, and organic compound authentication and identification. GC has been utilized to differentiate wines from same regions [14]. Volatile compounds such as 1-propanol, 2-methyl-1-propanol, 2-propen-1-ol, and 3-methyl-1-butanol in wine were measured and quantified by GC or GCMS, providing 30 physicochemical parameters usable for pattern classification [25]. In GC methods, the main components targeted are the volatile and semi volatile compounds present in the specimen, and the sensitivity of detection is high. Detection of adulteration of olive oil samples from sunflower seed oil was achieved using GC estimation of methyl esters of the oil components, steroids, and fatty acids. The technique could diagnose the contamination of olive oil by sunflower seed oil at least at 1% level with >95% certainty in bottling plants. Fatty acid as the purity indicator suggests the use of linolenic acid as the parameter for the detection of virgin olive oil with 5% of soybean oil [174].

5.6.3 Adulterant Authentication Through Spectroscopic Method

Various spectroscopic analytical techniques used in the food industry include FTIR, Fourier transform near-infrared (FT-NIR), Raman, Hyperspectral Imaging (HSI) [75], and Nuclear Magnetic Resonance (NMR). Yellow chalk powder in turmeric can be detected using terahertz spectroscopy.

Inexpensive method of food authentication relies on vibrational spectroscopy for both assessment of food quality and safety. Near-infrared and mid-infrared data with multivariate analysis provide data on analysis of food components. Raman spectroscopy is another upcoming methodology that can be used to specific functional group and provides information on sample molecule. Raman and infrared spectroscopy provides non-destructive analysis of samples. Milk and soy bean adulteration, adulteration of olive oil by vegetable oil, and Saffron adulteration have been determined by novel spectroscopic method. Corn starch in ginger, garlic, and onion can be detected by using FTIR and Raman spectroscopy.

5.6.4 Adulterant Authentication Through Ambient Mass Spectroscopy Techniques

Various ambient mass spectroscopic methods are used to detect food adulteration. Desorption electrospray ionization technique is used to authenticate meat. DESI MS has been used to identify triglycerides (TGs) in edible oils and margarine and the differentiation of postharvest methods of coffee beans. Though DESI MS has been used it is not much effective comparing other spectroscopic techniques. DART-MS, along with other AMS techniques such as DAPCI-MS and low temperature plasma-mass spectrometry (LTP-MS), is a technique which has been utilized to detect the presence of melamine in milk powder.

5.6.5 Adulterant Authentication Through Nuclear Magnetic Resonance Technique

NMR technique is a high-throughput spectroscopic method which helps in detection of adulterants with structural identification of the adulterant. It holds the metabolic profiles that can be used for food authentication. Peanut adulteration can be detected by NMR spectroscopy. Solid ingredients in juices can be found using proton NMR spectroscopy. NMR analysis can be used to identify adulteration, such as red wine adulteration with anthocyanins, synthetic flavors sold as natural, and addition of cane or corn sugar to maple syrup. Metabolic profile for adulteration cases can be done for saffron, wines, coffee, honey, fish, vinegar, and spirits. Table 5.3 suggest various analytical methods for detecting food adulterants.

Table 5.3 Analytical methods used for adulterant authentication.

Chemical/Bio-Chemical method	Food adulterant	Advantages	Disadvantage	References
High-performance liquid chromatography (HPLC): Solvent forced under high pressures of up to 400 atmospheres. Separation based on difference in the relative affinities of different molecules	Olive oil with hazelnut oil	Quick automated and highly accurate, highly reproducible	Costly, not suitable for all samples	[17]
	Quince jams with apple and pear puree			[18]
	Citrus juice with flavones glycosides and poly methoxylated flavones			[19]
	Black tea liquors containing phenolic pigments			[20]
	Wines containing proline isomers and amino acids			[21]
	Saffron adulteration			[22]
	Benzopyrene in meat products			[23]

(*Continued*)

Table 5.3 Analytical methods used for adulterant authentication. (*Continued*)

Chemical/Bio-Chemical method	Food adulterant	Advantages	Disadvantage	References
Liquid chromatography: Use liquid as mobile phase, separation is based on the difference in molecular structure	Residue analysis of toxic chemicals, pesticide, veterinary drug, and food additive	Broad application due to excellent separation and sensitivity	Poor qualitative data, use of toxic reagents, and more quantity of solvent is required	[24]
Gas chromatography: Uses gas as mobile phase, separation is based on the difference in molecular structure	Volatile compounds-1-propanol, 2-methyl-1-propanol 3-methyl-1-butanol in wine	High sensitivity and resolution with less separating time	Used for small no. of analysis, complex sample preparation	[25]
Gas chromatography with mass spectroscopy (GC-MS): Combines the separation properties of gas-liquid chromatography with the detection feature of mass spectrometry to identify different complex volatile substances within a test sample	Volatile compounds-1-propanol, 2-methyl-1-propanol 3-methyl-1-butanol in wine	Simple, rapid, sensitive, accurate, applied well to pesticide residue	High cost and trained man power	[25]
	Honey adulteration with syrups			[26]
	Tea with cashew husk			[27]
	pesticide (36) residue in tea sample			[28]

(*Continued*)

Table 5.3 Analytical methods used for adulterant authentication. (*Continued*)

Chemical/Bio-Chemical method	Food adulterant	Advantages	Disadvantage	References
Near-infrared spectroscopy (NIR): Electromagnetic spectrum generates vibrational absorption bands of a molecule and multivariate statistical analysis performed to extract desired chemical information	Soya-based product used as animal feed	Rapid, nondestructive	Unable to identify adulterant	[29]
	Black pepper adulteration			[1]
Mass spectroscopy (MS) Sample ions determined according to their mass by charge(m/z) ratio	Method cannot be used alone in food adulteration	High sensitivity, low sample requirement, rapid analysis	High purity of sample is required, analysis of result is complicated, expensive equipment	[1]
Nuclear Magnetic Resonance Spectroscopy (NMR): The application of external magnetic field to randomly moving nuclei make nuclei orientation aligned and then detected	Beer, juices, infant formula, milk adulteration	Detect adulterant with structural identification of contaminant, minimal sample preparation, high accuracy and repeatability	Expensive equipment, complex analysis of testing data	[1]

(*Continued*)

Table 5.3 Analytical methods used for adulterant authentication. (Continued)

Chemical/Bio-Chemical method	Food adulterant	Advantages	Disadvantage	References
Fourier transform infrared spectroscopy (FTIR): Molecular bonds present in sample absorb long wave infrared radiation with the use of Fourier transform algorithm converted to frequency peak which is related to presence of chemical component	Milk adulteration	Detailed spectral inspection differentiate adulteration	Unable to identify adulterant	[30, 31]
	Juices		Difference in ground state and excited state of element at particular wavelength can be detected.	[32]
Liquid chromatography time-of-flight mass spectroscopy: Ions accelerated by electric field, difference between the mass/charge ratio and time to reach ion to the detector is measured	Sugar adulteration in apple juice, analysis of pesticides in water and food	High resolving power, provides useful information about elemental compositions	Low abundance of ions difficult to identify, susceptible for contamination	[33]

(Continued)

Table 5.3 Analytical methods used for adulterant authentication. (*Continued*)

Chemical/Bio-Chemical method	Food adulterant	Advantages	Disadvantage	References
Isotope ratio mass spectroscopy (IRMS): Mass difference in different isotope of same element can be done by analyzing number of neutron present in sample	Differentiation of natural vs. synthetic compound like vanillin or CO_2 in champagne. Geographic origin of wine, seafood, coffee, meat, etc.	Mechanically simple, Measures relatively abundant stable isotope	Difficult to calibrate, Individual analysis takes time	[34, 35]
Fluorescence spectroscopy: Some aromatic compounds absorb light at specific wavelength and emission of photon at another wavelength that can be useful for qualitative and quantitative analysis	Extra virgin olive oil with seed oil	Simple, fast, accurate and non-invasive	High cost, difficult to interpret florescence spectra	[36]
Raman spectroscopy: Vibrating molecule in sample absorb or and lose energy from incident light. The energy shift provides the information about molecular composition and crystal structure	Olive oil adulteration with sunflower seed oil, soybean oil, corn oil, and rapeseed oil	Rapid and non-destructive analysis of sample	Low detection limits specific molecule	[37]

(*Continued*)

Table 5.3 Analytical methods used for adulterant authentication. (*Continued*)

Chemical/Bio-Chemical method	Food adulterant	Advantages	Disadvantage	References
Surface enhanced Raman spectroscopy: Metal surface or particles are used to enhance the vibrational spectrum of molecules. This is an advanced version of Raman spectroscopy	Vegetable oil authentication	Rapid, non-destructive and Sensitive	Complexity of output signal, costly, not suitable for quantitative analysis	[38]
Electrophoresis: Migration of charged molecules through a gel when an electric current is passed across it	Additional whey in dairy beverages	Accuracy, versatility in identification and low cost	Measurements are not precise, limited sample analysis	[39]
	Goat milk adulterated with cow milk			[40]
	Basmati rice adulteration (capillary electrophoresis done for microsatellite marker)			[41]

(*Continued*)

Table 5.3 Analytical methods used for adulterant authentication. (*Continued*)

Chemical/Bio-Chemical method	Food adulterant	Advantages	Disadvantage	References
Polyacrylamide gel electrophoresis (PAGE): Migration of protein under electric field due to difference in protein structure and charge	Goat milk adulterated with Cow milk	Sensitive for detecting different species	Not ideal for close genetic relationship	[42]
Isoenzyme electrophoresis: Migration of enzyme under electric field due to difference in enzyme structure and charge	Fish	Reproducible, stable, easy, and more applicability	Poor polymorphism, few enzymes are available	[43]
Enzyme Linked Immuno Sorbent Assay (ELISA): Uses antibodies and color change to identify a substance	Bovine milk adulteration in goats milk	High sensitivity, rapid, low cost, reliable	Standards are rare	[44]
	Melamine detection			[45]
	Meat with cheaper meat			[46]

5.7 Adulterant Identification by Molecular Techniques

Molecular techniques are much useful for the detection of source material, précised component of interest, additives, objectionable constituents, presence of microorganisms, and pesticides [47]. Information concerning the presence of a species, breed, or cultivar in a food product necessarily involves a DNA-based genetic analysis as it appears as the ideal tracking tool to prevent food frauds. Molecular methods are based on detection of nucleic acid (DNA and RNA) and that is why these methods are rapid and reproducible. Molecular techniques based on DNA identification are most suitable to identify biological adulterant in traded food commodities [3]. High stability of DNA helps in analyzing adulterant in processed food products as well as trace contaminants. Molecular technique involves isolation or extraction of DNA, amplification of isolated DNA with appropriate primer, and detection of amplified product. The resulting sequences then compared with the adequate control or references. The most promising potential use of molecular techniques is that age, physiological condition, and tissue type of material are not interfere in DNA identification and thus make it more sensitive, accurate, and precise. Various molecular authentication techniques depicted in Figure 5.3.

Restriction fragment length polymorphism (RFLP) and species-specific polymerase chain reaction (PCR) techniques are significantly used to identify microbial or any distinct source of contaminant present in foods that spoils quality and safety [48]. PCR-based, sequencing-based, and Hybridization-based (DNA bar coding to study small standard DNA sequences) techniques can be exploited for food adulterant authentication [49].

Identification of species or cultivated varieties based on DNA analysis use several markers such as mitochondrial DNA markers CO1 proven to be effective for the Identification and characterization of animal species in

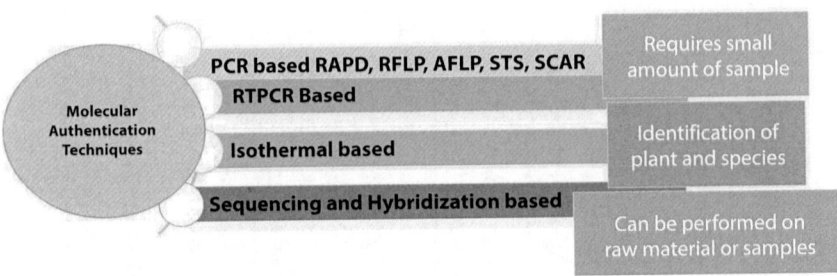

Figure 5.3 Molecular authentication techniques.

food products. This DNA barcode was employed for the discrimination of a high number of bovine breeds, for the analysis of fresh and degraded meat substrates and for the identification of mislabeled fishery products. Similarly, in plants, plastidial DNA and barcodes, such as rbcL+ matK, may allow to discriminate plants for food uses as, for example, *Ocimum* species or poison plants of the *Solanum* genus.

5.7.1 Polymerase Chain Reaction–Based Techniques for Adulterant Identification

PCR-based techniques have high potential in adulterant detection and authentication due to its rapidity and specific detection at low cost. PCR is used to amplify single or few copies of particular segment of DNA into thousand to millions of copies with the help of specific primer against the target. PCR-based adulterant detection and authentication consists of specific primer that produces high degree of individual specific banding profile of PCR products between individuals/strain/different species/subspecies or cultivar which can be analyzed by gel electrophoresis. Molecular techniques frequently adopted in authentication are RFLP, Random Amplified Polymorphic DNA (RAPD), Amplified Fragment Length Polymorphism (AFLP), Simple Sequence Repeat (SSR), Inter-Simple Sequence Repeat (ISSR), Sequence Tagged sites (STS), Single-Nucleotide Polymorphisms (SNPs), sequence characterized amplified regions (SCAR), etc. [1].

RFLP was used as the first markers used to analyze inter and intra-species genetic variability at genomic level [15]. DNA variations can be highlighted by comparing the specific restriction sites recognized by restriction enzymes and obtained DNA pattern allows identification of species or variety in different samples. Specific restriction sites recognized by the restriction enzymes RFLP markers have been widely employed for several purposes, as they are reproducible and co-dominant, such as the construction of linkage maps in several species including *Solanum lycopersicum* and *Zea mays* and the authentication of seafood products. Noncoding rDNA spacer segments also act as useful nuclear marker to discriminate closely related *Scombrid* fish species by RFLP. Despite RFLP popularity, it suffers from limitation that presence of mutation at restriction site, presence of mixed species, and chemical modification in DNA during canning interfere with authentication. The detection of RFLPs is an expensive, laborious, and time-consuming process.

Among the various molecular techniques (Table 5.4), RAPD is widely used and preferable technique for the detection of adulterant in commercial food samples due to its rapidness and no prior sequence information is required. RAPD analysis is extensively useful for species

Table 5.4 PCR-based molecular techniques for food authentication.

PCR-based techniques	Application	Primer/Target gene	References
Restriction fragment length polymorphism (RFLP): The length of PCR products cleaved with restriction enzymes differ due to presences or absence of restriction site which can be seen by gel electrophoresis and blotting procedures	Beef, carabeef, chevon, chicken, donkey, horse, mutton and pork meats	*cytochrome-b gene*	[51, 52]
	Spelt flour	ϒ-gliadin gene, GAG56D specific primer	[53]
	Meat, milks and milk derivative	Cytochrome b	[54]
	Mozzarella cheese	Cytochrome b	[55]
Random amplified polymorphic DNA (RAPD): A short random primer sequence is used to amplify the DNA randomly present in genome and generate species specific DNA fingerprints	Fish, fruit, mussel, vegetable foodstuff	Cytochrome b, 18S r DNA, ITS1	[56–58]
	Black pepper	OPC-1, OPC-4, OPC-6, OPC-7, OPJ09	[59]
	Saffron	OPA-14, MG11, MG12, AJ05	[60]
	Tea	OPF14	[61]
	Mediterranean Oregano	OPAG-06, OPAG-14, OPAG-18	[62]
	Herbal Medicine	OP-13B, OP-5A	[63]
	Virgin Olive oil	PLT253	[64, 65]

(Continued)

Table 5.4 PCR-based molecular techniques for food authentication. (*Continued*)

PCR-based techniques	Application	Primer/Target gene	References
	Turmeric powder	NR	[66]
	Chilli powder	NR	[67]
	Chinese drug (herba elephantopi)	OPC-06	[16, 168]
Amplified fragment length polymorphism (AFLP): First restriction enzyme used to digest genomic DNA, then synthetic adaptors ligated to the sticky ends of restriction fragment DNA and followed by PCR amplification is done to create fingerprint	Fish, olive oil, pasta	-	[69–72]
Inter-simple Sequence repeat (ISSR): Microsatellite sequences are used as primer for PCR amplification and resulted polymorphism in the number of repeats and length are used to create unique fingerprint	Olive oil, pig meat products, wine	-	[73, 74]

(*Continued*)

Table 5.4 PCR-based molecular techniques for food authentication. (*Continued*)

PCR-based techniques	Application	Primer/Target gene	References
Sequence characterized amplified region (SCAR): RAPD bands cloned and sequenced to make specific primers which amplify the genomic DNA under stringent condition for species differentiation	Fruit products,	NR	[75]
	Black pepper	P1, P2	[76]
	Saffron	SAF-L40, SAF-L70, SAF-L4	[77, 78]
		ScCo390	[79]
	Olive oil	NR	[80]
	Turmeric Powder	OPA01, OPE18	[81]
	Ginger	P3	[82]
	Herbal medicine	JG14	[83]
	Medicinal Echinacea species	OPA 20, OPA 10, OPA 11, OPA 17	[84]
Multiplex PCR: In a single PCR reaction multiple primer sets are used to amplify DNA at specific regions simultaneously for rapid analysis	Ginseng decoction and processed fish products	NR	[85, 86]

(*Continued*)

Table 5.4 PCR-based molecular techniques for food authentication. (*Continued*)

PCR-based techniques	Application	Primer/Target gene	References
Single stranded conformational polymorphism (SSCR): The amplified PCR products under non-denaturing conditions allowed to make defined conformation, followed by polyacrylamide gel electrophoresis to produce specific fingerprint	Cheese products, fish, meat products	-	[87–89]
Species-specific PCR: Identification of closely related species done by the PCR followed by electrophoresis	Fish and sea food, pig		[90, 91]
	Bread and pasta	Gliadin, glutanin	[92]
	Food allergen-hazelnut	Cor a 1.0401	[93]
	Food allergen-celery	Mannitol dehydrogenase	[94]
	Goats and mixture cheese	D-loop	[95]
	Mozerella cheese	Cytochrome oxidase I	[96]
		12s RNA	[97]
	Raw and heat treated milk	12s RNA	[98]

differentiation based on taxonomic and phylogenetic studies and detects genetic variations based on phylogeographic patterns. However, its applicability is limited due to poor reproducibility of results as slight variation in amplifying conditions affects the DNA pattern and makes it less reliable.

5.7.2 Application of Real-Time PCR in Adulterant Authentication

After the revealing of horsemeat scandal in the year 2013, DNA-based techniques get more attention to trace back the food supply chain. After this breakthrough, several PCR-based methods have been developed. Nowadays, real-time PCR (RTPCR) is used as most acknowledged analytical tool in food industry for analysis of food adulterant detection due to accuracy and reliability, as it can amplify highly fragmented DNA from processed sample [101]. Contrary to PCR, targeted DNA segment can be amplified as well as quantified by quantitative competitive PCR (QC-PCR) and RTPCR. This method is more reliable since it is highly suitable for the specific DNA detection needed for ingredient authentication. Different strategies for RTPCR diagnostic have developed including nonspecific detection independent of the target sequence using fluorescent dyes such as SYBR Green or by sequence-specific fluorescent oligonucleotide probes such as TaqMan probes or molecular beacons. Mostly mitochondrial gene and conserved sequences are the choice of target as it helps in species identification and available in many copies in cell. This method is rapid since it allows detection in early stages of amplification. This method is unique to unmask minced meat in vegetarian foods [99]. TaqMan-based RTPCR is another format where species-specific probes are used for the detection purpose. These techniques are used for the detection of raw and heat-treated meat mixture of chicken and turkey [100]. Application of RTPCR techniques used in various adulterant authentications is summarized in Table 5.5.

5.7.3 Isothermal Amplification Methods for Adulterant Identification

Loop-mediated isothermal amplification developed by Notomi and co-workers in 2000 is becoming increasingly popular because of its immense potential to use in field, easy to execute, and highly sensitive.

Table 5.5 Use of RTPCR techniques for adulterant authentication.

Application	Food/Product	Adulterant	Gene/Target	Limit of Detection (LOD)	References
Adulteration detection of basmati rice with non-basmati rice	Basmati Rice	Non-Basmati Rice	BAD2	1%	[102]
Simultaneous detection of wheat, and barley in food	wheat and barley	Gluten	PKABA1	NR	[103]
Adulteration detection of soft wheat (Triticum aestivum) with durum wheat (Triticum turgidum L.var. durum) and foodstuffs-based durum wheat	durum wheat	soft wheat	Microsatellite DNA	2.5%	[104]
Detection of allergenic almond protein in cumin	Cumin spice product	Prunus mahaleb (almond protein)	ITS*	1%	[105]
Adulteration detection of cereal (wheat, barley rye, oats) in gluten free foods	gluten free foods	wheat, barley rye, oats	ω-glidin (wheat); ω-secalin (rye), Hordey (barley); avenin (oat)	50 pg DNA/0.01–0.1%	[106]

(Continued)

Table 5.5 Use of RTPCR techniques for adulterant authentication. (Continued)

Application	Food/Product	Adulterant	Gene/Target	Limit of Detection (LOD)	References
Detection of potentially allergenic hazelnut residues (Corylus spp.) in food stuffs.	Food allergen	Hazelnut	hsp1	13 pg DNA/0.01% limit of detection	[107]
Detection of potentially allergenic peanut (Arachis hypogaea) in foods	peanut	Hazelnut	Cor a 1.04	50 pg DNA	[108]
		Allergenic peanut	Ara h1	Not reported	[109]
Detection of walnut residues in food	Food	Walnut residue	Ara h2	<0.001%	[110, 111]
			Ara h3	Not reported	[112]
			Jug r2	0.24ng DNA/0.01%	[113]
Detection of allergenic celery (Apium graveolens) in food	Food	Allergenic celery	Mannitol dehydrogenase	0.0005–0.001% and 0.01–0.001%	[114]
Origin and authenticity testing of virgin olive oil	Olive oil	Olea europeae cultivars	Plasma intrinsic protein	Not reported	[115]

(Continued)

Table 5.5 Use of RTPCR techniques for adulterant authentication. (*Continued*)

Application	Food/Product	Adulterant	Gene/Target	Limit of Detection (LOD)	References
Discrimination of *Euphorbia humifusa* and *E. maculate* (medicinal plants) from their adulterant	Euphoria spp.	Not specified	r DNA ITS*1	Not reported	[116]
Detection of adulteration in Dairy products	Mozzarella cheese	Cow, buffalo	Cytochrome b /nuclear growth hormone	0.1 ng	[117]
	Raw and heat treated milks	Cow, sheep	12S rRNA	0.5–10%	[118]
	Raw and heat treated milks	Bovine, buffalo	Cyt B,16S	220 ng/sample	[119]
	Bovine meat products (marinated, processed, fermented, dry-cured, halal)	Pork	12Sr RNA	0.1%,0.01 ng DNA	[120]

(*Continued*)

Table 5.5 Use of RTPCR techniques for adulterant authentication. (*Continued*)

Application	Food/Product	Adulterant	Gene/Target	Limit of Detection (LOD)	References
Detection of seafood adulteration	Foie gras	Mule duck, goose	12Sr RNA	0.01%DNA	[121]
	Cooked meat products	Beef, lamb, pork, chicken, turkey	Cytochrome b	0.1–0.05%DNA	[122]
	Cooked meat products Raw frozen canned	Cow, lamb, pork, chicken, turkey, ostrich	18 s RNA	0.03–0.80ng DNA	[123]
		Cow, pork, horse, wallaroo	ND6	<0.4ng DNA	[124]
		Cow, pork	Phospho-diesterase/rynodin/myostatin	<0.1%DNA	[125]
		Tuna	16S r RNA	NA	[126]

Contrary to PCR, amplification of specific target DNA performed under single temperature with the set of four to six distinct primers and a special DNA polymerase with high strand displacement activity. LAMP is advantageous over other PCR-based methods as it bypasses use of sophisticated and expensive thermal cycler, Limited risk of contamination, tolerance to some inhibitory material, easy to use and require less time to perform and the results can be seen directly in tube and more over field amenable [1]. Additionally, the detection limit of lamp is much lower and only a few copies are needed.

Other than LAMP, other isothermal amplification methods such as Helicase Dependent Amplification (HDA), Recombinase Polymerase Amplification (RPA), and Strand displacement amplification (SDA) have been reviewed [127].

5.7.4 Sequencing and Hybridization-Based Methods in Adulterant Identification

Each species has its own DNA sequence that can be used as a unique identifier and are commonly designated as a "DNA barcode". Insertion, deletion, or transfusion in mitochondrial or chloroplast gene regions acts as a key for differentiating and detecting biological adulterant from original foods and provides information on variation in species region [3]. Thus, variation in nucleotide sequence at specific region forms the basis for sequencing-based method. Adulterant detection can be done at the same time from variety of possible species by sequencing known DNA relative to adulterated one by hybridization method [128]. Next-generation sequencing is high-throughput sequencing (alignment) technique that provides extensive data of parallel short read per run. Among various NGS techniques, 454 sequencing employs pyro sequencing technique used for processed food material like fish feed, dairy, and dairy products [129]. Use of DNA bar coding for food authenticity has been adopted as the official regulatory test for seafood in US [130]. Authentication of herbal products [131] and identification of agricultural pests are also done by using hybridization method [132]. Accurate classification and identification of pest is critical and crucial when the species is quarantined [133].

Molecular method such as sequencing and hybridization-based method are irresistible for biological adulterant detection but requires prior information and knowledge of specific sequence for designing primer [134]. Sequencing and hybridization method requires large amount of DNA which is time consuming, laborious, and require stringent trials as compared to PCR-based methods [135]. NGS is powerful tool to analyze various sources DNA in single analysis and trace quantity of DNA [136]; and occurrence of amplification bias may cause inaccurate estimation of species composition and misleading results false negative or false positive [137]. NGS requires well-developed DNA reference library for sequence matching [138]. Selected examples of next-generation sequencing-based technique application in food authentication are given in Table 5.6.

5.8 Limitation in Use of Molecular-Based Methods for Adulterant Authentication

Molecular-based methods are considered as the best technique for the authenticity of food. However, some technical problems have to be taken in consideration to develop PCR, for amplifying DNA extracted from food samples. Food processing usually involves various treatments such as cooking, steaming, roasting, frying, and pasteurization, and due to this high heat treatment, DNA degrades into small fragments [150]. With the increase duration of heating and temperature, the DNA fragment size in meat and soybean product (836 bp to 162 bp) was progressively reduced into smaller fragment. Although DNA is reasonably stable in aqueous solution as well as under high pH, but if pH value very low (pH 3), then nicks will form in DNA due to depurination, which will lead to PCR failure [151]. DNA purity is essential for successful PCR amplification than DNA yield [152]. Food products consist of various PCR inhibitors like polyphenol, tannic acid, and secondary metabolites in the extracts of food products, oil, fats, animal tissues, meat, and seafood; hence, PCR amplification may fail if that has not been removed. Polysaccharides, polyphenolic compound and secondary metabolites, lipid, and casein interfere with the DNA purification. These compound bind to DNA and make it unavailable for the further process. Further high-lipid content in food and food products (butter and cheese) such as EDTA, silica, talc, iron oxide, calcium, and tannic acid are also limiting factors which affect the DNA extraction and PCR amplification. Thus, it is necessary to eliminate/reduce PCR inhibitor by optimizing DNA extraction procedure before the analysis [153].

Table 5.6 Application of next-generation sequencing (NGS) techniques in authentication of adulterants in processed food.

Techniques	Types of processed food	Target gene	Remark of techniques	References
Illumina MiSeq	Artisan cheese factory and cheese samples	16S rDNA gene	Facility-specific "house" microbiota and their role in shaping site specific characteristics in products	[139]
Roche 454 GS Junior	Spoiled retail foodstuffs	16S rDNA gene	Detection and characterization of spoilage microorganism in food	[140]
Hi Seq 2000 sequencer (Illumina)	Various meat products	Not available	Identification and quantification of heterogeneous meat products	[141]
454 Pyrosequencing Technology (Roche Diagnostics)	Milk and dairy products	16S rRNA gene	NGS reveals the changes in composition of milk microbiota which affect the nutritional value of milk products	[142]

(Continued)

Table 5.6 Application of next-generation sequencing (NGS) techniques in authentication of adulterants in processed food. (Continued)

Techniques	Types of processed food	Target gene	Remark of techniques	References
454 Pyrosequencing Technology (Roche Diagnostics)	Milk and kefir grains	16S rRNA gene	Identify the composition of different symbiotic microorganism present in milk kefir grain	[143]
Ion Personal Genome Machine System (Thermo Fisher Scientific)	Meat products	12S and 16S rRNA gene	multiple primer pairs used to avoid PCR amplification competition and inhibition	[144]
Roche 454 GS-FLX Titanium	Pig musculature	16S r DNA gene	Microorganism present in pork sample was dominated by psychrophilic spoilers; *E. coli* and can be used as marker in pork contamination cases	[145]
454 Pyrosequencing Technology (Roche Diagnostics, Basel, Switzerland	Fish feed	COI gene	NGS can differentiate composition of fish in feeds	[146]

(Continued)

Table 5.6 Application of next-generation sequencing (NGS) techniques in authentication of adulterants in processed food. (Continued)

Techniques	Types of processed food	Target gene	Remark of techniques	References
Ion Personal Genome Machine System (Thermo Fisher Scientific)	Oil-free tablets and gelatin capsule	ITS2 and *rbc*L region	Poor DNA recovery from plant extracts can cause NGS failure	[147]
Hi Seq 2000 sequencer (Illumina, san Diego, CA, USA)	GMO rice and noodles	Not available	Developed for routine analysis of GMO crops	[148]
Ion Personal Genome Machine System (Thermofischer Scientific, Walthem, MA, USA)	Candies	16S rRNA gene	Result consistent to PCR-cloning –sequencing, time and cost/sample can be reduced	[149]

5.9 Conclusion

Food adulteration is the major issue which is responsible for the economic losses of one's own country. It also affects health upon intake. Currently, various techniques are available which involve the modern use of instruments with full or partial automation. Most techniques undergo less time and simple procedures. The development of newer techniques with less time and cost is much needed for the authentication of foods. Ambient mass spectrometry is one of the techniques which provides higher authenticity and traceability. Various ionization techniques such as DART, DAPCI, EASI, and LESA coupled with mass spectrometry help the way to get better and accurate results. PCR and RT-PCR considered as gold standard for molecular-based techniques. With increasing importance on food origin and authentication, use of RT-PCR increased, as it is reliable, fast, and extremely sensitive. DNA from two different species in processed samples can be quantified by RT-PCR technique [154]. Advancement of RT-PCR analysis by newly developed Cy0 method allows reliable and precise quantification for PCR data by linear regression even in the presence of PCR inhibitors [155]. DNA quality is an important factor for molecular-based authentication. DNA often deteriorated by heat, irradiation, contamination, and presence of inhibitor or removed during processing; therefore, it is essential to develop quick protocol to extract DNA with more efficiency without damaging quality. With the advancement of novel molecular techniques like isothermal amplification, next-generation sequencing, genome wide studies, and abundant molecular data available in Gene Bank database provide invaluable insights for advancement of quick protocols for on-site evaluation of processed products for authentication and quantification.

References

1. Bansal, S., Singh, A., Mangal, M., Mangal, A., Kumar, S., Food Adulteration: Sources, Health Risks and Detection Methods. *Crit. Rev. Food Sci. Nutr.*, 57, 6, 1174–1189, 2015.
2. *Global food safety forum*, 2017, http://globalfoodsafetyforumorg/2017/11/eu-and-china-jointly-tackle-food-fraud/27 Nov2017.
3. Dhanya, K. and Sasilkumar, B., Molecular marker based adulteration detection in traded food and agriculture commodities of plant origin with special reference to spices. *Curr. Trends Biotechnol. Pharm.*, 4, 1, 454–489, 2010.
4. *Monthly Summary of Articles on Food Fraud and Adulteration*, European commission, 2020, RC tool Medisys, http://medisys.newsbrief.eu/.

5. Gonzalez, M.I., Hernández, H., González, C.J., Use of NIRS technology with a remote reflectance fibre-optic probe for predicting mineral composition (Ca, K, P, Fe, Mn, Na, Zn), protein and moisture in alfalfa. *Anal. Bioanal. Chem.*, 387, 2199–2205, 2007.
6. FSSAI, *Manual of Methods of Analysis of Foods (Spices and Condiments)*, Lab manual 10, Government of India, New Delhi, 2012.
7. Pruthi, J.S. and Kulkarni, B.M., A simple technique for the rapid and easy detection of papaya seeds in black pepper berries. *Indian Food Packer*, 23, 51–52, 1969.
8. Pruthi, J.S., *Spices and condiments chemistry, microbiology, technology*, p. 449, Academic Press Inc., New York, 1980.
9. Pearson, D., *The chemical analysis of foods*, p. 575, Churchill and Livingstone, New York, 1976.
10. Louveaux, J., Maurizio, A., Vorwohl, G., Methods of melissopalynology. *Bee World*, 59, 4, 139–157, 1978.
11. Dustmann, J.H. and Ohe, K., *Scanning electron microscopic studies on pollen from honey. IV. Surface pattern of pollen of Sapium sebiferum and Euphorbia spp (Euphorbiaceae)*, Apidologie, Springer Verlag, 24(1), 59–66, 1993, https://hal.archives-ouvertes.fr/hal-00891057.
12. Joshi, V.C., Pullala, V.S., Khan, I.A., Rapid and easy identification of Illicium verum Hook. f. and its adulterant Illicium anisatum Linn. by fluorescent microscopy and gas chromatography. *J. Assoc. Off. Anal. Chem. Int.*, 88, 703–706, 2005.
13. Cserháti, T., Chromatography in authenticity and traceability tests of vegetable oils and dairy products: a review. *Biomed. Chromatogr.*, 19, 3, 183–190, 2005.
14. Wang, M., Li, R., Zou, S., Determination of carbofuran residue in aquatic products by gas chromatography. *Chin. J. Chromatogr.*, 26, 6, 775–777, 2008.
15. Wu, Y. and Chang, C., Biochemical genetic analysis of isozymes in three populations of Misgurnusanguilli caudatus. *J. Northwest A&F Univ. (Natural Sci. Edition)*, 37, 11, 37–42, 2009.
16. Shaw, P.C., Ngan, F.N., But, P.P.H., Wang, J., Molecular markers in Chinese medicinal materials, in: *Authentication of Chinese medicinal materials by DNA technology*, Singapore, pp. 1–23, World Scientific Publishing Co, 2002. https://www.worldscientific.com/doi/abs/10.1142/9789812706591_0001
17. Blanch, G.P., Mar Caja, M., Castillo, M.L., Herraiz, M., Comparison of different methods for the evaluation of the authenticity of olive oil and hazelnut oil. *J. Agric. Food Chem.*, 46, 3153–3157, 1998.
18. Silva, B.M., Andrade, P.B., Mendes, G.C., Valentao, P., Seabra, R.M., Ferreira, M.A., Analysis of phenolic compounds in the evaluation of commercial Quince Jam Authenticity. *J. Agric. Food Chem.*, 48, 2853–2857, 2000.
19. Mouly, P., Gaydou, E.M., Auffray, A., Simultaneous separation of flavone glycosides and polymethoxylated flavones in citrus juices using liquid chromatography. *J. Chromatogr.*, A800, 171–179, 1998.

20. McDowell, I., Taylor, S., Gay, C., The Phenolic Pigment Composition of Black Tea Liquors Part I: Predicting Quality. *J. Agric. Food Chem.*, 69, 467–474, 1995.
21. Calabrese, M., Stancher, B., Riccobon, P., High-Performance Liquid Chromatography Determination of Proline Isomers in Italian Wines. *J. Agric. Food Chem.*, 69, 361–366, 1995.
22. Haghighi, B., Feizy, J., Kakhk, A.H., LC Determination of Adulterated Saffron Prepared by Adding Styles Colored with Some Natural Colorants. *Chromatographia*, 66, 5/6, 325–332, 2007.
23. Hao, X., Hu, J.Z., Zhong, H., Detection of Benzo (a) pyrene in meat food by high pressure liquid chromatography (HPLC). *Food Sci. Technol.*, 7, 219–221, 2007.
24. Li, Y., Zheng, F., Wang, M., Rapid screening and confirmation of 156 pesticide residues in concentrated fruit and vegetable juices using liquid chromatography-tandem mass spectrometry. *Chin. J. Chromatogr.*, 27, 2, 127–137, 2009.
25. Nogueira, J.M.F. and Nascimento, A.M.D., Analytical characterization of Madeira wine. *J. Agric. Food Chem.*, 47, 566–575, 1999.
26. Matute, A.I.R., Soria, A.C., Castro, I.M., Sanz, M.L., A new methodology based on GC-MS to detect honey adulteration with commercial syrups. *J. Agric. Food Chem.*, 55, 18, 7264–9, 2007.
27. Dhiman, B. and Singh, M., Molecular detection of cashew husk (Anacardium occidentale) adulteration in market samples of dry tea (Camelliasinensis). *Planta Med.*, 69, 882–884, 2003.
28. Jiang, J.S., Zhao, B., Zhou, L., Simultaneous determination of 36 pesticide residues in tea by GC-MS. *Food Sci.*, 30, 14, 276–330, 2009.
29. Haughey, S.A., Graham, S.F., Cancouet, E., Elliott, C.T., The application of Near- Infrared Reflectance Spectroscopy (NIRS) to detect melamine adulteration of soya bean meal. *Food Chem.*, 136, 3–4, 1557–1561, 2012.
30. Ozen, B.F. and Mauer, L.J., Detection of hazelnut oil adulteration using FTIR spectroscopy. *J. Agric. Food Chem.*, 50, 3898–3901, 2002.
31. Nicolaou, N., Xu, Y., Goodacr, R., Fourier transform infrared spectroscopy and multivariate analysis for the detection and quantification of different milk species. *J. Dairy Sci.*, 93, 5651–5660, 2010.
32. Kelly, J.F. and Downey, G., Detection of sugar adulterants in apple juice using fourier transform infrared spectroscopy and chemometrics. *J. Agric. Food Chem.*, 53, 9, 3281–3286, 2005.
33. Di Stefano, V., Avellone, G., Bongiorno, D., Cunsolo, V., Muccilli, V., Sforza, S., Dossena, A., Drahos, L., Vékey, K., Applications of liquid chromatography–mass spectrometry for food analysis. *J. Chromatogr.*, 1259, 74–85, 2012.
34. Hohmann, M., Differentiation of Organically and Conventionally Grown Tomatoes by Chemometric Analysis of Combined Data from Proton Nuclear Magnetic Resonance and Mid-Infrared Spectroscopy and Stable Isotope Analysis. *J. Agric. Food Chem.*, 63, 43, 9666–9675, 2015.

35. Drivelos, S.A. and Georgiou, C.A., Multi-element and multi-isotope-ratio analysis to determine the geographical origin of foods in the European Union. *TrAC - Trends Anal. Chem.*, 40, 38–51, 2012.
36. Poulli, K.I., Mousdis, G.A., Georgiou, C.A., Rapid synchronous fluorescence method for virgin olive oil adulteration assessment. *Food Chem.*, 105, 1, 369–375, 2007.
37. Qi, X., Zou, M., Liu, F., Rapid Authentication of Olive Oil by Raman Spectroscopy Using Principal Component Analysis. *Analytical Lett. J.*, 44, 2209–2220, 2011.
38. Danezis, G.P., Tsagkaris, A.S., Federica, C., Vladimir, B., Georgiou, C.A., Food authentication: Techniques, trends and emerging approaches. *Trends Anal. Chem.*, 85, 123–132, 2016, http://dx.doi.org/doi: 10.1016/j.trac.2016.02.026.
39. De-Souza, E.M.T., Arruda, S.F., Brandao, P.O., Siqueira, E.M.A., Electrophoretic analysis to detect and quantify additional whey in milk and dairy beverages. *Ciênc. Tecnol.*, 20, 3, 314–317, 2000.
40. Cartoni, G., Coccioli, F., Jasionowska, R., Masci, M., Determination of cows' milk in goats' milk and cheese by capillary electrophoresis of the whey protein fractions. *J. Chromatogr. A*, 846, 1–2, 135–141, 1999.
41. Vemireddy, L.R., Archak, S., Nagaraju, J., Capillary electrophoresis is essential for microsatellite marker based detection and quantification of adulteration of Basmati rice (Oryza sativa). *J. Agric. Food Chem.*, 55, 8112–8117, 2007.
42. Zhang, J., Zhang, X., Lorena, Victor, C., Review of the current application of fingerprinting allowing detection of food adulteration and fraud in China. *Food Control*, 22, 1126–1135 Elsevier, 2011.
43. Zhihua, L., Shouju, J., Yinghui, D., Xueliang, C., Rong-mao, L., Yongpu, Z., The isozymes in different populations of clam Meretrix. *J. Dalian Fish. Univ.*, 6, 525–530, 2009.
44. Xue, H., Hu, W., Son, H., Han, Y., Yang, Z., Indirect ELISA for detection and quantification of bovine milk in goat milk. *J. Food Sci. Technol.*, 31, 24, 370–373, 2010.
45. FSSAI, *Manual of Methods of Analysis of Foods (Spices and Condiments)*, GOI, New Delhi, India, 2012.
46. Hernandez, P.E., Martín, R., García, T., Morales, P., Anguita, G., Haza, A.I., González, Sanz, B., Antibody based analytical methods for meat species determination and detecting adulteration of milk. *Food Agric. Immunol.*, 6, 1, 95–104, 1994.
47. Casale, M., Oliveri, P., Armanino, C., NIR and UV Vis spectroscopy, artificial nose and tongue: comparison of four fingerprinting techniques for the characterization of Italian red wines. *Anal. Chim. Acta*, 668, 143–148, 2010.
48. Singh, V.P., Pathak, V., Nayak, N.K., Verma, A.K., Umaraw, P., Recent developments in meat species speciation – a review. *J. Livest. Sci.*, 5, 49–64, 2014.
49. Yip, P.Y., Chau, C.F., Mak, C.Y., Kwan, H.S., DNA methods for identification of Chinese medicinal materials. *Chin. Med.*, 2, 9, 2007.

50. Marieschi, M., Torelli, A., Poli, F., Sacchetti, G., Bruni, R., RAPD-based method for the quality control of mediterranean oregano and its contribution to pharmacognostic techniques. *J. Agric. Food Chem.*, 57, 5, 1835–40, 2009.
51. Doosti, A., Ghasemi, D.P., Rahimi, E., Molecular assay to fraud identification of meat products. *J. Food Sci. Technol.*, 51, 1, 148–152, 2014.
52. Kumar, D., Singh, S.P., Karabasanavar, N.S., Singh, R., Umapathi, V., Authentication of beef, carabeef, chevon, mutton and pork by a PCR-RFLP assay of mitochondrial cytb gene. *J. Food Sci. Technol.*, 51, 11, 3458–3463, 2014.
53. Buren, M., Stadler, M., Luthy, J., Detection of wheat adulteration of spelt flour and products by PCR. *Eur. Food Res. Technol.*, 212, 234–239, 2001.
54. Lanzilao, I., Burgalassi, F., Fancelli, S., Settimelli, M., Fani, R., Polymerase chain reaction-restriction fragment length polymorphism analysis of mitochondrial cyt-b gene from species of dairy interest. *AOAC Int.*, 88, 128–135, 2005.
55. Rea, S., Chikuni, K., Branciari, R., Sangamayya, R.S., Ranucci, D., Avellini, P., Use of duplex polymerase chain reaction (duplex-PCR) technique to identify bovine and water-buffalo milk used in making mozzarella cheese. *J. Dairy Res.*, 68, 689–698, 2001.
56. Ramella, M.S., Kroth, M.A., Tagliari, C., Arisi, A.C.M., Optimization of random amplified polymorphic DNA protocol for molecular identification of Lophius gastrophysus. *Food Sci. Technol. (Campinas)*, 25, 733–735, 2005.
57. Rego, I., Martinez, A., Gonzalez, T.A., Vieites, J., Leira, F., Mendez, J., PCR technique for identification of mussel species. *J. Agric. Food Chem.*, 50, 7, 1780–1784, 2002.
58. Spychaj, A., Mozdziak, P.E., Pospiech, E., PCR methods in meat species identification as a tool for the verification of regional and traditional meat products. *Acta Sci. Pol. Technol. Aliment.*, 8, 2, 5–20, 2009.
59. Khan, S.K., Mirza, J., Anwar, F., Abdin, M.Z., Development of RAPD marker for authentication of Piper nigrum (L). *Environ. We Int. J. Sci. Tech.*, 5, 47–56, 2010.
60. Babaei, S., Talebi, M., Bahar, M., Developing an SCAR and ITS reliable multiplex PCR-based assay for safflower adulterant detection in saffron samples. *Food Control*, 35, 1, 323–328, 2013.
61. Mneney, E.E., Review of the use of molecular marker technologies for genetic improvement of cashew (Anacardium occidentale L) in Tanzania. *Proceedings of Second International Cashew Conference*, 2010.
62. Marieschi, M., Torelli, A., Poli, F., Sacchetti, G., Bruni, R., RAPD-based method for the quality control of mediterranean oregano and its contribution to pharmacognostic techniques. *J. Agric. Food Chem.*, 57, 5, 1835–40, 2009.
63. Shim, Y.H., Park, C.D., Kim, D.H., Cho, J.H., Cho, M.H., Kim, H.J., Identification of Panax species in the herbal medicine preparations using gradient PCR method. *Biol. Pharm. Bull.*, 28, 671–676, 2005.
64. Busconi, M., Foroni, C., Corradi, M., Bongiorni, C., Cattapan, F., Fogher, C., DNA extraction from olive oil and its use in the identification of the production cultivar. *Food Chem.*, 83, 127–134, 2003.

65. Muzzolupo, I. and Peri, E., Recovery and characterisation of DNA from virgin olive oil. *Eur. Food Res. Technol.*, 214, 528–531, 2002.
66. Sasikumar, B., Syamkumar, S., Remya, R., John, Z.T., PCR based detection of adulteration in the market samples of turmeric powder. *Food Biotechnol.*, 18, 299–306, 2005.
67. Dhanya, K., Syamkumar, S., Jaleel, K., Sasikumar, B., Random amplified polymorphic DNA technique for the detection of plant based adulterants in chilli powder (Capsicum annuum). *J. Spices Aromat. Crops*, 17, 75–81, 2008.
68. Cao, H., But, P.P., Shaw, P.C., Authentication of the Chinese drug "Kudidan" (herba Elephantopi) and its substitutes using random-primed polymerase chain reaction (PCR). *Acta Pharm. Sin.*, 31, 543–553, 1996.
69. Miggiano, E., De Innocentiis, S., Ungaro, A., Sola, L., Crosetti, D., AFLP and microsatellites as genetic tags to identify cultured gilthead seabream escapees: Data from a simulated floating cage breaking event. *Aquac. Int.*, 13, 1, 137, 2005.
70. Montemurro, C., Pasqualone, A., Simeone, R., Sabetta, W., Blanco, A., AFLP molecular markers to identify virgin olive oils from single Italian cultivars. *Eur. Food Res. Technol.*, 226, 6, 1439, 2007.
71. Pafundo, S., Agrimonti, C., Marmiroli, N., Traceability of plant contribution in olive oil by amplified fragment length polymorphisms. *J. Agric. Food Chem.*, 53, 18, 6995–7002, 2005.
72. Terzi, V., Morcia, C., Giovanardi, D., D'Egidio, M.G., Stanca, A.M., Faccioli, P., DNA-based analysis for authenticity assessment of monovarietal pasta. *Eur. Food Res. Technol.*, 219, 4, 428–431, 2004.
73. Martins-Lopes, P., Gome, S., Santos, E., Guedes-Pinto, H., DNA markers for Portuguese olive oil fingerprinting. *J. Agric. Food Chem.*, 56, 24, 11786–1179, 2008.
74. Pereira, L., Martins-Lopes, P., Batista, C., Zanol, G.C., Clímaco, P., Brazão, J., Molecular markers for assessing must varietal origin. *Food Anal. Methods*, 5, 6, 1252–1259, 2012.
75. Marieschi, M., Torelli, A., Beghe, D., Bruni, R., Authentication of Punica granatum L: Development of SCAR markers for the detection of 10 fruits potentially used in economically motivated adulteration. *Food Chem.*, 202, 438–444, 2016.
76. Dhanya, K., *Detection of probable plant based adulterants in selected powdered market samples of spices using molecular techniques* PhD thesis, p. 251, Mangalore University, Mangalore, India, 2009.
77. Javanmardi, N., Bagheri, A., Moshtaghi, N., Sharifi, A., Kakhki, A.H., Identification of Safflower as a fraud in commercial Saffron using RAPD/SCAR Marker. *Cell Mol. Res.*, 3, 1, 31–37, 2011.
78. Marieschi, M., Torelli, A., Bruni, R., Quality control of saffron (Crocus sativus L) Development of SCAR markers for the detection of plant adulterants used as bulking agents. *J. Agric. Food Chem.*, 60, 44, 10998–1004, 2012.

79. Torelli, A., Mureischi, M., Bruni, R., Authentication of saffron in different processed retail products by means of SCAR markers. *Food Control*, 36, 126–131, 2014.
80. Pafundo, S., Agrimonti, C., Maestri, E., Marmiroli, N., Applicability of SCAR markers to food genomics: olive oil traceability. *J. Agric. Food Chem.*, 55, 6052–6059, 2007.
81. Dhanya, K., Syamkumar, S., Siju, S., Sasikumar, B., Sequence characterized amplified region markers: A reliable tool for adulterant detection in turmeric powder. *Food Res. Int.*, 44, 9, 2889–2895, 2011.
82. Chavan, P., Warude, D., Joshi, K., Patwardhan, B., Development of SCAR (sequence characterized amplified region) markers as a complementary tool for identification of ginger Zingiber officinale Roscoe) from crude drugs and multi-component formulations. *Biotechnol. Appl. Biochem.*, 50, 61–69, 2008.
83. Choi, Y.E., Ahn, C.H., Kim, B.B., Yoon, E.S., Development of species specific AFLP-derived SCAR marker for authentication of Panax japonicas. *Biol. Pharm. Bull.*, 31, 135–138, 2008.
84. Nieri, P., Adinolfi, B., Morelli, I., Breschi, M.C., Simoni, G., Martinotti, E., Genetic characterization of the three medicinal Echinacea species using RAPD analysis. *Planta Med.*, 69, 7, 685–6, 2003.
85. Catanese, G., Manchado, M., Fernández-Trujillo, A., Infante, C., A multiplex-PCR assay for the authentication of mackerels of the genus Scomber in processed fish products. *Food Chem.*, 122, 1, 319–326, 2010.
86. Lo, Y.T., Li, M., Shaw, P.C., Identification of constituent herbs in ginseng decoctions by DNA markers. *Chin. Med.*, 10, 1, 1, 2015.
87. Asensio, L., Gonzalez, I., Fernandez, A., Rodriguez, M.A., Hernandez, P.E., Garcia, T., PCR-SSCP: A simple method for the authentication of grouper (Epinephelus guaza), wreck fish (Polyprion americanus), and Nile perch (Lates niloticus) fillets. *J. Agric. Food Chem.*, 49, 4, 1720–1723, 2001.
88. Csikós, Á., Tisza, Á., Simon, Á., Gulyás, G., Jávor, A., Czeglédi, L., Species identification in meat and cheese products by PCR-single strand conformation polymorphism (PCRSSCP) and DNA sequencing. *Anim. Welf. Ethol. Hous. Syst.*, 11, 2, 78–83, 2015.
89. Schiefenhovel, K. and Rehbein, H., Differentiation of Sparidae species by DNA sequence analysis, PCR-SSCP and IEF of sarcoplasmic proteins. *Food Chem.*, 138, 1, 154–160, 2013.
90. Rasmussen, R.S., Morrissey, M.T., Hebert, P.D.N., DNA-Based Methods for the Identification of Commercial Fish and Seafood Species. *J. Agric. Food Chem.*, 57, 8379–8385, http://10.1021/jf901618z, 2009.
91. Arun, K., Kumar, R.R., Sharma, B.D., Mendiratta, S.K., Sharma, D., Gokulakrishnan, P., Species-specific polymerase chain reaction (PCR) assay for identification of pig (Sus domesticus) meat. *Afr. J. Biotechnol.*, 11, 89, 15590–15595, 2012.
92. Terzi, V., Malnati, M., Barbanera, M., Stanca, A.M., Faccioli, P., Development of analytical systems based on real-time PCR for Triticum species specific

detection and quantitation of bread wheat contamination in semolina and pasta. *J. Cereal Sci.*, 38, 87–94, 2003.
93. Holzhauser, T., Wangorsch, A., Vieths, S., Polymerase chain reaction (PCR) for detection of potentially allergenic hazelnut residues in complex food matrixes. *Eur. Food Res. Technol.*, 211, 360–365, 2000.
94. Dovicovicova, L., Olexová, L., Pangallo, D., Siekel, P., Polymerase chain reaction (PCR) for the detection of celery (Apium graveolens) in food. *Eur. Food Res. Technol.*, 218, 493–495, 2004.
95. Maudet, C. and Taberlet, P., Detection of cows' milk in goats' cheeses inferred from mitochondrial DNA polymorphism. *J. Dairy Res.*, 68, 229–235, 2001.
96. Feligini, M., Bonizzi, I., Curik, V.C., Parma, P., Greppi, G.F., Enne, G., Detection of adulteration in Italian mozzarella cheese using mitochondrial DNA templates as biomarkers. *Food Technol. Biotechnol.*, 43, 91–95, 2005.
97. López-Calleja, I., Alonso, I.G., Fajardo, V., Rodríguez, M.A., Hernández, P.E., García, T., Martín, R., PCR detection of cows' milk in water buffalo milk and mozzarella cheese. *Int. Dairy J.*, 15, 1122–1129, 2005a.
98. López-Calleja, I., González, I., Fajardo, V., Martín, I., Hernández, P.E., García, T., Martín, R., Application of polymerase chain reaction to detect adulteration of sheep's milk with goats' milk. *J. Dairy Sci.*, 88, 3115–3120, 2005b.
99. Cheng, C.Y., Shi, Y.C., Lin, S.R., Use of real-time PCR to detect surimi adulteration in vegetarian foods. *J. Mar. Sci. Technol.*, 20, 5, 570–574, 2012.
100. Kesmen, Z., Yetiman, A.E., Sahin, F., Yetim, H., Detection of Chicken and Turkey Meat in Meat Mixtures by Using Real-Time PCR Assays. *J. Food Sci.*, 77, 2, C167–173, 2012.
101. Soares, S., Amaral, J.S., Mafra, I., Oliveira, M.B., Quantitative detection of poultry meat adulteration with pork by a duplex PCR assay. *Meat Sci.*, 85, 3, 531–536, 2010.
102. Lopez, S.J., TaqMan based real time PCR method for quantitative detection of basmati rice adulteration with non-basmati rice. *Eur. Food Res. Technol.*, 227, 619–622, 2008.
103. Ronning, S.B., Berdal, K.G., Andersen, C.B., Holst-Jensen, A., Novel reference gene PKABA1, used in a duplex real-time polymerase chain reaction for detection and quantitation of wheat- and barley-derived DNA. *J. Agric. Food Chem.*, 54, 682–697, 2006.
104. Pasqualone, A., Montemurro, C., Grinn- Gofron, A., Sonnante, G., Blanco, A., Detection of soft wheat in semolina and durum wheat bread by analysis of DNA microsatellites. *J. Agric. Food Chem.*, 55, 3312–3318, 2007.
105. Malcolm, B., Michael, W., Timothy, W., Laurie, H., Kirstin, G., Gavin, N., Development of a Real-Time PCR Approach for the Specific Detection of Prunus mahaleb. *Food Nutr. Sci.*, 7, 703–710, 2016.
106. Sandberg, M., Lundberg, L., Ferm, M., Yman, I.M., Real time PCR for the detection and discrimination of cereal contamination in gluten free foods. *Eur. Food Res. Technol.*, 217, 344–349, 2003.

107. Piknova, L., Pangallo, D., Kuchta, T., A novel real-time polymerase chain reaction (PCR) method for the detection of hazelnuts in food. *Eur. Food Res. Technol.*, 226, 1155–1158, 2008.
108. Arlorio, M., Cereti, E., Coisson, J.D., Travaglia, F., Martelli, A., Detection of hazelnut (Corylusspp) in processed foods using real-time PCR. *Food Control*, 18, 140–148, 2007.
109. Zhang, W.J., Qin, C.X., Guan, Q.C., Analytical Methods, Detection of peanut (Arachis hypogaea) allergen by Real-time PCR method with internal amplification control. *Food Chem.*, 174, 547–552, 2015.
110. Stephan, O. and Vieths, S., Development of a real-time PCR and a sandwich ELISA for detection of potentially allergenic trace amounts of peanut (Arachis hypogaea) in processed foods. *J. Agric. Food Chem.*, 52, 3754–3760, 2004.
111. Hird, H., Lloyd, J., Goodier, R., Brown, J., Reece, P., Detection of peanut using real-time polymerase chain reaction. *Eur. Food Res. Technol.*, 217, 265–268, 2005.
112. Scaravelli, E., Brohee, M., Marchelli, R., Hengel, A.J., Development of three real-time PCR assays to detect peanut allergen residue in processed food products. *Eur. Food Res. Technol.*, 227, 857–869, 2008.
113. Brezna, B., Hudecova, L., Kuchta, T., A novel real-time polymerase chain reaction (PCR) method for the detection of walnuts in food. *Eur. Food Res. Technol.*, 223, 373–377, 2006.
114. Mustorp, S., Axelsson, C.E., Svensson, U., Holck, A., Detection of celery (Apium graveolens), mustard (Sinapis alba, Brassica juncea, Brassica nigra) and sesame (Sesamum indicum) in food by real-time PCR. *Eur. Food Res. Technol.*, 226, 771–778, 2008.
115. Wu, Y., Chen, Y., Wang, Y.G.J., Xu, B., Huang, W., Yuan, F., Detection of olive oil using the Evagreen real-time PCR method. *Eur. Food Res. Technol.*, 227, 1117–1124, 2008.
116. Xue, H.G., Wang, H., Li, D.Z., Xue, C.Y., Wang, Q.Z., Differentiation of the traditional Chinese medicinal plants Euphorbia humifusa and E maculate from adulterants by TaqMan real-timepolymerase chain reaction. *Planta Med.*, 74, 302–304, 2008.
117. Lopparelli, R.M., Cardazzo, B., Balzan, S., Giaccone, V., Novelli, E., Real time TaqMan polymerase chain reaction detection and quantification of cow DNA in pure water buffalo Mozzarella cheese: Method validation and its application on commercial samples. *J. Agric. Food Chem.*, 55, 3429–3434, 2007.
118. López-Calleja, I., González, I., Fajardo, V., Martín, I., Hernández, P.E., García, T., Martín, R., Real-time TaqMan PCR for quantitative detection of cows' milk in ewes' milk mixtures. *Int. Dairy J.*, 17, 729–736, 2007.
119. Drummond, M.G., Brasil, B.S.A.F., Dalsecco, L.S., Brasil, R.S.A.F., Teixeira, L.V., Oliveira, D.A.A., A versatile real-time PCR method to quantify bovine contamination in buffalo products. *Food Control*, 29, 131–137, 2013.
120. Rodriguez, M.A., Garcia, T., Gonzalez, I., Hernandez, P.E., Martin, R., TaqMan real time PCR for the detection quantification of pork in meat mixtures. *Meat Sci.*, 70, 113–120, 2005.

121. Rodríguez, M.A., García, T., González, I., Asensio, L., Hernández, P.E., Martín, R., Quantification of mule duck in goose foie gras using TaqMan real time PCR. *J. Agric. Food Chem.*, 52, 1478–1483, 2004.
122. Dooley, J.J., Paine, K.E., Garrett, S.D., Brown, H.M., Detection of meat species using TaqMan real-time PCR assays. *Meat Sci.*, 68, 431–438, 2004.
123. Lopez-Andreo, M., Lugo, L., Garrido-Pertierra, A., Prieto, M.I., Puyet, A., Identification and quantitation of species in complex DNA mixtures by real-time polymerase chain reaction. *Anal. Biochem.*, 339, 73–82, 2005.
124. López-Andreo, M., Garrido-Pertierra, A., Puyet, A., Evaluation of post polymerase chain reaction melting tempreture analysis for meat species identification in mixed DNA samples. *J. Agric. Food Chem.*, 54, 7973–7978, 2006.
125. Laube, I., Spiegelberg, A., Butschke, A., Zagon, J., Schauzu, M., Kroh, L., Broll, H., Methods for the detection of beef and pork in foods using real-time polymerase chain reaction. *Int. J. Food Sci. Technol.*, 38, 111–118, 2003.
126. Lopez, I. and Pardo, M.A., Application of relative quantification TaqMan real-time polymerase chain reaction technology for the identification and quantification of Thunnus alalunga and Thunnus albacares. *J. Agric. Food Chem.*, 53, 4554–4560, 2005.
127. Deng, H. and Gao, Z., Bioanalytical applications of isothermal nucleic acid amplification techniques. *Anal. Chim. Acta*, 853, 30–45, 2015.
128. Carles, M., Cheung, M.K., Moganti, S., Dong, T.T., Tsim, K.W., Ip, N.Y., Sucher, N.J., A DNA microarray for the authentication of toxic traditional Chinese medicinal plants. *Planta Med.*, 71, 580–584, 2005.
129. Lo, Y.T. and Shaw, P.C., DNA-based techniques for authentication of processed food and food supplements. *Food Chem.*, 240, 767–774, 2018.
130. Handy, S.M., Deeds, J.R., Ivanova, N.V., Hebert, P.D.N., Hanner, R., Ormos, A., Weigt, L.A., Moore, M., Yancy, H.F., A single laboratory validated method for the generation of DNA barcodes for the identification of fish for regulatory compliance. *J. AOAC Int.*, 94, 201–210, 2011.
131. Newmaster, S.G., Grguric, M., Shanmughanandhan, D., Ramalingam, S., Ragupathy, S., DNA detects contamination and substitution in North American herbal products. *BMC Med.*, 11, (1), 1–13, 2013.
132. Frewin, A., Scott, D.C., Hanner, R., DNA barcoding for plant protection: applications and summary of available data for arthropod pests. *CAB Rev.*, 8, 1–13, 2013.
133. Naaum, A.M.N., The University of Guelph, Ontario, Canada, *Methods of Species and Product Authenticity and Traceability Testing Using DNA Analysis for Food and Agricultural Applications*. (Thesis), The University of Guelph, Ontario, Canada 2014, Naaum_amanda_201404_PhDpdf.
134. Lockley, A.K. and Bardsley, R.G., DNA-based methods for food authentication. *Trends Food Sci. Technol.*, 11, 67–77, 2000.
135. Zammatteo, N., Lockman, L., Brasseur, F., De, P.E., Lurquin, C., Lobert, P.E., Hamels, S., Boon, T., Remacle, J., DNA microarray to monitor the expression of MAGE-A genes. *Clin. Chem.*, 48, 25–34, 2002.

136. Burns, M., Wiseman, G., Knight, A., Bramley, P., Foster, L., Rollinson, S., Measurement issues associated with quantitative molecular biology analysis of complex food matrices for the detection of food fraud. *Analyst*, 141, 1, 45–61, 2016.
137. Thudi, M., Li, Y., Jackson, S.A., May, G.D., Varshney, R.K., Current state-of art of sequencing technologies for plant genomics research. *Briefings Funct. Genomics*, 11, 1, 3–11, 2012.
138. Egan, A.N., Schlueter, J., Spooner, D.M., Applications of next-generation sequencing in plant biology. *Am. J. Bot.*, 99, 2, 175–185, 2012.
139. Bokulich, N.A. and Mills, D.A., Facility-specific "house" microbiome drives microbial landscapes of artisan cheese making plants. *Appl. Environ. Microbiol.*, 79, 5214–5223, 2013, 101128/AEM00934-13.
140. Pothakos, V., Taminiau, B., Huys, G., Nezer, C., Daube, G., Devlieghere, F., Psychrotrophic lactic acid bacteria associated with production batch recalls and sporadic cases of early spoilage in Belgium between 2010 and 2014. *Int. J. Food Microbiol.*, 191, 157–163, 2014, 101016/jijfoodmicro201409013.
141. Ripp, F., Krombholz, C.F., Liu, Y., Weber, M., Schafer, A., Schmidt, B., All-Food-Seq (AFS): A quantifiable screen for species in biological samples by deep DNA sequencing. *BMC Genomics*, 15, 639, 2014.
142. Zhang, R., Huo, W., Zhu, W., Mao, S., Characterization of bacterial community of raw milk from dairy cows during subacute ruminal acidosis challenge by highthroughput sequencing. *J. Sci. Food Agric.*, 95, 5, 1072–1079, 2015.
143. Garofalo, C., Osimani, A., Milanovic, V., Aquilanti, L., De, F.F., Stellato, G., Bacteria and yeast microbiota in milk kefir grains from different Italian regions. *Food Microbiol.*, 49, 123–133, 2015.
144. Bertolini, F., Ghionda, M.C., D'Alessandro, E., Geraci, C., Chiofalo, V., Fontanesi, L., A next generation semiconductor based sequencing approach for the identification of meat species in DNA mixtures. *PLoS One*, 10, 4, 0121701, 2015.
145. Mann, E., Wetzels, S.U., Pinior, B., Metzler-Zebeli, B.U., Wagner, M., Schmitz-Esser, S., Psychrophile spoilers dominate the bacterial microbiome in musculature samples of slaughter pigs. *Meat Sci.*, 117, 36–40, 2016, 101016/jmeatsci201602034.
146. Khallaf, G.A., Osman, A.G.M., Carleos, C.E., Garcia, V.E., Borrell, Y.J., A case study for assessing fish traceability in Egyptian aqua feed formulations using pyrosequencing and metabarcoding. *Fish. Res.*, 174, 143–150, 2016.
147. Ivanova, N.V., Kuzmina, M.L., Braukmann, T.W., Borisenko, A.V., Zakharov, E.V., Authentication of herbal supplements using next-generation sequencing. *PLoS One*, 11, 5, 0156426, 2016.
148. Willems, S., Fraiture, M.A., Deforce, D., De, K.S.C., De, L.M., Ruttink, T., Statistical framework for detection of genetically modified organisms based on Next Generation Sequencing. *Food Chem.*, 192, 788–798, 2016.

149. Colmenero, M.M., Martinez, J.L., Roca, A., Garcia, V.E., NGS tools for traceability in candies as high processed food products: Ion Torrent PGM versus conventional PCR-cloning. *Food Chem.*, 214, 631–636, 2017.
150. Novak, J., Grausgruber, G.S., Lukas, B., DNA-based authentication of plant extracts. *Food Res. Int.*, 40, 3, 388–392, 2007.
151. Bauer, T., Weller, P., Hammes, W.P., Hertel, C., The effect of processing parameters on DNA degradation in food. *Eur. Food Res. Technol.*, 217, 4, 338–343, 2003.
152. Sarkinen, T., Staats, M., Richardson, J.E., Cowan, R.S., Bakker, F.T., How to open the treasure chest? Optimising DNA extraction from herbarium specimens. *PLoS One*, 7, 8, 43808, 2012.
153. Di Pinto, A., Forte, V.T., Guastadisegni, M.C., Martino, C., Schena, F.P., Tantillo, G., A comparison of DNA extraction methods for food analysis. *Food Control*, 18, 76–80, 2007.
154. Deagle, B.E., Eveson, J.P., Jarman, S.N., Quantification of damage in DNA recovered from highly degraded samples – A case study on DNA in faeces. *Front. Zool.*, 3, 11, 2006.
155. Guescini, M., Sisti, D., Rocchi, M.B., Stocchi, L., Stocchi, V., A new real-time PCR method to overcome significant quantitative inaccuracy due to slight amplification inhibition. *BMC Bioinf.*, 9, 326, 2008.
156. Chau, F.T., Chan, T.P., Wang, J., TLCQA: Quantitative study of thin-layer chromatography, in: *Bioinformatics*, 1998, https://doi.org/10.1093/bioinformatics/14.6.540.
157. Mohamad, I., K.S.S., Shakeel, W., Rapid Detection of Adulteration in Indigenous Saffron of Kashmir Valley, India. *Res. J. Forensic Sci.*, 3(3), 7–11, 2015.
158. Dar, M.M., Detection of Sudan Dyes in Red Chilli Powder by Thin Layer Chromatography. *J. Allergy Ther.*, 2(1), 1–3, 2012, https://doi.org/10.4172/scientificreports.586.
159. Banti, M., Food Adulteration and Some Methods of Detection, Review. *International Journal of Research studies in Science, Engineering and Technology (IJRSSET)*, 9, 3, 86–94, 2020, https://doi.org/10.11648/j.ijnfs.20200903.13.
160. Cheah, W.L. and Fang, M., HPLC-Based Chemometric Analysis for Coffee Adulteration Foods, 9(7), 880, 2020.
161. Taylor, P., Ehling, S., Tefera, S., Ho, I.P., High-performance liquid chromatographic method for the simultaneous detection of the adulteration of cereal flours with melamine and related triazine by-products ammeline, ammelide, and cyanuric acid. *Food Addit. Contam.*, 24, 37–41, October 2014, https://doi.org/10.1080/02652030701673422.
162. Jablonski, J.E. and Harnly, J.M., Nontargeted Detection of Adulteration of Skim Milk Powder with Foreign Proteins Using UHPLC – UV, *J. Agric. Food Chem.* 62, (22), 5198–206, 2014.
163. Chen, R., Chang, L., Chung, Y., Lee, M., Ling, Y., Quantification of cow milk adulteration in goat milk using high-performance liquid chromatography with electrospray ionization mass spectrometry, Rapid Communications

in Mass Spectrometry, 18(10), 1167–1171, 2020, May 2004, https://doi.org/10.1002/rcm.1460.
164. Jabeur, H., Zribi, A., Rebai, A., Detection of Chemlali Extra-Virgin Olive Oil Adulteration Mixed with Soybean Oil, Corn Oil, and Sunflower Oil by using GC and HPLC, *J. Agric. Food Chem.*, 62(21), 4893–4904, 2014, May, https://doi.org/10.1021/jf500571n.
165. Parvathy, V.A., Swetha, V.P., Sheeja, T.E., Leela, N.K., Chempakam, B., Sasikumar, B., DNA Barcoding to Detect Chilli Adulteration in Traded Black Pepper Powder. *Food Biotechnol.*, 28, 25–40, 2014, https://doi.org/10.1080/08905436.2013.870078.
166. Lage, M. and Cantrell, C.L., Quantification of saffron (Crocus sativus L.) metabolites crocins, picrocrocin and safranal for quality determination of the spice grown under different environmental Moroccan conditions. *Sci. Hortic.*, 121(3), 366–373, 2009, https://doi.org/10.1016/j.scienta.2009.02.017.
167. Sproll, C., Ruge, W., Andlauer, C., Godelmann, R., Lachenmeier, D.W., HPLC analysis and safety assessment of coumarin in foodsle. *Food Chem.*, 109, 462–469, 2008.
168. Blahova, J. and Svobodova, Z., Assessment of coumarin levels in ground cinnamon available in the Czech retail market. *Sci. World J.*, vol 2012, 1–4, 2012.
169. Tateo, F. and Bononi, M., Fast determination of Sudan I by HPLC/APCI-MS in hot chilli spices, and oven-baked foods. *J. Agric. Food Chem.*, 52, 655–658, 2004.
170. Kamboj, A., Analytical evaluation of herbal drugs, in: *Drug Discovery Research in Pharmacognosy*, pp. 33–34, 2012.
171. Abrantes, M.R., De Oliveira, A.R.M., De Oliveira Cabral Rocha, M., De Souza, G.O., Telles, E.O., Sakamoto, S.M., Da Silva, J.B.A., Detection of bovine milk contaminants in adulterated milk and curd goat cheese. *Acta Sci. Vet.*, 42, 1213, 2014.
172. Romero, C., Perez-Andújar, O., Olmedo, A., Jiménez, S., Detection of cow's milk in ewe's or goat's milk by HPLC. *Chromatographia*, 42, 181–184, 1996, https://doi.org/10.1007/BF02269650.
173. Wang, C.Z., Ni, M., Sun, S., Li, X.L., He, H., Mehendale, S.R., Yuan, C.S., Detection of adulteration of notoginseng root extract with other Panax species by quantitative HPLC coupled with PCA. *J. Agric. Food Chem.*, 57, 6, 2363–2367, 2009, https://doi.org/10.1021/jf803320d.
174. Gamazo-Vázquez, J., García-Falcón, M.S., Simal-Gándara, J., Control of contamination of olive oil by sunflower seed oil in bottling plants by GC-MS of fatty acid methyl esters. *Food Control*, 14(7), 463–467, 2003, https://doi.org/10.1016/S0956-7135(02)00102-0.
175. Pospiech, M., Lukaskova, Z.R., Tremlova, B., Randulova, Z., Bartl, P., Microscopic methods in food analysis. *Maso Int. Brno*, 1, 27–34, 2011.

6

Trends of Food Adulteration in Developing Countries and Its Remedies

Satyam Chachan[1], Anand Kishore[1*], Khushbu Kumari[2] and Arun Sharma[1]

[1]*National Institute of Food Technology and Entrepreneurship and Management, Sonepat, India*
[2]*National Dairy Research Institute, Karnal, India*

Abstract

Food adulteration is a threat to mankind since the beginning of civilization; it not only degrades the quality of food products but also has negative impact on human health and economy. Adulteration is the act of deterioration of quality of food products by incidental or intentional means through the addition of chemicals, extraneous matter, non-edible, and toxic substances or removal of vital substance. It is widespread all over the world but most common fraud practices are in developing countries where there is a technological gap between all stakeholders such as manufacturers, raw material provider, supply chain providers, and consumers. Tractability of food products through manufacturing, distribution channels, and the retail system play vital role to control this threat. The possible remedies to food adulteration can be technological interventions for development of supply chain infrastructure, hands-on training to food processors, and distribution partners on regular intervals to mitigate food adulteration. Availability of low-cost advance adulteration detection tools and techniques at consumer level empowers customers to monitor food adulteration. Multidirectional approach to ensure safety and quality of food products is indispensable to contain food adulteration and includes effective implementation of food safety laws and government initiatives to spread awareness about food adulteration among all stakeholders, scientific, and stringent food safety standards.

Keywords: Food, adulteration, food fraud

---

*Corresponding author: anand.iitkgp14@gmail.com

Mousumi Sen (ed.) *Food Chemistry: The Role of Additives, Preservatives and Adulteration*, (165–188) © 2022 Scrivener Publishing LLC

6.1 Introduction

Food adulteration is a threat which is described as food products that do not comply with legal standards. It is a malpractice which poses serious health risks to many people. Adulteration is a menace which could be found in every product, whether it be daily groceries or our life saving medicines. Nowadays, infants' milk products are also being tempered which has negative impact on children health and, in some cases, leads to deformities. It reflects the inhuman behavior, insensitivity, and lack of awareness among all stakeholders ranging from food producers, processors, to consumers. Adulterated food is perilous as it can affect health and also it could deprive consumers from vital nutrients required for physical and mental development of human beings. Some of the adulterants can result in life threatening diseases, especially in elderly, children, pregnant women, and immune compromised subjects. The lack of supply chain infrastructure, poor implementation of the existing food safety laws, lack of awareness among food processors and consumers about the harmful effects of adulterants, infrastructure deficit for accurate and rapid testing methods and unorganized food processing sector are major factors responsible for food adulteration in developing countries. There are various conventional perspectives followed in developing countries, but those approaches had not been the most effectual alternative to combat against food adulteration. Therefore, effective remediation of the food adulteration has become a crucial requirement for ensuring safety and quality of food products. The objective of this chapter is to identify common adulteration malpractices of food adulteration in developing countries, evaluating the existing laws, technology, findings of gaps for prevention, and remedial approach to counter food adulteration.

6.2 Food Fraud in Developing Countries

Food adulteration is highly prevalent in many developing countries, such as India, Bangladesh, China, Pakistan, Afghanistan, Ethiopia, and others. It is expected that around 22% of foods are tempered every year and consumption of these adulterated and contaminated foods affected the health of about 57% of population internationally [1]. Maintaining food quality has been a great challenge in various developing countries. The rampant food tempering is due to following augmenting factors:

- Increased food demand for a rapidly growing population: It has been forecasted by Food and Agriculture Organization that food production needs to increase by 70% if the population reaches 9.1 billion by 2050, placing pressure on food quality standards. Since the demand is increasing and natural resources is depleting day by day, so the demand is met by declining the production standard and raising the yield.
- Lack of knowledge of consumers/producers/manufacturers: Knowledge on food quality and safety is important in minimizing the risk of foodborne illnesses during food handling and consumption. Food licensing and certification systems were developed to build trust and validate the concepts of food safety, food integrity, and food quality which is useful for all the personnel involved in food business as well as consumers. But due to lack of awareness, this malfunction is spreading all over the world.
- Practiced as a part of the business strategy: In the prevailing business environment, main motive is to make maximum profit from lesser investments. Thus, food fraud is done as part of business strategy by various stakeholders whether it be local vendors or some established multinational brands, putting public health at risk.
- Unethical practices by cultivators, manufacturers, importers, and traders: Most of the food manufacturers, processors, restaurant owners, and so forth are involved in one way or another in unethical practice of adulteration. Many injurious products are promoted by them with misleading or false information. Also, unfair prices and unfair target marketing are happening nowadays.
- Scarcity of advanced equipment and proper staff: Food testing laboratories are not equipped with high caliber testing equipment and also lack trained staff who can successfully conduct the adulteration test. Digital devices should be used for detecting adulterated foods and there should be additional features incorporated in the gadgets which are consumer friendly.
- The food regulatory authorities do not take strict actions to control the unlawful act like food impurity, poor quality of food, lack of hygiene, misleading information on

labels, selling expired products, lack of standard infrastructure, and fake license. Occasionally, the regulatory bodies become active and if found guilty then sellers/producers are punished for selling adulterated food products but they are functional only for a short period of time.
- Bio-terrorism and food safety: Intentional contamination of food products by terrorists or some antisocial groups has become a matter of concern in recent year due to rising incidents of terrorist attacks globally. Food terrorism involves the tempering of food products during harvesting, storage, processing, manufacturing, or distribution. It was stated by WHO that food terrorism can be prevented by strengthening food safety national policy, having proper food monitoring and surveillance system, food inspection should be conducted regularly, adequate training should be provided at regular interval, and proper resources should be available to detect food adulteration.

All above factors thus lead to food fraud which ultimately results that the consumer is either cheated or often become victim of diseases.

6.2.1 Impact of Adulteration

Impacts on farmers/producers: Farmers and producers are affected to great extent due adulteration. They lose the confidence of the buyer and sometimes the farmers who do get involved in the fraud are affected as they cannot compete with the economically motivated food frauds.

Impacts on processors: Food processors are impacted by increased risk of legal enforcements from government and regulatory agencies. Many a times, consumers lose confidence in their product, enterprise has to bear the loss occurred due to recall of product from the market, destruction of the recalled adulterated product, expenses occurred in complaint redressal, and other legal formalities. An enterprise will not only suffer regulatory legal implications but also suffer huge economic loss due to the retracted product and have to face the lost sale due to loss of consumer trust in the brand. If any particular food product of a company or brand is found with adulteration, then the public perception goes against other product range of such enterprise and thus other competitors can outnumber such enterprise in terms of market share. The recovery from such situation will require significant time and enterprise has to again expense a huge amount for image branding as consumers will shift to other company.

Impacts on consumers: Consumers are impacted by deterioration in health and loss of nutrition even after spending money for the same. Excessive or prolonged use of adulteration causes mild to severe health associated with diarrhea, abdominal pain, nausea, vomiting, eyesight problem, headache, cancer, anemia, insomnia, muscular paralysis and brain damage, stomach disorder, digestive system disorder, kidney failure, glaucoma carcinogenic effects, cardiac arrest, respiratory distress, edema, dropsy, and various other diseases [2–4]. Various harmful chemicals are used in fruits and vegetables which has a great impact on consumer's health. For example, calcium carbide, copper sulfate, and different other chemicals are used on large scale to ripen the fruits like mango and bananas; oxytocin hormone used to increase the size or production of vegetables like pumpkin, cucumber, and brinjal affects significantly to the consumer's health. Coating on apples and other fruits with harmful wax and dyes to make it shiner and look fresh for long period of time also has a bad impact on human body. Metallic lead is applied on leafy vegetables to give fresh or brighter green color.

To meet the increasing demand of fruits and vegetables and to increase the production capacity, the producers use the pesticides and herbicides excessively which has adverse impact on consumers. Consumption of such chemical laden vegetables and fruits results in various diseases mainly the digestive disorder [5]. As a matter of fact and case studies from India on food adulteration borne diseases:

- 1998 Kolkata (West Bengal), after consuming adulterated rapeseed oil with tricresyl phosphate (generally used in hydraulic fluid and varnishes), about 600 persons suffered from paralysis in fore limbs.
- 1998 Delhi, 2000 Gwalior (Madhya Pradesh), 2002 Kannauj (Uttar Pradesh), and 2005 Lucknow (Uttar Pradesh), more than 100 people suffered from dropsy after consuming the mustard oil adulterated with Mexican prickly poppy/argemone (Argemone mexicana) oil [6].

6.3 Classification of Food Adulteration

There are two types of food adulteration in general.

(i) Intentional Adulteration
The food is contaminated deliberately with intention to increase profit that cause illness or death on a large scale. For example, marble chips, sand,

chalk powder, and stones are added to food products. The following cases of deliberate adulteration gained the attention recent years.

- In 2016, about 33 causalities have been reported from Layyah district of Punjab province in Pakistan where consumers ate sweets deliberately mixed with pesticide.
- In 2018, Australian-grown strawberries (Queensland and Western Australia) deliberately contaminated with sewing needles [21].

(ii) Incidental/Unintentional Adulteration

Food being tempered is usually attributed to negligence, ignorance, and carelessness scarcity of proper infrastructure for maintaining standard food quality. Food products also get contaminated during growth, harvesting, storage, processing, transporting, and marketing. Some examples of incidental or unintentional adulteration are pesticide residues as a result of poor agricultural practices, insect larvae, or bird droppings due to poor hygienic practices at processing level. In brassica fields, argemone mexicana is frequently grows as an unwanted plant, and at the time of cultivation, if proper care is not taken, then seeds of argemone mexicana get mixed with seeds of brassica and the mustard oil expressed with that batch of seed will also contains argemone oil thus adulterated. Presence of argemone oil in edible mustard oil is not good for human health; consumption of contaminated oil may lead to dropsy. Metallic contamination with lead, arsenic, and mercury while processing of food products can also occur incidentally. Although these metallic adulterations can happen unintentionally while processing but they have severe impact on human health. Lead is very toxic element, it brings pathological changes in liver, kidneys, and arteries. Consumption of lead contaminated foods may lead to nausea, anemia, abdominal pain, insomnia, brain damage, and muscular paralysis. Most of the water bodies are getting polluted from various sources which lead to contamination of fish and other sea food with various chemicals. For example, fish get contaminated with mercuric salts due to which the concentration of mercury gets increased in it. Some of the organic mercury compound like methyl mercury has neurological effect on human body. It affects the brain, which could make consumers blind and deaf and can paralyze various muscles. Pesticides and fertilizers used in field are most common unintentional adulterants nowadays. Due to poor agriculture practices and lack of good infrastructure of storage and transportation, agro produce get contaminated with the pesticides and fertilizers used in the field, which have a severe impact on human body if accumulated in high quantity. Some of them are Dichloro diphenyl trichloroethane (DDT), malathion, pyrethrum, sulfates, and nitrates. The maximum permissible limit is 3 ppm for Malathion and DDT and 10 ppm for pyrethrum.

6.3.1 Intentional Adulteration

More alarming type of food adulteration is the intentional one where the low quality product is not added mistakenly during processing, or poor sanitation between products but added due to purposeful act of food tempering. The following are the different ways which are generally accompanied to adulterate food products intentionally.

(i) Replacement: Food constituent specially valued pure ingredients are completely or partially replaced with a less expensive component that is generally carried off through dilution or addition with various adulterants. Few examples: Artificially increasing protein content of milk by adding melamine; fraudulently increase the titratable acidity of lemon juice by adding water and citric; and over treating frozen fish with ice (which adds extra water).

These types of deceptive practices are done with intention to get financial benefit. This fallacious activity is also carried out by substituting one ingredient with other when there is short supply of particular food ingredient or product. For example, when supply of beef and other meat products get reduced in European countries, horse meat was found in prepared meals.

Replacement type of food adulteration also includes false declaration and claim which are as follows:

> Misleading consumers by falsely declaring botanical, species, geographic, or varietal origin: Examples are substituting milk fat with low-cost vegetable oil, imitating less expensive cow milk with sheep or goat milk, and substitution of synthetically produced vanillin for botanically derived natural vanillin.
> False declaration of origin of product to evade taxes/tariffs (e.g., shipment of products from China origin, like shrimp and honey, to different Asian countries get relief from anti-dumping duties and to hide potentially unsafe product;

U.S. imports of catfish from Vietnam branded as grouper get rebate on anti-dumping duties).

Misrepresentation of production and manufacturing processes (e.g., deceitful labeling of artificially produced flavored chemicals as natural; fraudulent claims of certifications as organically or naturally produced).

(ii) Addition: In order to disguise low quality ingredients, non-authentic substances are added into food products. For example, addition of color additive like sudan red dyes to enhance the color of poor quality paprika.

(iii) Removal: Intentional removal of original, naturally available, and valuable components without explicitly disclosing to consumers. One example is the removal of non-polar constituents from paprika (e.g., lipids and flavor compounds) to produce paprika-derived flavoring extracts, or "defatted" paprika, which lacks valuable flavoring compounds, as normal paprika [22]. Another example is removal of pollen and other geographical origin indicating components from honey so that consumers or technical evaluators find difficulty in identify original geographical source.

6.4 Common Food Adulterants

An adulterant is a substance that is incorporated in any food item to diminish its quality. For example, when water is added in milk, the water is called as an adulterant. There are a number of adulterants, such as water, starch, caustic soda, cane sugar, urea, chalk powder, sodium chloride, skimmed milk, fructose, formalin, ammonium sulfate, lead chromate, tamarind seeds, papaya seed, brick powder, fat, washing powder, argemone seed, hydrogen peroxide, metanil yellow, potassium sulfate, saw dust, stone powder, melamine, methanol, talc powder, maltodextrin, coal tar dye, saccharin, palm oil, fertilizers, sudan red, ammonium chloride, boric acid, saw, dust, and cassia bark are added in various types of foods [7, 24]. Some common foods susceptible to fraud and their adulterants are summarized in Table 6.1. The following are the common categories of food and their adulterants.

(a) Milk and Dairy Products: Milk is considered as an ideal food because it provide complete nutrients being required by both adults and infants,

Table 6.1 Common adulterant for food product [3, 19].

Food commodity	Common adulterants
Milk	Water, Chalk, Urea, Starch, Formalin, Urea, Hydrogen peroxide, and skimmed milk
Coffee Beans	Tamarind seeds, Mustard seeds, and Chicori
Honey	Starch syrup, Inverted syrup, Mollases
Dal	Metanil yellow
Mustard seeds Edible oils and fats	Argemone seeds Argemone oil
Tea	Exhausted tea leaves or Foreign leaves, Saw, Dust artificially colored
Turmeric powdered or whole, and mixed spices	Lead Chromate
Cottonseed flour	Gossypol
Jaggery	Metanil Yellow, Chalk powder, Washing soda
Wheat, rice, barley	Dust, Stone, Pebble, Damaged grain
Salt	White powder, Urea
Meat and bone meal	Leather meal, Sand, Rock phosphate, Blood meal, Formalin
Chilli and Coriander powder	Redbrick powder, Red lead, Rhodamine B Dye, Water-soluble synthetic colors, Soluble salts, Dung powder, and other Common salts
Jam, Juice and Candies	Non-permitted dyes which includes metanil yellow and other artificial food dyes
Butter and cream	Palm oil, Sunflower oil and soybean oil
Ice Cream	Washing powder
Fruits and vegetables	Wax, Oxytocin sachharin, Copper sulfate, and Calcium carbide

but unfortunately, it is very easily adulterated in almost all the countries. It became the global concern after the melamine adulteration scandal in infant milk products, China 2008. Milk is mainly adulterated for economically stimulated fraud by adding milk from different species, vegetable protein, whey, and dilution of water into milk [17]. These adulterants do not have any severe impact on human health but some of the adulterants are too harmful to be ignored such as formalin, boric acid, ammonium sulfate, urea, caustic soda, hydrogen peroxide, melamine, salicylic acid, and sugars, which are threat to humankind as it may causes life threatening disease. When we talk about to milk adulteration, water is the most common adulterants, because it is the easiest way and cheapest source for impregnation in milk. Milk generally contains 87% of water and rest 13% of solid part, but milk with extra water is not good because it will degrade the nutrient profile of milk and will also promote the addition of chemicals to compensate its density and other quality parameters. Some adulterants are mainly added to alter the parameters like fat percentage, SNF (Solid not Fat), and protein content which are being checked to evaluate the quality of milk, to increase the milk quality in unethical way. For example to increase the SNF, starch, cane sugar, urea, and sulfate salts are added to milk. Excessive undigested starch in the colon may leads to diarrhea. In a similar way, to increase the non-protein nitrogen content in milk urea is added which puts an extra pressure on human kidney as it has to filter extra urea from the human body leads to kidney failure. Apart from this, urea also causes acidity, indigestion, ulcers, and cancers and has a significant harmful effect on heart and liver too. To increase the protein content in milk, melamine is added [8] but its consumption above the prescribed level may lead to renal failure and death in infants [9]; to increase the density, ammonium sulfate is added; for increasing the shelf life, preservatives like salicylic acid, formalin, hydrogen peroxide, and benzoic acid are added [10] and detergents are added which emulsify and dissolve oil in water to give more froth to milk [11] which can cause gastro-intestinal complications, inflammation of the intestine, and gastritis. Apart from the intentional economically motivated adulteration, sometime due to carelessness of farmers and due to lack of good infrastructure and hygienic conditions of processing, storing, and transportation at farm level, milk get adulterated in developing countries. Apart from milk, some milk-based products are also adulterated like paneer and khoya is adulterated with starch. Similarly ghee is adulterated with vanaspati or margarine [26].

(b) Grains and Pulses: Grains and pulses is being adulterated by replacing the low quality with superior grade ones or by addition of inedible components to increase the weight of the product for the financial benefits or unintentionally get contaminated at farm side with pesticides residues, dropping of rodents, larva, etc., because of lack of proper hygienic conditions at farm level. Furthermore, husk, synthetic materials, sand, stones, marble chips, and filths are generally added to increase product bulk. In developing countries, grains are mainly adulterated by damaged grains like kernel or pieces of kernels sprouted or internally damaged, kernel burnt grains, weevil led grains (kernels that are partially or wholly bored by insects), or by low quality food grains like arhar dhal is adulterated with kesari dhal (Lathyrus Sativus), which is considered as staple food for low-income population in some parts of central India. Continuous consumption of the kesari dhal for 2–3 months will lead to lathyrism (progressive spastic paralysis of lower limbs). Apart from this, pulses are also adulterated with some harmful colors like metanil yellow and tartazinc dye to improve the color appearance of the old stock which are carcinogenic in nature and can cause stomach disorder and testicular degeneration in the male if consumed for long period.

(c) Juice: Fruits juice is mainly used as a healthier beverage which helps in providing important micro and macro nutrients. They are rich in antioxidants, polyphenols, micronutrients, etc., thus posing many health benefits. However, increasing popularity and demand of fruit juices makes them an easy target for adulteration. Other reasons of juice adulteration are the high price of fruit pulp and lack of infrastructure for storing and transportation of pulp under frozen state. Fruit juices are adulterated by adding components of lower grades, sugars or diluting it with water or pulp wash substituting the juice with cheaper or rotten juices, and addition of different preservative and dyes. Sugar syrups such as inverted cane syrup, high fructose corn syrup (HFCS), and synthetic solution of fructose, glucose, and sucrose are used as adulterants in fruit juice industry.

(d) Herbs and Spices: Herb and spice adulteration takes many forms. These items can be adulterated with artificial dyes, fillers, and cheaper cousins and synthetics. Each of these fraudulent methods serves to reduce the cost associated with herbs and spices by replacing some of their weight or volume with a less expensive alternative, often with additional measures to conceal the addition. A common example being oregano tempered with myrtle leaves and olive. Some of the common spices adulterants are dirt, sand, earth gritty matter, artificial color, chalk powder, starch, stone, bark, horse dung, saw dust, papaya seed, and lead chromate. Corn flour, yellow color talk, and saw dust are some foreign material used as adulterant in turmeric powder to increase its volume, while lead chromate and metanil yellow are artificial dyes which are used as to provide bright yellow tinge to turmeric powder. Similarly, saw dust and brick powder are used as adulterants in red chili powder and sudan red is the artificial color used to give desired red color to red chili powder. While in case of coriander, powder husk and ash are used to increase the volume [12].

(e) Edible oil and Fats: Some of the more expensive edible oil and fats such as milk fat, olive oil, oil from various nuts and seeds, and cocoa butter are adulterated with low price vegetable oils and fats in order to earn more profit. Basically, there are two types of fraud in edible oil and fats: one is replacing the more expensive oil and fats with cheaper one and other is mixing refined oil with cold pressed oil. Adulteration with vegetable oil is basically adding inferior, harmful, cheaper, and unnecessary ingredients to degrade the nature and quality of oil. A very common example is mustard oil adulterated with argemone oil. Other sources of oil adulterants include olive pomace oil, rapeseed oil, corn oil, sunflower oil, and soybean oil [23].

(f) Honey: Honey is a high-valued product produced by bees from nectar of different plants. Honey is mainly used as sweetening agent as well as for some health benefit purposes as it has some pharmacological properties such as anti-inflammatory, antioxidant, anti-cancer activities against breast, cervical and prostate cancer, and osteosarcoma. Honey plays an important role in treatment of osteoporosis, laryngitis, gastrointestinal ulcers, insomnia, anorexia, and constipation [13]. Honey adulteration has a significant economic impact and is mainly done by adding foreign materials mainly syrups which is cheaper and also lacks medicinal properties that honey has. Honey adulteration is done by mainly two ways: first direct adulteration in which sugar syrups from different source like sugar beet, fructose sugar syrup, maltose sugar syrup, or industrial sugar syrup (mainly obtain by thermal, enzymatic, or acidic treatment of starch) are added directly in harvested honey in a fixed proportion with purpose of increasing its sweetness and second is indirect adulteration in which sugar syrups are used for overfeeding the bees to increase honey yield in hives [14]. Apart from these two methods, honey is also adulterated by blending pure or rare honey (high quality and nutrient profile) with low-quality or cheap honey, and sometimes, honey is also mixed with synthetic honey [15]. Fructose corn syrup (HFCS), inverted sugar syrup (ISS), high fructose inulin syrup, corn sugar syrup (COSS), and cane sugar syrup (CASS) are the most used sugar syrups for honey adulteration purpose [16]. Other common adulterants in honey are water, molasses, liquid glucose, maize syrup, banana, and rice syrup. Some of biological and chemical properties like specific compound content, enzymatic activity and electrical conductivity of honey are affected because of these adulterants.

(g) Meat, Fish, and Seafood: The intentional addition of low-cost and low-quality meat into high quality and expensive meat products not only reduces the nutritional value but also increases risk of contamination and increase microbial activity in adulterated food.

(h) Tea and Coffee: Tea and coffee are two most common beverages in almost all developed and developing countries. Both tea and coffee are adulterated with harmful chemicals for economic benefits. Tea is mainly adulterated with harmful colors to give more attractive appearance. Many cases have been reported in which sub-standards tea leaves are adulterated or colored with different harmful dyes like potassium blue, coal tar dye, azo dye bismarck brown, indigo, turmeric, and plumbago (black lead mainly used in pencil manufacturing) to impart attractive color and glossiness to the product. The consumption of these dyes and chemical leads to liver infection and many gastrointestinal diseases. Coffee beans are mainly adulterated to counter the high price in the market and to reduce the production cost in an unethical way with motive of economic benefits. Coffee stem and husks, twigs, roots, and some roasted grains like barley, wheat middlings, brown sugar, corn, oat, soy, and rice are the main adulterants used for the purpose. The main reason behind using these adulterants is that they are relatively cheaper than coffee beans and do not alter the physiochemical and organoleptic properties of coffee beans to that extent. In some countries like England, selling coffee mixed with other materials like figs (Viennese coffee) or chicory (French coffee) is illegal. Coffee seeds are also adulterated with tamarind seed and mustard seed.

6.5 Adulteration Remedy Strategies

6.5.1 Government and Regulatory Agency Initiative

  I. Food Defense: The government and regulatory agency are required to formulate legal frameworks and strategies to monitor, identification, control, and mitigate strategies for food adulteration.

a. Vulnerability assessment: The severity of the hazard depends upon numerous factors that might affect public health. The impacts will depend upon the volume of product, the number of servings the consumer consumes in a day which is calculated as the total exposure limit to the consumer. Moreover, the adversity of hazard will depend that how fast it moves inside the body, consumed, and affects the health of the consumer also calculated as the LD50 limit which is the lethal dose to kill 50% of the population. The hazard before reaching the customer has to pass various barriers like the packaging layers, lids, and seals which it has to cross. The product contamination will vary according to the situation.

b. Mitigation strategies: The implementation strategies have to follow a trickle down approach where the top management has to be the most motivated to implement food safety and should integrate it into its business. The organization should take up various steps to have a safe, unadulterated, and non-mislabeled product through the implementation of HACCP, VACCP, and TACCP. The mitigation strategies should be real time to take prompt corrective and preventive decisions. As technological advancements are penetrating into field of on-site testing, adulteration should be encompassed into umbrella of technological driven, fast, reliable, and sensitive mitigation strategies.

c. Strategies for mitigation of food adulteration: The important aspects of strategies for mitigation of food adulteration include 1) monitoring of adulteration, 2) corrective actions, and 3) verification of appropriate corrective actions.

d. Training and record keeping: The staff employed in mitigation strategies should be properly trained about various aspects revolving around mitigation strategies ranging from information about regulatory and preventive measures to be taken at individual and collective level along with technical professional competent to conduct sampling and tests for determination of adulteration in food products.

II. Skill development training programs: Training is required to be arranged by government and regulatory bodies for all the stake holders including food inspectors, local food and health authorities, and food analyst and chemists.

III. Sophisticated equipments: Governments are supposed to facilitate latest and reliable testing facilities to food testing laboratories so that at least each state is appropriately equipped with advanced techniques. Additionally efforts are being made to keep these laboratories functional by providing them consumables and repairing the damaged equipment on regular basis and providing proper training to the analysts/chemists to use the equipment.
IV. Educating the masses: The government has also taken initiative to educate consumers on the harmful effects caused due to consumption of adulterated food through electronic and print media regularly. Consumers are significant stake holders and they should be kept updated and well informed about their right and responsibilities toward safe and hygienic food practices. So, once consumer become aware about the menace caused by contaminated food they will become conscious about their health as well society and end up buying faulty food products.
V. Remedies on food adulteration by FSSAI: FSSAI has started a scheme to develop robust food testing laboratory network throughout India. The scheme consist of six initiative for strengthening of Food Testing Laboratory [25]:

   a. Strengthening of state food testing laboratory
   b. Strengthening of Referral food testing laboratory
   c. Support mobile food labs
   d. Capacity building of food testing personnel
   e. Incentivize state to use the facilities available in FSSAI notified private labs
   f. Develop food testing culture in college/school

6.5.2 Loopholes in Existing Method of Eliminating Adulteration

The regulatory agency's approach has not been profoundly effective in eliminating food fraud. Loopholes in law have emboldened erring manufacturers, and lack of consumer awareness has made things worse. The following are the few loopholes that should be corrected for prevention of food fraud.

- Although consumers are informed about the standards set by regulatory bodies through different modes, most of them remain ignorant about it. They are not aware of Prevention of Food Adulteration Act, and they prefer to buy low-grade cheaper food products by compromising on quality which affects their health. Consumers do not make an effort to check the content provided on the label and even if they found spurious and substandard food products, expired food products, or some necessary information missing on the label, they will not take responsibility to lodge a complaint. Consumers have to be made aware and empowered to bring end to adulteration.

- Generally, people do not have fear of law involved in food counterfeiting as no one is put behind bar for this deed. Additionally, the penalty is very small amount for selling tempered food products which every trader can pay easily. Every single person must be made accountable including food inspectors, police officer, food analyst, and drug inspector if any incidence of adulteration is found in their jurisdiction area. If the officials and analyst will not be involved in the food fraud, then will also make sure that the person engaged in such practices are punished.
- If any person is caught even once with any type of malpractice in food, then there should be a law to cancel the certificate and they should not be allowed to sell anything in the market. This heinous crime will end up to some extent if the food inspector, officials and judiciary give deterrent punishment to the sinners.
- One of the probable causes of adulteration is indifferent behavior of general population toward the sanitary conditions of food products. Also, law takes its own time so until the general mass does not understand the importance of hygienic value and protest against the food fraud then such crimes will prevail. The only way to remain healthy and combat adulteration is self-realization, and therefore whenever and wherever someone comes across the food counterfeit, they should bring the crime to officials and jurisdiction. A social awakening can only help to end such crime. We should be responsible for what we consume as unless we care for our health, nothing can be changed.
- Conventional analytical, morphological, and organoleptic techniques have good resolution to easily detect synthetic adulterants present in the food products [18]. But biological adulterants cannot be detected by the conventional

techniques because of morphological similarity of actual product to the adulterant. New-generation adulteration detection methods are not made available to the analyst and also there is lack of cost-efficient detection method that consumers can directly access.
- There is lack of sophisticated equipment in state and central laboratory. Furthermore, there is scarcity of technical and trained manpower to use the advanced equipment.

Recommendations:

To the government: Rigorous testing procedure should be adopted, and law should be implemented strictly. Low-grade food product must not be allowed to be sold in the market.

To the Manufacturers: Manufacturers and processors largely focus on increasing their quantity rather than quality. They care least about the well-being of the consumers which should be avoided, and they must work on the principles involving business ethics.

To the Wholesalers: It is compulsory for the wholesalers to pick standard quality products and store them in proper hygienic condition.

To the Retailers: Since retailers are directly linked to the consumer, they should choose wisely and purchase the best quality products.

To the Consumers: Consumers being the ultimate users of the food products should be fully aware of the adulteration practices that are prevalent nowadays. They should educate themselves and keep themselves updated with food standards set by the regulatory bodies.

6.5.3 Process and Product Verification

Product and process verification both are critical for assessment of food standards. Verification plays an important role in almost every stage, from initial product development to production and up-scaling. Quality of food products cannot be assured just by testing or inspecting the finished product or raw material. Each and every step of manufacturing process needs to be controlled and should be under strict observation so that all the product specification is met. Proper documentation, evidence, and records need to be maintained so that any food counterfeit can be verified easily. Figure 6.1 shows that the potential sources of contamination into the food.

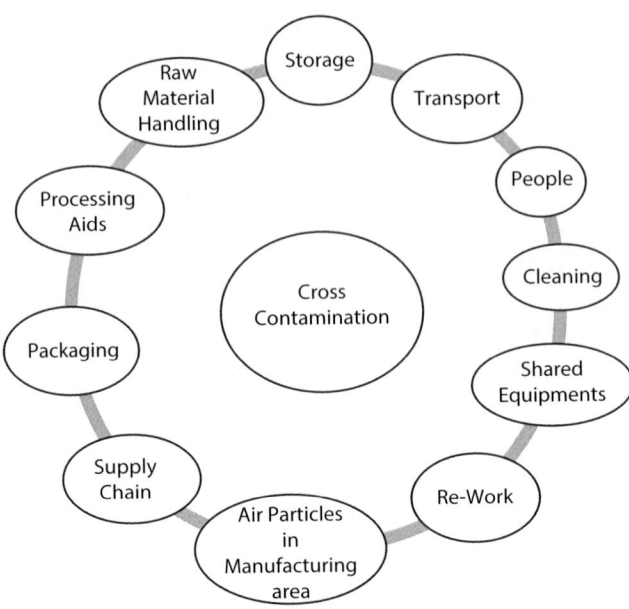

Figure 6.1 Potential sources/processes for adulterant entry/contamination into the food.

Random Sampling: In order to obtain descriptive analysis of adulteration in food, random sampling vigil by regularly drawing food samples from all potential sources of food products subjected to adulteration, *viz.*, manufacturers, wholesalers, and retailers, is advised to government and regulatory authorities so that strict control can be enforced to protect interest of all stakeholders including customers.

6.5.4 Higher Levels of Transparency/Traceability in Supply Chain

Transparency in a supply chains is very crucial to generate high level information of the product processed and hence tracing the accurate path become very easy. The product component captured like name of suppliers, certification, location of facilities, and regulatory requirements is accurate and specific. Pictures or scan able QR codes mentioned on packaging nowadays give the details information and journey about any product.

6.5.5 Use of Novel Technology

The adulterants can be identified by using various sophisticated equipment and novel technology. There are various techniques like chemical/

biochemical, physical, and molecular techniques used to detect the adulteration depending on the type of adulterants. DNA-based technique and molecular method helps in rapid and reliable detection of various types of contamination in food products and helps in assessing the authenticity of food products like milk, cereals, pulses, and meat. Recently, large numbers of sophisticated technique have been developed to curb food fraud and detect adulterant. DNA-based technology follows three main strategies: PCR-based, sequencing-based, and hybridization-based that could be an easy, rapid, and universal method for detecting food adulteration [20]. Currently, a range of vibrational spectroscopic techniques combined with chemo metrics has been proved valuable tool for fraud detection. These are very fast and sensitive techniques to define the food authentication of wide range of food product. Advantageously, these techniques are non-destructive and are comparatively low cost.

6.5.6 Training

The food vendors should be trained and well informed about the ill effects of food contamination. They must be made aware about the harm it causes to the consumers and also their own family. FOSTAC training is given to personnel involved where are informed about the measures taken to ensure food safety and about the penalties they have to pay if found guilty, and therefore, this encourage them to carry out fair trade practices.

6.5.7 Awareness

"Consumer Awareness" is the first step to eradicate food adulteration. Consumers have to be aware of their rights. The health department should create awareness through texts, advertisements through television and social media at major centers of the city and educate the consumers about the methods of identifying adulterated food from its odor, texture, density, taste, etc. Awareness and motivation will lead to empowerment and will help to end the menace of food adulteration.

The following concerted efforts might be beneficial to curb food adulteration:

- Mass awareness of the impact and consequences of the food adulteration on the human body and long-term health among the people.
- People involved in food adulteration must be punished severely. There should be exemplary punishment to the food

contaminators depending upon the ultimate effect of adulterated food.
- Food inspection service throughout the country should be strengthened with the help of skilled manpower and analytical instruments.
- The safety limit of adulterants in food material which are potentially toxic to humans and can have both acute and chronic health effects should be reduced to match the international guidelines.
- Food adulteration should be included in the academic curriculum and made compulsory for school and college students in order to acquaint them lethal effect of contaminated foods.
- Farmers and producers should be encouraged to use safe chemicals by the agriculture department and inform the about side effects of harmful chemical being used in farms.
- All the food business leaders must be motivated to develop and sustain a personal moral value and a strong understanding of the damage that can be caused by food fraud.
- Relevant international food safety policy and regulations should be enforced by the government through food ministry department.
- Government should involve scientists from the institutions to verify the food products on regular time interval through laboratory analysis by independent research.
- Government should give some initiative and awards if anyone notify to them about food fraud cases.
- There should be proper communication between all the stakeholders engaged in the supply chain. This will be helpful in tracing the counterfeit products and can decreases potential food frauds.
- Adulteration awareness centers can be arranged by the government, which will allow the people to report cases of food adulteration directly. The raids and surprise visits can be arranged by the government to the go down or places where food is stored to ensure whether the conditions are suitable or not.
- Development of rapid and real-time robust system for detection of adulteration.

6.6 Conclusion

Food is considered as one of the most important and basic needs for a human life. For the survival, every human being on earth has to consume food as it helps in nourishment of the body but, if the food which we are consuming is toxic, then instead of nourishing the body, it will harm it. Food adulteration is a socio-economic crime and is harmful for human beings, and source of profits is for businessmen who are involved in adulteration activities. Food adulteration is rampant in developing countries due to lack of knowledge of strict laws and punishments against food adulteration and ignorance of their duties and rights toward food safety. It is necessary to understand the main reasons behind food adulteration practices, if we want to stop it. In most the cases, money is the motive behind the ill practices of food adulteration. The intention of making more money and extra profit in the business mainly motivates the food business operators to adulterate foods in most of the cases. Food business operators in various tropical countries mix or add various harmful chemicals to increase the shelf life or add cheap food products of similar physical attributes in food for making extra profit. They are adding harmful textile dye in food to make it look more attractive to the costumer and to make more profit. Another main reason for food adulteration is to compete with other traders in the market. Nowadays, consumers are price-oriented end up buying low-grade food products, and as a result, it becomes difficult for the honest manufacturer to compete with the manufacture engaged in food fraud. For a healthy nation and bright future, food safety and security has to be ensured and to achieve that every government should stop and control the food adulteration practices in their country. For adulteration controls, integrated approach through statutory and regulatory authorities, industry, scientific community, consumer guidance, voluntary agencies, proper counseling, and IECT (Information, Education, Communication and Training) can play a vital role. Regular surveillance, monitoring, and random sampling should be done to prevent food adulteration. It is very crucial to impart health education to the food producers and consumers about the harmful effects of different adulterants in foods. People are advised not to purchase loose milk and other food items from unauthorized sources. It is imperative that the regular quality control tests must be conducted to assure adulterant free food for human consumption.

Researchers should be encouraged to develop composite technology driven systems with minimal analytical technology for rapid, nondestructive, highly efficient, and economic food-adulteration detection maneuvers that may be used in the field/at point of use by less trained personnel to generate data with significant reproducibility. Regular monitoring by government and regulatory agencies along with awareness among all stakeholders including food producers, processors, supply chain providers, wholesalers, retailers, and consumers can mitigate food adulteration.

References

1. Devrani, M. and Pal, M., *How to Detect Adulteration of Maltodextrin in Milk? Food and Beverage Processing*, vol. 5, pp. 22–23, 2018. https://www.researchgate.net/publication/328262332_How_to_Detect_Adulteration_of_Maltodextrin_in_Milk
2. Banti, M., Food Adulteration and Some Methods of Detection, Review. *Int. J. Nutr. Food Sci.*, 9, 3, 86–94, 2020.
3. Narayan, D., Food Adulteration: Types, worldwide laws and futures. Biotech Articles, Healthcare, 2014. https://www.biotecharticles.com/Healthcare-Article/Food-Adulteration-Types-Worldwide-Laws-Future-3165.html
4. Lakshmi, V., LABS R.V, Guntur, Pradesh, A., Food adulteration. *IJSIT*, 1, 2, 106–113, 2012.
5. Mebdoua, S., Pesticide residues in fruits and vegetables, Bioactive molecules in food, Reference series in Phytochemistry, pp. 1–39, Springer, Cham, 2018.
6. Food Safety and Standards (Contaminants, Toxins and Residues) Regulation, FSSAI, 2011, https://www.fssai.gov.in/upload/uploadfiles/files/Compendium_Contaminants_Regulations_20_08_2020.pdf
7. Food safety and Standard Act, 2006. https://fssai.gov.in/dart/, 2021.
8. Liu, Y., Todd, E.E.D., Zhang, Q., Shi, J.-R., Liu, X.-J., Recent developments in the detection of melamine. *J. Zhejiang Univ. Sci. B*, 13, 7, 525–532, 2012.
9. Domingo, E., Tirelli, A.A., Nunes, C.A., Guerreiro, M.C., Pinto, S.M., Melamine detection in milk using vibrational spectroscopy and chemometrics analysis: a review. *Food Res. Int.*, 60, 131–139, 2014.
10. Singh, P. and Gandhi, N., Milk preservatives and adulterants: Processing, regulatory and safety issues. *Food Rev. Int.*, 31, 236–261, 2015.
11. Singuluri, H., Sukumaran, MK., Milk Adulteration in Hyderabad, India – A Comparative Study on the Levels of Different Adulterants Present in Milk. *J. Chromatogr. Sep. Techn.*, 5, 212, 2014.
12. Sudhabindu, K. and Samal, K., Common adulteration in spices and Do-athome tests to ensure the purity of spices. *Food Sci. Rep.*, 1, 9, 66–68, 2020. https://foodandscientificreports.com/details/seed-production-during-summer-reduces-proportion-of-hard-seeds-in-mung-bean-vigna-radiata-l.html
13. Pasupuleti, R.V., Sammugam, L., Ramesh, N. and Gan, S., Honey, Propolis, and Royal Jelly: A comprehensive review of their biological actions and health benefits. *Oxid. Med. Cell. Longev.*, 2017, 1–21, 2017.
14. Zábrodská, B. and Vorlová, L., Adulteration of honey and available methods for detection: A review. *Acta Vet. Brno*, 83, 10, 85–102, 2015.

15. Cordella, C.B.Y., Militão, J., Clément, M.-C., Drajnudel, P., Cabrol-Bass, D., Detection and quantification of honey adulteration via direct incorporation of sugar syrups or bee-feeding: Preliminary study using high-performance anion exchange chromatography with pulsed amperometric detection (HPAEC-PAD) and chemometrics. *Anal. Chim. Acta*, 531, 239–248, 2005.
16. Se, K.W., Ghoshal, S.K., Wahab, R.A., Ibrahim, R.K.R., Lani, M.N., A simple approach for rapid detection and quantification of adulterants in stingless bees (Heterotrigona itama) honey. *Food Res. Int.*, 105, 453–460, 2018.
17. Asrat, A. and Zelalem, Y., Patterns of milk and milk products adulteration in Boditti town and its surrounding, South Ethiopia. *Sch. J. Agric. Sci.*, 4, 10, 512–516, 2014.
18. Awasthi, S., Jain, K., Das, A., Alam, R., Surti, G., Kishan, N., Analysis of food quality and food adulterants from different departmental & local grocery stores by qualitative analysis for food Safety. *IOSR J. Environ. Sci. Toxicol. Food Technol.*, 8, 2, 22–26, 2014.
19. Pardeshi, S., Food adulteration: Injurious adulterants and contaminants in foods and their health effects and its safety measures in India. *Int. J. Sci. Eng. Dev. Res.*, 4, 6, 229–236, June 2019.
20. El Sheikha, A.F., DNA Foil: Novel technology for the rapid detection of food adulteration. *Trends Food Sci. Technol.*, 86, 544–552, 2019.
21. https://www.dw.com/en/australias-needle-in-strawberry-scare-widens/a-45525110, 2018.
22. Moore, C., Spink, J., Lipp, M., Development and application of a database of food Ingredient fraud and economically motivated adulteration from 1980 to 2010. *J. Food Sci.*, 77, 4, 118–126, 2012.
23. Banti, M., Food Adulteration and Some Methods of Detection, Review. *Int. J. Nutr. Food Sci.*, 9, 3, 86–94, 2020.
24. Pal, M. and Mahinder, M., Food adulteration: A global public health concern. *Food and Drink Industry*, 1, 3, 38–40, 2020.
25. Pathan, F. and Pawar, M., Current Situation of Food Adulteration: Laws, policy and governance in India and remedies to the problems. *Int. J. Sci. Eng. Manag.*, 2, 4, 2017.
26. Aun, P.J.A., Food adulteration in developing countries. *J. Assoc. Public Anal.*, 17, 115–120, 1979.

7

Food Adulteration and Its Impacts on Our Health/Balanced Nutrition

Suka Thangaraju, Nikitha Modupalli and Venkatachalapathy Natarajan*

Department of Food Engineering, National Institute of Food Technology Entrepreneurship and Management (formerly Indian Institute of Food Processing Technology - IIFPT), Thanjavur, Tamilnadu, India

Abstract

Adulteration of food is a common issue, and it's a big concern since the beginning of civilization. Because, it decreases the quality of food either by the addition of low-quality material or by subtracting/extracting the valuable products from the food. Food adulteration can be intentional/deliberate (for the financial lift to producers, processors, retailers) or unintentional/incidental (happens during production, handling, and storage). Food products prone to adulteration include: juices, fats and oils, honey, pepper, cereal products, dairy products, etc., and these adulterants have hazardous health effects. It is estimated that approximately 57% of people have developed health issues because of indigested adulterants and contaminants, and about 22% of foods are adulterated every year globally. If people are educated about common contaminants, then it is possible to prevent adulteration. There should be quick, precise, easy, and cheap source techniques and sources to detect/identify adulterants in various foods. Regulations that castigate illegal traders and dishonest producers that adulterate different food products in various places should be applied. This chapter is aimed to cover food adulteration, why it is done, its types, its effect on human health, and balanced nutrition.

Keywords: Adulteration, intentional, unintentional adulterants, hazards, health impacts

---

*Corresponding author: venkat@iifpt.edu.in

Mousumi Sen (ed.) Food Chemistry: The Role of Additives, Preservatives and Adulteration, (189–216) © 2022 Scrivener Publishing LLC

7.1 Introduction

Food is any substance from a plant or animal source composed of carbohydrates, water, proteins, fat, and several micro and macronutrients ate or drunk by a human for their growth and maintenance of health and well-being vitality. Food is a basic requirement for life. The advent of increase in requirement of convenience foods by the consumers has directly increased the market and production of packed foods. The food industry's growing globalization and the consequent isolation between suppliers and consumers have raised the likelihood of food adulteration. Adulterant use has been common in societies with few legal controls on food quality and therefore poor or even non-existent monitoring by the authorities. Humans have altered the condition of food during prehistoric times to extend its longevity or enhance its flavor. Before 300,000 years ago, humans used fire to cook and preserve meat. Later, they found that salt can be used to preserve meat without cooking. The wine was mixed with honey, spices, herb, saltwater, lead, or chalk which acts both as a sweetener and a preservative in ancient Rome and Greece. As time emerges, the act of adulterating food for economic gain has grown up. Imported and high valued products had high demand and limited supply. The producer and processing unit started to mix foreign materials with the original product, and accordingly, contemporary accounts adulteration date from the 1850s to the present day. Adulteration was first investigated in 1820 by Frederick Accum, a German chemist. He investigated the presence of toxic metals and colors in foods and drinks. In early 1850, physician Arthur Hill Hassall carried out extensive research on adulteration, leading to the 1860 food adulteration act and further legislation [1]. Mostly, adulteration of foods lead to food fraud and adulteration and create health hazards to the consumer. Food fraud means deliberate substitution or addition, tampering, misrepresenting food ingredients, packaging, or information about the food product for economic gain.

Food safety and quality and the factors affecting them are major concern areas for the food supply chain. Food adulteration can be visualized in different ways, like mixing or substituting low-quality substances with the original or natural product or removing valuable constituents from the given natural food product [2]. Adulterated in legal terms means that the food product fails to meet standards. Adulteration could also be adding a non-food item to the food product to increase the quality, preservation, and improvement of appearance. It can also be any poisonous or deleterious food substance that leads to health effects. According to Food and

Drug Administration (FDA), any food can be acknowledged to be adulterated if

(i) added with any foreign material that can be injurious to health
(ii) added with same material of inferior quality
(iii) the product quality not meeting the quality standards
(iv) added with foreign materials to alter the appearance or consumer appeal of the original product
(v) added with any substance to manipulate the product weight or density
(vi) removal of any important or characteristic compound from the product [3]

Certain special cases of food poisoning, food spoilage, and food contamination can also be categorized into food adulteration. Further, false and misleading labeling of food products to manipulate the consumer choice is termed as misbranding and is as dangerous as direct adulteration of foods. Consumers are either the sufferers of being cheated or the victims of diseases because of the adulteration of food products. So, the public needs to know common adulterants and the mostly adulterated food and their health implications. This makes product authenticity and quality testing, issues, or utmost of importance among the food manufacturers and industries.

The result of food adulteration is that the consumer is deprived of the information about the actual quality of product that can affect their purchase choice. Food fraud and economically motivated adulteration (EMA) are most prevalently reported in current scenario due to the increase in globalization of the food market and development of complex supply chains from farm to fork. Food fraud and EMA are growing concern for reasons such as jeopardy to the public health, supply chain control, transparency, and loss in consumer trust. Most cases of intentional adulteration in the food market are the product of the motivation to gain some sort of economic incentives through it. Such acts generally involve adulterating the targeted products without using ant materials that pose serious or immediate health risks. The reason for this is that any health risk developed can expose the fraud and can be catastrophic. Due to this, many occurrences of EMA are not likely to be reported or detected [4]. The most common reasons for opportunity to carry on adulteration of food products are as follows:

a. complex supply chains that extend to many countries across the globe,
b. raise in demand for imported ingredients among the consumers, and
c. gap in the food safety and quality laws in different countries, that serves as a loophole for EMA [5].

7.2 Types of Adulteration

Adulteration is the act of devaluing the food product to duplicate a natural commodity or addition of low-quality material into the genuine product to gain illegal profit. Adulteration of food can be four ways: 1. addition of extraneous material (e.g., addition of water to milk and the addition of papaya seeds into pepper); 2. mixing low-quality material with the real or natural one (e.g., mixing used tea leaves with the fresh tea leaves); 3. using prohibited preservatives, colors, or flavoring agents (coloring of spices); and 4. extraction of valuable components from the native food (e.g., oils from spices). Food has also been adulterated for increasing the volume. Generally, adulterations can be categorized into intentional and incidental adulteration [6, 7].

7.2.1 Intentional Adulteration

Intentional adulteration is the purposefully performed type of adulteration. It is usually done for financial gain for the food producers. Intentional adulteration is also termed sometimes as food fraud [8]. It is the inclusion of inferior substances with similar physical properties to the food they are added. This adulterant can be physical or biological. The most common example of intentional adulteration is the addition of colors. Some of the common examples of intentional adulteration are the addition of synthetic made milk or water into natural milk, dried papaya seeds into black pepper, jaggery into honey, washing powder into ice cream, brick powder into red chilli powder, chicory into coffee, and argemone seeds into mustard seeds [9]. The seven foods most likely to be the target for the intentional or deliberate use of adulterants are olive oil, milk, honey, saffron, coffee, orange, and apple juices [10]. Calcium carbide in mango and banana, use of oxytocin for faster growth of pumpkin, watermelon, gourds, and cucumber, and external application of wax to increase the shine of apple and pear are also considered as intentional adulteration [5].

7.2.2 Incidental Adulteration

Incidental or unintentional adulteration is the outcome of ignorance or the absence of food safety maintenance facilities. Accidental food adulteration, as the name implies, occurs incidentally or accidentally without our knowledge. Leakage out from pesticides and fertilizers, droppings of rodents, larvae in food, tin from cans, lead from water, and mercury from effluents are examples of incidental adulteration. This type of adulteration is because of food products' improper hygienic conditions starting from the production site to the consumer's table. In this type of adulteration, manufacturers or traders are not in a position to incorporate various forms of adulterants, but how the goods are made, manufactured, processed, packaged, shipped, and marketed could be areas where they have been tainted or adulterated, because any material without its origin is alien to the commodity and is thus assumed to be adulterated [7, 10, 11]. Table 7.1 elaborates on the differences between intentional and accidental adulteration with suitable examples.

7.2.3 Other Types of Adulteration

7.2.3.1 Natural Contamination

Natural adulteration is another type of impurity that can occur in foods due to various chemicals, organic compounds, or radicals that naturally occur in foods that are detrimental to health and are not purposely or accidentally added to foods. Any examples of various forms of natural adulteration include poisonous varieties of pulses, mushrooms, green and other vegetables, fish, and shrimp, which can in some cases be considered anti-nutrients. Several types of marine fish species are known to be poisonous; many are edible species [12, 13].

7.2.3.2 Metallic Contamination

Metallic contaminants are present in the food in trace amounts and enter through the environment, or during the food production process. Tin from cans, lead, and mercury from water and other effluents,

Table 7.1 Type of adulterants/contaminants with their examples.

Type of contaminants/ adulterants	Examples
Intentional adulterants • Physical • Biological	Sand, stone, mud, marble chips, chalk powder, water, mineral oil, and other filth Papaya seeds in black pepper, argemone seeds in mustard, animal fat in ghee, chicory in coffee, etc
Accidental adulterants • Natural • Non-natural • Metal	Toxic varieties of fish, sea food, green vegetables, pulses, mushroom Pesticide and fungicide residues, tin and zinc from can, larvae, droppings of animals, etc Arsenic, barium, cadmium, cobalt, copper, lead, mercury, tin, and zinc
Microbial contaminants • Bacteria • Fungi • Parasites	*Bacillus cereus, Clostridium botulinum, Clostridium perfringens, Salmonella, Shigella sonnei, Staphylococcus aureus, Streptococcus pyogenes* *Aspergillus flavus, Aspergillus versicolour, Aspergillus nidulans, Aspergillus bipolaris, Fusarium sporotrichioides, Penicillium inslandicum, Penicillium atricum, Penicillium citreovirede, Penicillium inslandicum, Penicillium atricum, Penicillium citreovirede* *Ascaris lumbricoides, Entamoeba histolytica*

and pesticide residues are common examples of metallic contaminants. Metallic contamination is considered the type of incidental or accidental adulteration [10, 11].

7.2.3.3 Microbial Contamination

Spoilage of food is due to the presence of microorganisms through various sources. Food can be contaminated by bacteria, fungi, or parasites at any time, in any environment from many sources during the processing like harvest, storage, processing, and distribution.

7.2.3.4 Adulteration in Organic Foods

If food is claimed to be organic, then it must be cultivated and processed to comply with the country's requirements in which it is sold. In organic farming, there is no use of synthetic pesticides in the area. Bt. pyrethrum and rotenone are the most common pesticides, approved for restricted use by most organic standards. Any food stated to be organic and not compliant with the guidelines is thus adulterated [14, 15].

7.2.3.5 Adulteration During Irradiation of Foods

Food irradiation is one of the preservation techniques by applying ionizing radiation that improves food products' safety and shelf life by killing the insects and harmful microorganisms. Ionizing radiation such as radionuclide Cobalt-60 is used in food to kill and control the multiplication of bacteria, viruses, or insects that may be found in food and prevent sprouts, delay in ripening, and increase rehydration. In ionizing radiation, the internal metabolism of cells, the division of DNA, and the production of free radicals are interrupted. The mechanism entails the destruction of chemical bonds, the potential production of resistant microorganisms. There is also an insufficient diagnostic method for the identification of food irradiation and public opposition. If this irradiation is used at a level that can be overdosed, then it is clear that there would be significant health risks. So, without the consent of the consumer, it can be considered adulterated [16].

7.2.3.6 Genetically Modified Foods

Organisms in which genetic materials (DNAs) are transformed in a way not found in nature are genetically modified organisms (GMOs). It is also termed gene technology. Genetic alteration is said to be the synthesis of genes from various species. Medicines, vaccines, nutritional and food additives, feed, and fibers are used in GM products. Although GM foods are produced and sold because either the manufacturer or user of these foods has a perceived advantage, safety concerns like allergens, the transmission of antibiotic resistance markers, and unknown consequences can occur. Therefore, unless proper information is conveyed to customers, genetically modified foods are also considered adulterated [17].

7.3 Adulteration in Foods

As revealed by the findings of numerous writers worldwide, there are different food goods and beverages that are vulnerable to adulteration. This ensures that it can be difficult to procure food products, such as flour, pulse, grease, meat, vegetables, milk, candy, spices, tea, coffee, honey, bakery, chocolate, fruit juice, and the likes, which are free of one or more types of adulterants. Also, animal feeds such as cakes as protein supplements for lactating animals are adulterated, accounting for around 90% of non-branded loose types, as shown in the literature [18]. Consumers are not always aware of the adulteration of pre-packaged foods and foods known as "loose" (e.g., without any branding or packaging). This is important to illiterate consumers who are frequently misled about approved additives' quality requirements and are victims of industry violations or malpractices. For example, in the open market, cereals and pulses are sold with sand, dirt, stones, earth, or talc. Good sorghum and corn grains are combined with fumonisin-toxin-containing mouldy grains [19, 20]. Bad quality cardamoms (from which essential oils have been extracted) are blended with high-quality green cardamoms, and red pepper powder is adulterated with colored sawdust on weekly markets. Papaya seeds, grass seeds (coated with charcoal dust), or mineral oil are adulterated with black pepper seeds. Pigeon pea (*Cajanus cajan*) or chickpea (*Cicer arietinum*) split grains or flour are adulterated with grass pea (Lathyrus sativus) when cooking snacks or dishes. Consumption records of wheat/millet grains mixed with buckwheat (*Polygonatum Fagopyrum*) flour or Crotalaria spp. seeds have been recorded. It contains poisonous alkaloids. Edible oils are combined with oils and fats which contain butylated hydroxyanisole or butylated hydroxytoluene. The mixture of rancid oil with edible oils kills vitamins A and E from a dietary perspective. The combination of oleomargarine (beef fat) in butter and gelatin and formaldehyde in milk is usually found adulterants [18, 21].

Asystemic fungicide (benomyl) is used in crops to suppress microorganisms' development and discourage spoilage. Wax (containing morpholine as a solvent and emulsifier) spreads to fruit to maintain moisture, avoid bursting and physical injury, increase the appearance, and prolong storage time/shelf life. However, the consumer samples' wax content is below the Appropriate Daily Intake (ADI) of 2.0–3.6 µg/kg body weight/day. Unripe fruits are chemically ripened with ethylene to maintain firmness and give a ripening appearance. Calcium carbonate powder containing traces of arsenic and phosphorus is added to fruit; fruit and vegetables are ingested by injection of the hormone "oxytocin" to maintain freshness; colored water is injected into the watermelon to add redness to the pulp [22]. Many non-permitted color food commodities (NPC) are related to human diseases. Fruit samples obtained from the market showed an NPC content of up to 730 ppm, which exceeded the given limit. Metanil yellow in parboiled rice, turmeric powder and split pulse grains, and rhodamine (10–95 ppm), orange II (135–560 ppm), or auramine (15–400 ppm) in sweets and roadside foods are reported. At present, flavor concentrates, natural colors (beta-carotene, riboflavin, caramel, annatto, saffron, and curcumin), and synthetic colors (brilliant blue FCF, carnosine, erythrosine, fast green FCF, ponceau 4R, sunset yellow FCF, indigo carmine, and tartrazine) are permitted up to100 ppm for all foods or 200 ppm for canned products. During trading, malachite green or copper sulfite to crucifers, green peppers, leafy vegetables and eggplant, and Congo red and Sudan dyes to red peppers are added for retaining natural appearance and freshness. Similarly, acidity regulators, anticoagulating agents, antifoaming agents, antioxidants, bulking agents, color retention agents, emulsifiers, flavors, and flavor enhancers are frequently utilized in the production and processing units. Additives that are most frequently mixed in the market yards include fat-soluble azo dye to mustard oil, monosodium glutamate to meat and Chinese foods, sulfur dioxide to preserved foods, artificial sweeteners/sugar substitutes to sweets; 3,4 benzopyrene to smoked foods; cottonseed oil containing gossypol to edible oils, all are found beyond the safe limit [23]. The adulterants commonly occurring in different food commodities have been listed and categorized in Table 7.2.

7.3.1 Global Food Environment

The understanding of global food market and reports of adulteration in different countries can improve the understanding of the risks of EMA and food fraud. Commercialization of food trade and growing complexity in supply chain has a profound effect on the incidences. Several countries have reported successful legal measures and prosecutions on incidents of food adulterations and fraud [87]. The European Union (EU) has reported incidence of horse meat and phenyl butazone drug in

Table 7.2 Different type of adulterants in food products.

Food	Type of adulterants			References
	Physical	Chemical	Biological	
Pepper Pepper powder	-	-	Papaya seeds Chilli powder, cassava starch, corn flour, millet, buck and whole wheat flour	[24, 25] [26–29]
Chilli	Brick powder, stone powder, talc powder	Metanil yellow, sudan dyes	Starch, saw dust, almond shell, tomato waste	[30, 31]
Coriander	-	-	Starch	[32]
Turmeric	Colored saw dust, chalk powder	Metanil yellow, sudan dye, methyl orange, red lead salt	Starch, wild curcumin	[33]
Cumin powder	Straw	-	Starch, almond, hazelnut, Brazil nut, pistachio shell powder, coriander powder	[34, 35]
Mustard seed and oil	-	-	Argemone seeds and oil, rapeseed, ragi	[36]
Onion, garlic powder	-	-	Starch, red pepper powder	[37–39]
Ginger	-	-	Crude Moroccan beans	[40]

(*Continued*)

Table 7.2 Different type of adulterants in food products. (*Continued*)

Food	Type of adulterants			References
	Physical	**Chemical**	**Biological**	
Saffron	-	Synthetic dyes- tartrazine, ponceau 2R, sunset yellow, amaranth, orange GG, methyl orange, eosin and Erythrosine oil; honey; glycerine; solutions of potassium or ammonium nitrate; sodium sulfate; magnesium sulfate; barium	Different parts of the saffron flower itself, dried petals of safflower and Scotch marigold, calendula, poppy, turmeric, annatto, pomegranate, Spanish oyster and maize, dyed corn silk, meat fiber, red sandal wood, turmeric powder, paprika	[41–46]
Coffee	-	-	Maize, wheat, corn, soybean, barley, rice, chickpea, roasted husks	[47–51]
Honey	-	-	Cane sugar syrup, jaggery, beet invert sugar, high fructose corn syrup, glucose syrup, saccharose syrup, rice syrup	[52–59]
Egg	-	Melamine, cyanuric acid	Turmeric, starch, soy protein, maltodextrin, water	[60–65]
Meat	Metals, glass, stones, bones	-	Other meats, soybean protein,	[66–71]

(*Continued*)

Table 7.2 Different type of adulterants in food products. (*Continued*)

Food	Type of adulterants			References
	Physical	Chemical	Biological	
Milk and milk powder	-	Water, urea, melamine	Other milk varieties, whey protein, allergenic buck wheat, soy milk	[72–79]
Rice	Dust, stones, straw, weed seeds, damaged grains, insects, rodent's hair and excreta	-	Other varieties of rice	[80–86]

about 4.7% and 0.51% of the meat samples analyzed in April 2013 [88]. Certain cases were reported to have presence of chalk, alum, and plaster in baker's floor and lead and copper in jellies and sweets to impart bright colors [89]. Another form of EMA is to declare non-organically grown produce and ingredients as organic foods, which comes under misleading labeling category. A report in 2012 on fish samples in United States revealed that 94% of white tuna products were tested to contain no tuna fish, with the fraud involving more smaller markets (40%) that larger establishments (12%) [90]. Melamine contamination in milk stood out to be a huge scandal in China in 2006, when the entire dairy industry of the country was prosecuted. Melamine is a nitrogen-rich compound and contains about 67% of nitrogen per unit mass, making it a good source of adulterant to increase the nitrogen content of milk, thus falsely projecting greater protein content. This has caused formation of kidney stone sin several infants [89]. WHO has quoted that at least 22 manufacturers of dairy products in China have been found to have melamine traces in their products [91]. The entire melamine adulteration incident was a classic example of the EMA to bridge the gap between demand and supply of a particular commodity. In order to fill the supply-demand gradient to stabilize the cost structure of the product, such EMA can occur at a large scale. Most specifically, the products with higher market value like oils, spices, and meats are prone to such adulteration. A study on processed meats and meat products other than carcass meats

in South Africa has inferred that about 68% of the samples contained species that were not declared in product labeling [92]. The adulteration of high valued commodities actually deprives the consumers of the experience obtained from the original product, along with fraud of extracting more value than the product quality. The substitution of highly expensive white sturgeon caviar with the caviar from beluga costs about five times more from the buyers [93]. In a survey of milk samples in India, adulteration was found in 68% of the samples collected, with detergent (8%), skimmed milk powder (45%), and glucose (27%) detected as the prime adulterants [94].

7.4 Effects of Food Adulteration

7.4.1 Health Effects

EMA and fraud have been recognized as a serious threat to public health by most governing bodies and organizations worldwide. Several instances of food fraud like melamine in milk products, heavy metal contents in processed food products, etc., have increased the consumer conscience about adulteration and its health effects [13]. Food contamination has also been a serious problem in the food business and distribution sector. Food contamination critically affects the health of the consumers if unattended and hence needs to be checked and controlled along with adulteration. Food contamination instances like aflatoxins in spice products or the formation of biogenic amines in processed foods like animal foods and fermented foods have been known to cause ill-health to the persons. Health risks of EMA can be manifested into immediate problems like diarrhoea, nausea, and vomiting. Food adulteration hazards include deterioration of consumers' health causing problems like carcinogenesis, ulcerations, digestive tract problems, anaemia, insomnia, paralysis, dropsy, muscle dystrophy, liver and kidney malfunctions, respiratory distress, and oedema [95]. Dropsy can manifest into partial paralysis of limbs of the body, whereas muscular dystrophy causes weakness in muscles, causing problems in muscle coordination and movement.

The use of sodium cyclamate as an adulterant has been reported widely, in sugar and its products as it imparts free-flowing characteristics to sugar. Sodium cyclamate is a carcinogenic chemical and can cause cancer and ulcerations in different parts of the body, especially children and vulnerable groups [96]. Formalin is another important additive added by the growers and vendors, especially to products like fish, meat, and milk, to extend its shelf stability and manipulate its protein content. Formalin is known to cause carcinogenesis, skin diseases, breathing and asthma troubles, etc. Food colors like erythrosine, Sudan dyes (I–IV), rhodamine, and tartrazine are some of the most frequently added prohibited additives to improve the food's aesthetic appeal products and affect the appeal of the food quality. Nitric acid and sulfides are commonly used additives in food to extend shelf-life. These compounds can cause disturbances in health like anaphylaxis and asthma, along with urinary tract problems [17]. Many edible oils can be adulterated with argemone oil, which can cause dropsy, due to compounds like sanguinarine and dihydro-sanguinarine [96]. The use of non-food substances like sawdust, brick powder, and chalk for adulteration of food products can cause severe health retardations, which can be irreversible at times. The pesticides and herbicides used on fresh produce can also be very dangerous to human health.

Melamine, a nitrogen-rich compound, used as an adulterant agent in milk, can create hepatic and renal stress and cause kidney stones. Melamine has also been reported to be an active carcinogenic agent [97]. Metanil yellow, which is used to impart good color to turmeric powder, etc., and coal-tar dye that is used in production of sweet meats have also been reported to cause cancer by several previous studies [98]. Detergent powders might be added in ice creams and confectionary products to improve the aeration and look glossy, thus improving consumer appeal. The detergents cause severe stomach problems like dysentery, nausea, ulcer, and abdominal cramps and can lead to cancer on a long run. Addition of wheat slurry or any plant-based milk extracts (groundnut milk, etc.) to milk to manipulate the thickness can cause discomfort to consumers with gluten sensitivity. Consumption of products adulterated with simple starches like sugar solution or fructose syrup in honey and glucose powder in milk can create a raise in immediate blood sugar levels of the consumers, which can be dangerous for critical diabetes patients. Use of preservatives such as benzoic acid, boric acid, formaldehyde, ammonium fluoride, and salts of sodium and potassium can cause various adverse health effects on prolonged usage [98, 99]. The health hazards that can occur by ingestion of different adulterants in several commodities have been elaborated in Table 7.3. The perils of consumption of adulterated foods can manifest on the human health over a period, rather than with immediate effect. The effect will be profound in vulnerable groups of population, like children, geriatrics, and pregnant women. This is one of the major problems in detecting the presence and levels of adulteration in food commodities.

Table 7.3 Types of adulterants and their health effects.

Adulterants or contaminants	Food items	Health hazards
Accidental adulterants		
Pesticide residues	All types of food	Acute poisoning that damages the nervous system, liver, kidney
Fluorine	Water, seafood	Fluorosis (mottling of teeth, and neurological disorders)
Polycyclic Aromatic Hydrocarbons	Smoked fish, meat, water, oils, shell-fish	Cancer
Physical adulterants		
Stone, sand, filth	Cereals, pulses, etc.	Damages the digestive system
Brick, chalk powder, sawdust, artificially colored leaves	Chilli powder, coriander powder, tea leaves	Cancer, injurious to health
Biological adulterants		
Rancid oils	All type of oils	Loss of vitamin A and E
Colored foreign seeds	Pepper, cumin, poppy, and mustard seeds	Injurious to health, cancer
Low-quality ingredients	Oregano, rice, olive oil, meat, milk	Low-quality standards affect health
Husk, shell dust, petals, stem, and fruits of other plant sources	Saffron, clove, chillies, cashew	Low-quality standards affect health
Microbial contaminant Bacteria	Cereals products, custard, pudding, sauce	Nausea, vomiting, diarrhoea, abdominal pain
Bacillus cereus	Milk, canned meat, fish, and gravy	Nausea, diarrhoea, abdominal pain, gas formation

(Continued)

Table 7.3 Types of adulterants and their health effects. (*Continued*)

Adulterants or contaminants	Food items	Health hazards
Clostridium perfringens	Meat and meat products, raw vegetables, shell-fish, egg products, salads	Salmonellosis (fever and chills)
Salmonella spp.	Dairy products, baked food, meat products, low acid frozen foods, salads, sauces, etc.	Salivation, vomiting, abdominal cramp, diarrhoea, severe thirst, cold sweat
Staphylococcus aureus, Enterotoxins – A, B, C, D or E	Milk, poultry, potato, beans, tuna, shrimp, mixed moist food	Shigellosis (bacterial dysentery)
Shigella	Groundnut, cottonseed Wheat, millet, oats, etc.	Liver cancer Alimentary toxic aleukia (ATA) (epidemic panmyelotoxicosis)
Fungi Aflatoxins Toxins from *Fusarium sporotrichioides*	Foodgrains	Carcinogenic and mutagenic. Kidney and liver damage. Skin and hepatic tumor.
Sterigmatocystin from *Aspergillus versicolour* *Aspergillus nidulans* *Aspergillus bipolaris* Toxins from *Penicillium inslandicum* *Penicillium atricum* *Penicillium citrovirede* *Fusarium, Rhizopus* *Aspergillus* **Paraciticus**	Yellow rice	Mouldy rice disease.
	Any food or water contaminated by human faces that contains human eggs	Ascariasis
Ascaris lumbricoides *Entamoeba histolytica*	Uncooked vegetables and fruits	Amoebic dysentery

(*Continued*)

Table 7.3 Types of adulterants and their health effects. (*Continued*)

Adulterants or contaminants	Food items	Health hazards
Metallic contaminants		
Arsenic	Water, apples sprayed with lead arsenate	Dizziness, chills, cramps, paralysis, death
Barium	Foods contaminated by rat poison	Violent peristalsis, arterial hypertension, muscular twitching, convulsions, cardiac disturbances
Cadmium	Fruit juices, soft drinks, water in contact with Cadmium equipment, shell-fish	Itai-Itai (ouch-ouch) disease, Increased salivation, acute gastritis, liver and kidney damage, prostate cancer
Cobalt	Water and liquors	Cardiac insufficiency and myocardial failure
Copper	Any food	Vomiting diarrhoea
Lead	Water, natural and preserved foods	Lead poisoning causing foot-drop, insomnia, anaemia, constipation
Mercury	Fish	Brain damage, paralysis, death
Tin and zinc	Any food	Vomiting

Source: [3]. Reprinted with permission of the publisher (Taylor & Francis Ltd, http://www.tandfonline.com).

7.4.2 Balanced Nutrition

The effect of adulterated and contaminated food substances falls seriously on the consumers' nutrition and a balanced diet. The addition of products like ergot and chalk in wheat flour reduces the amount of nutrition obtained from the flour. Similarly, the addition of foreign chemicals like formalin or melamine can cause severe gastrointestinal distress, leading to symptoms like nausea, dysentery, and diarrhoea, causing improper digestion and absorption. Food adulterants can also cause food poisoning and metabolic dysfunction, directly affecting individuals' overall health and nutritional status [17].

7.5 Measures to Mitigate Food Adulteration

7.5.1 Producer's or Manufacturer's End

At the farmers' end, contamination and unintentional adulteration can be avoided by following Good Agricultural Practices (GAP), along with integrated storage solutions and pest management. Avoidance of indiscriminate use of fertilizers and pesticides can be followed by adapting to the need-basis application. Extreme care needs to be taken for pest management to avoid over-application of chemicals, which may cause accumulation in the final product. Alternatively, natural, and sustainable measures like spraying of water extracts of plants like turmeric and neem can be used in initial and less-intense infestation stages.

In the manufacturer's end, strict Good Manufacturing Practices (GMP) must be followed throughout the processing line, from procurement, processing, packaging, storage, distribution, to waste disposal and management. The management needs to be apprehensive at all times and regularly monitor the production process. Critical control points to evaluate the system (HACCP system) can be a very effective production control method. There should also be a timely check on different sources of contamination, waste disposal management, following of standards, labeling, etc. Compliance with standards set by regulatory bodies like Codex and FSSAI needs to be followed to minimize the risk of any contamination or adulteration by several folds. Novel process evaluation systems like Threat Assessment and Critical Control Points (TACCP) and Vulnerability Assessment and Critical Control Points (VACCP) can be very helpful in keeping the incidence of EMA on check. TACCP critically focuses on food defence and tampering. Though it is similar to HACCP in many ways, it requires more personnel than HACCP. VACCP focuses on prevention of potential food adulteration in a systemic fashion by identifying and managing the vulnerable points in the supply chain [99].

7.5.2 Consumer's End

Consumers need to be apprehensive during the purchase, storage, and preparation of food products. It is advisable to purchase quality products packed properly and certified by the competent authority. Visual inspection of the products before the purchase can help avoid visible food adulteration. The package label should be carefully read for information like date of manufacture, expiry date, and ingredients. The consumers should be aware of the most commonly adulterated food products and their detection techniques for cross-verification at periodic intervals. In case of detection of adulteration in any food products, the consumers need to complain to the manufacturer of the product for suitable inquiry and action. The fresh foods should be thoroughly washed and rested before usage to avoid agricultural chemical residues. The foods should be consumed with immediate effect rather than storage for longer periods to avoid spoilage due to microorganisms and aflatoxins' growth. Opting for seasonal foods can help avoid the consumption of highly processed foods, which have a higher risk of exposure to contamination and adulteration.

7.5.3 Government and Regulatory Agencies

Government and regulatory bodies like Codex Alimentarius Commission (CAC), US-Food and Drug Authority (US-FDA), European Union (EU), Food Safety and Standards Authority of India (FSSAI), and Bureau of Indian Standards (BIS) have been some of the regulatory bodies for checking the food quality and safety. These bodies control and inspect the food supply chain and play a key role in the policy-making process. The food supply chain, being complex, adaptive, and highly dynamic, needs these authorities to coordinate the entire supply chain to ensure its smooth functioning. Food safety measures are of utmost importance to food producers, retailers, public authorities, and health inspectors. Nowadays, food safety programs focus on a "farm to table/fork" approach as an effective means of reducing foodborne diseases. These programs are managed by the regulatory bodies that help ensure food products' safety by adapting suitable protocols like periodic market surveys, testing, and inspection.

The HACCP system helps identify, evaluate, and control the potential hazards in the food processing and production lines. The Food Safety Management System (FSMS) is also a very useful compliance system to ensure food safety throughout the supply chain, i.e., farm to fork. For detecting contamination of the foodborne diseases caused by microorganisms, modern laboratory facilities with pollution control measures, food quality inspection, and monitoring and management (including treatment, packaging, and preservation) have been established by most regulatory bodies country. The modern day requirement of labeling and declaring the particulars of the food commodities and its contents has increased the need for the legislator and regulatory measures to curb food fraud and EMA. The supply chain of food products, being highly complex and adaptive, has certain drawbacks like weak infrastructure, lack of supervised logistics, unorganized food market, and developing economies, which can havoc the occurrence of adulteration such vulnerable points. The regulatory systems like HACCP, TACCP, and VACCP critically look after such vulnerable points and avoid such occurrences. The system of "farm to table" has been drawing the primary focus of the regulatory organizations in developing a holistic approach for food control. The main focus of the general food industry and international regulatory agencies has been on adulterant substances because they are a commonly identified a public health threat, and the response is usually led by a public health agency. The policy making for the sake of safeguarding public health is generally proceeded in stages:

a. Problem identification: The need for formulation of policy and legislature is identified by governmental or nongovernmental organizations, giving more stress on public health and food safety. It can also be focused on the economic fraud and scam caused due to EMA.
b. Agenda setting: The identified problem would be investigated and objectives of the solution can be formulated. This step includes personnel as the investigation involves market surveys, sample analysis, forums, and questionnaires to evaluate the seriousness of the identified concerns.
c. Policy formulation: The development of the outline of the policies and legislation by the interested organizations is based on the intensity of the situation and the reach of the organization. This can be backed up by the evidences collected during the objective formulation for more elaboration.
d. Policy legitimation: The formulated policy would be evaluated by experienced panel to understand its effectiveness in safeguarding the public health by avoiding the instances of food fraud. The expert committee can also allow required changes in the policy statements to suit the situation. However, this differs for different countries as well as various organizations.
e. Policy implementation: The formulated legislations are executed in to force by the governing bodies. This could include clarifying where non-governmental or industry activities align with current regulations including the explanation of

compliance expectations. The policy implementation phase also includes propagation and advertising about the policy and its benefits to gain the public interest and educating them about the public health measures. This also creates good understanding about the food adulteration and its ill health effects to the common public.

f. Policy evaluation: The executed policy needs to be periodically evaluated for effectiveness and significance of the outcomes. Based on the evaluation, certain changes will be needed to be made to the policy statements yet again to suit mass applicability and better enforcement. The evaluation step is to monitor and continually adjust food fraud public policy.

The food fraud and adulteration is an emerging concern for public health and an economic scam against the consumers. Formulation of a holistic, comprehensive, and harmonized legislator provision for checking the incidences of food adulteration which is poised to be a severe threat to human health can help to achieve the same. Food, being the primary requirement of the life, needs to be safe and quality for the value offered by the consumers.

References

1. Manasha, S. and Janani, M., Food adulteration and its problems [Intentional, Accidental and natural food adulteration]. *Int. J. Res. Finance Mark.*, 6, 4, 131–40, 2016.
2. Ayalew, H., Birhanu, A., Asrade, B., Review on food safety system: Ethiopian perspective. *Afr. J. Food Sci.*, 7, 12, 431–40, 2013.
3. Bansal, S., Singh, A., Mangal, M., Mangal, A.K., Kumar, S., Food adulteration: Sources, health risks, and detection methods. *Crit. Rev. Food Sci. Nutr.*, 57, 1174–1189, 2017.
4. Johnson, R., *Food fraud and economically motivated adulteration of food and food ingredients*, Congressional research service, CRS Report, Washington DC, 2014.
5. Everstine, K., Supply Chain Complexity and Economically Motivated Adulteration [Internet], in: *Food Protection and Security: Preventing and Mitigating Contamination during Food Processing and Production*, pp. 1–14, Wood Head Publishing, UK, 2017. Available from: http://dx.doi.org/10.1016/B978-1-78242-251-8.00001-1.
6. El-Loly, M.M., Ibrahim, A., Mansour, A., Ahmed, R.O., Evaluation of Raw Milk for Common Commercial Additives and Heat Treatments Internet.

J. Food Safety, 15, 10, 2013. Available from: https://www.researchgate.net/publication/329885369.
7. Ayza, A. and Yilma, Z., Patterns of milk and milk products adulteration in Boditti town and its surrounding, South Ethiopia. *Sch. J. Agric. Sci.* [Internet], 4, 10, 512–6, 2014, Available from: http://www.scholarly-journals.com/SJAS.
8. Karoui, R., *Food authenticity and fraud*, Second Edi, pp. 579–608. Available from: http://dx.doi.org/10.1016/B978-0-12-813266-1/00013-9, Chemical Analysis of Food. Elsevier Inc., US, 2020.
9. Spink, J. and Moyer, D.C., Defining the Public Health Threat of Food Fraud. *J. Food Sci.*, 76, 9, 2011, https://doi.org/10.1111/j.1750-3841.2011.02417.x
10. World Health Organization, *Ten chemicals of major public health concern*, World Health Organization, 2010. Available on http://www.who.int/ipcs/assessment/public_health/chemicals_phc/en/
11. Choudhary, A., Gupta, N., Hameed, F., Choton, S., An overview of food adulteration: Concept, sources, impact, challenges and detection. *Int. J. Chem. Stud.*, 8, 1, 2564–2573, 2020. https://doi.org/10.22271/chemi.2020.v8.i1am.8655
12. Das, M., *Food Contamination and Adulteration*, 2007.
13. Khan, M., Food Adulteration and its Effect on Health. *Community Based Med. J.*, 2, 2, 1–3, 2013.
14. Hong, E., Lee, S.Y., Jeong, J.Y., Park, J.M., Kim, B.H., Kwon, K. *et al.*, Modern analytical methods for the detection of food fraud and adulteration by food category. *J. Sci. Food Agric.*, 97, 12, 3877–96, 2017.
15. Spink, J., Vincent Hegarty, P., Fortin, N.D., Elliott, C.T., Moyer, D.C., The application of public policy theory to the emerging food fraud risk: Next steps. *Trends Food Sci. Technol.* [Internet], 85[September 2017], 116–28, 2019. Available from: https://doi.org/10.1016/j.tifs.2019.01.002.
16. Mahindru, S.N., *Food preservation and irradiation*, APH Publishing, New Delhi, 2013.
17. Gahukar, R.T., Are Indian foods from genetically modified crops safe? *J. Food Agric. Environ.*, 2, 11–3, 2004.
18. Alauddin, S., Food adulteration and society. *Glob. Res. Anal.*, 1, 7, 3–5, 2012.
19. Janardhana, G.R., Raveesha, K.A., Shetty, H.S., Mycotoxin contamination of maize grains grown in Karnataka [India]. *Food Chem. Toxicol.*, 37, 8, 863–8, 1999.
20. Bhatt, R.V. and Mathur, P., Food Safety Evaluation-National and International Perspectives. *NFI Bull.*, 18, 6–8, 1997.
21. Majumdar, S., Food hazards and food security. *Everyman's Sci.*, 64, 348–55, 2010.
22. Siddiqui, M.W., and Dhua, R.S., Eating artificially ripened fruits is harmful. *Curr. Sci.*, 99, 12, 1664–1668, 2010.
23. Sudershan, R.V., Pratima, R., Kalpagam, P., Food safety research in India: a review. *Asian J. Food Ag.-Ind.*, 2, 3, 412–33, 2009.

24. Wilde, A.S., Haughey, S.A., Galvin-King, P., Elliott, C.T., The feasibility of applying NIR and FT-IR fingerprinting to detect adulteration in black pepper. *Food Control*, 100, 1–7, 2019 Jun 1.
25. Orrillo, I., Cruz-Tirado, J.P., Cardenas, A., Oruna, M., Carnero, A., Barbin, D.F. *et al.*, Hyperspectral imaging as a powerful tool for identification of papaya seeds in black pepper. *Food Control* [Internet], 101, December 2018, 45–52, 2019. Available from: https://doi.org/10.1016/j.foodcont.2019.02.036.
26. Parvathy, V.A., Swetha, V.P., Sheeja, T.E., Leela, N.K., Chempakam, B., Sasikumar, B., DNA barcoding to detect chilli adulteration in traded black pepper powder. *Food Biotechnol.*, 28, 1, 25–40, 2014.
27. Bhattacharjee, P., Singhal, R.S., Gholap, A.S., Supercritical carbon dioxide extraction for identification of adulteration of black pepper with papaya seeds. *J. Sci. Food Agric.*, 83, 8, 783–6, 2003.
28. McGoverin, C.M., September, D.J.F., Geladi, P., Manley, M., Near infrared and mid-infrared spectroscopy for the quantification of adulterants in ground black pepper. *J. Near Infrared Spectrosc.*, 20, 5, 521–8, 2012.
29. Li, X., Lu, R., Wang, Z., Wang, P., Zhang, L., Jia, P., Detection of corn and whole wheat adulteration in white pepper powder by near infrared spectroscopy. *Am. J. Food Sci. Technol.*, 6, 114–7, 2018.
30. Dhanya, K., Syamkumar, S., Jaleel, K., Sasikumar, B., Random amplified polymorphic DNA technique for detection of plant based adulterants in chilli powder [Capsicum annuum]. *J. Spices Aromat. Crops*, 17, 2, 75–81, 2008.
31. Cornet, V., Govaert, Y., Moens, G., van Loco, J., Degroodt, J.-M., Development of a fast analytical method for the determination of sudan dyes in chili- and curry-containing foodstuffs by high-performance liquid chromatography–photodiode array detection. *J. Agric. Food Chem.*, 54, 3, 639–44, 2006.
32. Sen, S., Mohanty, P.S., Suneetha, V., Detection of food adulterants in chilli, turmeric and coriander powders by physical and chemical methods. *Res. J. Pharm. Technol.*, 10, 9, 3057–60, 2017.
33. Sasikumar, B., Syamkumar, S., Remya, R., John Zachariah, T., PCR based detection of adulteration in the market samples of turmeric powder. *Food Biotechnol.*, 18, 3, 299–306, 2004.
34. Tahri, K., Tiebe, C., el Bari, N., Hübert, T., Bouchikhi, B., Geographical provenience differentiation and adulteration detection of cumin by means of electronic sensing systems and SPME-GC-MS in combination with different chemometric approaches. *Anal. Methods*, 8, 42, 7638–49, 2016.
35. Arslan, F.N., Akin, G., Elmas, ŞNK, Yilmaz, I., Janssen, H.-G., Kenar, A., Rapid detection of authenticity and adulteration of cold pressed black cumin seed oil: A comparative study of ATR–FTIR spectroscopy and synchronous fluorescence with multivariate data analysis. *Food Control*, 98, 323–32, 2019.
36. Dhanya, K. and Sasikumar, B., Molecular marker based adulteration detection in traded food and agricultural commodities of plant origin with special reference to spices. *Curr. Trends Biotechnol. Pharm.*, 4, 1, 1, 454–489, 2010.

37. Lee, S., Lohumi, S., Lim, H., Gotoh, T., Cho, B., Kim, M.S. et al., Development of a detection method for adulterated onion powder using Raman spectroscopy. *J. Fac. Agric. Kyushu Univ.*, 60, 1, 151–6, 2015.
38. Lohumi, S., Lee, S., Lee, W.-H., Kim, M.S., Mo, C., Bae, H. et al., Detection of starch adulteration in onion powder by FT-NIR and FT-IR spectroscopy. *J. Agric. Food Chem.*, 62, 38, 9246–51, 2014.
39. Kim, J.-H. and Baik, S.-H., Molecular identification of economically motivated adulteration of red pepper powder by species-specific PCR of nuclear rDNA-ITS regions in garlic and onion. *Food Anal. Methods*, 9, 12, 3287–97, 2016.
40. Terouzi, W. and Oussama, A., Evaluation of Ginger adulteration with beans using Attenuated Total Reflectance Fourier Transform Infrared spectroscopy and multivariate analysis. *Int. J. Eng. Appl. Sci.*, 3, 7, 257627, 2016.
41. Heidarbeigi, K., Mohtasebi, S.S., Foroughirad, A., Ghasemi-Varnamkhasti, M., Rafiee, S., Rezaei, K., Detection of adulteration in saffron samples using electronic nose. *Int. J. Food Prop.*, 18, 7, 1391–401, 2015.
42. Er, S.V., Eksi-Kocak, H., Yetim, H., Boyaci, I.H., Novel spectroscopic method for determination and quantification of saffron adulteration. *Food Anal. Methods*, 10, 5, 1547–55, 2017.
43. Petrakis, E.A., Cagliani, L.R., Tarantilis, P.A., Polissiou, M.G., Consonni, R., Sudan dyes in adulterated saffron [Crocus sativus L.]: Identification and quantification by 1H NMR. *Food Chem.*, 217, 418–24, 2017.
44. Soffritti, G., Busconi, M., Sánchez, R.A., Thiercelin, J.-M., Polissiou, M., Roldán, M. et al., Genetic and epigenetic approaches for the possible detection of adulteration and auto-adulteration in saffron [Crocus sativus L.] spice. *Molecules*, 21, 3, 343, 2016.
45. Soffritti, G., Busconi, M., Sánchez, R.A., Thiercelin, J.-M., Polissiou, M., Roldán, M. et al., Genetic and epigenetic approaches for the possible detection of adulteration and auto-adulteration in saffron [Crocus sativus L.] spice. *Molecules*, 21, 3, 343, 2016.
46. Petrakis, E.A. and Polissiou, M.G., Assessing saffron [Crocus sativus L.] adulteration with plant-derived adulterants by diffuse reflectance infrared Fourier transform spectroscopy coupled with chemometrics. *Talanta* [Internet], 162, 558–66, 2017. Available from: http://dx.doi.org/10.1016/j.talanta.2016.10.072.
47. Nogueira, T. and do Lago, C.L., Detection of adulterations in processed coffee with cereals and coffee husks using capillary zone electrophoresis. *J. Sep. Sci.*, 32, 20, 3507–11, 2009.
48. Sezer, B., Apaydin, H., Bilge, G., Boyaci, I.H., Coffee arabica adulteration: Detection of wheat, corn and chickpea. *Food Chem.*, 264, 142–8, 2018.
49. Cai, T., Ting, H., Jin-Lan, Z., Novel identification strategy for ground coffee adulteration based on UPLC–HRMS oligosaccharide profiling. *Food Chem.*, 190, 1046–9, 2016.

50. Ebrahimi-Najafabadi, H., Leardi, R., Oliveri, P., Casolino, M.C., Jalali-Heravi, M., Lanteri, S., Detection of addition of barley to coffee using near infrared spectroscopy and chemometric techniques. *Talanta*, 99, 175–9, 2012.
51. Jham, G.N., Winkler, J.K., Berhow, M.A., Vaughn, S.F., γ-Tocopherol as a marker of Brazilian coffee [Coffea arabica L.] adulteration by corn. *J. Agric. Food Chem.*, 55, 15, 5995–9, 2007.
52. El-Bialee, N.M. and Sorour, M.A., Effect of adulteration on honey properties. *Int. J. Appl.*, 1, 6, 122–32, 2011.
53. Siddiqui, A.J., Musharraf, S.G., Choudhary, M.I., Application of analytical methods in authentication and adulteration of honey. *Food Chem.*, 217, 687–98, 2017.
54. Sivakesava, S. and Irudayaraj, J., Detection of inverted beet sugar adulteration of honey by FTIR spectroscopy. *J. Sci. Food Agric.*, 81, 8, 683–90, 2001.
55. Strayer, S.E., Everstine, K., Kennedy, S., Economically motivated adulteration of honey: quality control vulnerabilities in the International honey market. *Food Prot. Trends*, 34, 1, 8–14, 2014.
56. Chen, L., Xue, X., Ye, Z., Zhou, J., Chen, F., Zhao, J., Determination of Chinese honey adulterated with high fructose corn syrup by near infrared spectroscopy. *Food Chem.*, 128, 4, 1110–4, 2011.
57. Wu, Z., Chen, L., Wu, L., Xue, X., Zhao, J., Li, Y. et al., Classification of Chinese honeys according to their floral origins using elemental and stable isotopic compositions. *J. Agric. Food Chem.*, 63, 22, 5388–94, 2015.
58. Tosun, M., Detection of adulteration in honey samples added various sugar syrups with 13C/12C isotope ratio analysis method. *Food Chem.*, 138, 2–3, 1629–32, 2013.
59. Xue, X., Wang, Q., Li, Y., Wu, L., Chen, L., Zhao, J. et al., 2-Acetylfuran-3-glucopyranoside as a novel marker for the detection of honey adulterated with rice syrup. *J. Agric. Food Chem.*, 61, 31, 7488–93, 2013.
60. Liu, P. and Ma, M.H., Application of hyperspectral technology for detecting adulterated whole egg powder. *Spectrosc. Spect. Anal.*, 38, 1, 246–52, 2018.
61. Agulló, E. and Gelós, B.S., Gas-liquid chromatographic determination of total and free cholesterol in egg pastas. *Food Res. Int.*, 29, 1, 77–80, 1996.
62. Biancolillo, A., Santoro, A., Firmani, P., Marini, F., Identification and quantification of turmeric adulteration in egg-pasta by near infrared spectroscopy and chemometrics. *Appl. Sci.*, 10, 8, 2647, 2020.
63. Zhang, W., Pan, L., Tu, S., Zhan, G., Tu, K., Non-destructive internal quality assessment of eggs using a synthesis of hyperspectral imaging and multivariate analysis. *J. Food Eng.* [Internet], 157, 41–8, Available from: http://dx.doi.org/10.1016/j.jfoodeng.2015.02.013, 2015.
64. Palazoğlu, T.K. and Miran, W., Computational investigation of the effect of orientation and rotation of shell egg on radio frequency heating rate and uniformity. *Innovative Food Sci. Emerg. Technol.*, 58, 102238, 2019.

65. Rodriguez Mondal, A.M., Desmarchelier, A., Konings, E., Acheson-Shalom, R., Delatour, T., Liquid chromatography– tandem mass spectrometry [LC–MS/MS] method extension to quantify simultaneously melamine and cyanuric acid in egg powder and soy protein in addition to milk products. *J. Agric. Food Chem.*, 58, 22, 11574–9, 2010.
66. Mane, B.G., Mendiratta, S.K., Tiwari, A.K., Bhilegaokar, K.N., Detection of adulteration of meat and meat products with buffalo meat employing polymerase chain reaction assay. *Food Anal. Methods*, 5, 2, 296–300, 2012.
67. Mandli, J., Fatimi, I.E.L., Seddaoui, N., Amine, A., Enzyme immunoassay [ELISA/immunosensor] for a sensitive detection of pork adulteration in meat. *Food Chem.*, 255, 380–9, 2018.
68. Li, T.T., Jalbani, Y.M., Zhang, G.L., Zhao, Z.Y., Wang, Z.Y., Zhao, X.Y. *et al.*, Detection of goat meat adulteration by real-time PCR based on a reference primer. *Food Chem.*, 277, 554–7, 2019.
69. Ren, J., Deng, T., Huang, W., Chen, Y., Ge, Y., A digital PCR method for identifying and quantifying adulteration of meat species in raw and processed food. *PLoS One*, 12, 3, e0173567, 2017.
70. Inbaraj, B.S. and Chen, B.H., Nanomaterial-based sensors for detection of foodborne bacterial pathogens and toxins as well as pork adulteration in meat products. *J. Food Drug Anal.*, 24, 1, 15–28, 2016.
71. Leitner, A., Castro-Rubio, F., Marina, M.L., Lindner, W., Identification of marker proteins for the adulteration of meat products with soybean proteins by multidimensional liquid Chromatography– Tandem mass spectrometry. *J. Proteome Res.*, 5, 9, 2424–30, 2006.
72. Swathi, J.K. and Kauser, N., A study on adulteration of milk and milk products from local vendors. *Int. J. Biomed. Adv. Res.*, 6, 09, 678–81, 2015.
73. Bania, J., Ugorski, M., Polanowski, A., Adamczyk, E., Application of polymerase chain reaction for detection of goats' milk adulteration by milk of cow. *J. Dairy Res.*, 68, 2, 333, 2001.
74. Ullah, R., Khan, S., Ali, H., Bilal, M., Potentiality of using front face fluorescence spectroscopy for quantitative analysis of cow milk adulteration in buffalo milk. *Spectrochim. Acta Part A: Mol. Biomol. Spectrosc.*, 225, 117518, 2020.
75. Sharma, R., Rajput, Y.S., Dogra, G., Tomar, S.K., Estimation of sugars in milk by HPLC and its application in detection of adulteration of milk with soymilk. *Int. J. Dairy Technol.*, 62, 4, 514–9, 2009.
76. Cheng, Y., Dong, Y., Wu, J., Yang, X., Bai, H., Zheng, H. *et al.*, Screening melamine adulterant in milk powder with laser Raman spectrometry. *J. Food Compos. Anal.*, 23, 2, 199–202, 2010.
77. Khan, K.M., Krishna, H., Majumder, S.K., Gupta, P.K., Detection of urea adulteration in milk using near-infrared Raman spectroscopy. *Food Anal. Methods*, 8, 1, 93–102, 2015.

78. Bilge, G., Sezer, B., Eseller, K.E., Berberoglu, H., Topcu, A., Boyaci, I.H., Determination of whey adulteration in milk powder by using laser induced breakdown spectroscopy. *Food Chem.*, 212, 183–8, 2016.
79. Fu, X., Kim, M.S., Chao, K., Qin, J., Lim, J., Lee, H. et al., Detection of melamine in milk powders based on NIR hyperspectral imaging and spectral similarity analyses. *J. Food Eng.* [Internet], 124, 97–104, Available from: http://dx.doi.org/10.1016/j.jfoodeng.2013.09.023, 2014.
80. Bligh, H.F.J., Detection of adulteration of Basmati rice with non-premium long-grain rice. *Int. J. Food Sci. Technol.*, 35, 3, 257–65, 2000.
81. Ibrahim, M., Rahim, R.A., Nordin, J.M., Nyzam, S.Z.A., Abdullah, M.T., Thermal distribution analysis of rice weevil disinfestation using microwave heating treatment. *Indones. J. Electr. Eng. Comput. Sci.*, 13, 2, 759–65, 2019.
82. Zhou, L., Ling, B., Zheng, A., Zhang, B., Wang, S., Developing radio frequency technology for postharvest insect control in milled rice. *J. Stored Prod. Res.*, 62, 22–31, 2015.
83. Siripatrawan, U. and Makino, Y., Monitoring fungal growth on brown rice grains using rapid and non-destructive hyperspectral imaging. *Int. J. Food Microbiol.* [Internet], 199, 93–100, Available from: http://dx.doi.org/10.1016/j.ijfoodmicro.2015.01.001, 2015.
84. Timsorn, K., Lorjaroenphon, Y., Wongchoosuk, C., Identification of adulteration in uncooked Jasmine rice by a portable low-cost artificial olfactory system. *Measurement.*, 108, 67–76, 2017.
85. Archak, S., Lakshminarayanareddy, V., Nagaraju, J., High-throughput multiplex microsatellite marker assay for detection and quantification of adulteration in Basmati rice [Oryza sativa]. *Electrophoresis*, 28, 14, 2396–405, 2007.
86. Vemireddy, L.R., Archak, S., Nagaraju, J., Capillary electrophoresis is essential for microsatellite marker based detection and quantification of adulteration of Basmati rice [Oryza sativa]. *J. Agric. Food Chem.*, 55, 20, 8112–7, 2007.
87. Accum, F.C., *A Treatise on Adulterations of Food and Culinary Poisons: Exhibiting the Fraudulent Sophistications of Bread, Beer, Wine, Spirituous Liquors, Tea, Coffee, Cream, Confectionery, Vinegar, Mustard, Pepper, Cheese, Olive Oil, Pickles and Other Articles Employed in Domestic Economy; and Methods of Detecting Them*, J. Mallett (Ed.), Longman, Hurst, Rees, Orme, and Brown, London, 1820.
88. EC, *Commission publishes European test results on horse DNA and Phenylbutazone: No Food Safety Issues but Tougher Penalties to Apply in the Future to Fraudulent Labelling*, European Commission, Brussels, 2013. Available from: http://europa.eu/rapid/press-release_IP-13-331_en.htm
89. Manning, L. and Soon, J.M., Developing systems to control food adulteration. *Food Policy.*, 49, 23–32, 2014.
90. Warner, K., Timme, W., Lowell, B., *Widespread seafood fraud found in New York City*, Oceana, New York, NY, 2012.

91. WHO, *Melamine-Contaminated Powdered Infant Formula in China*, World Health Organization, Geneva, 2008. Available from: http:// www.who.int/csr/don/2008_09_19/en/, accessed 26.03.13.
92. Cawthorn, D.-M., Steinman, H.A., Hoffman, L.C., A high incidence of species substitution and mislabelling detected in meat products sold in South Africa. *Food Control*, 32, 2, 440–9, 2013.
93. Cohen, A., Sturgeon poaching and black market caviar: a case study. *Environ. Biol. Fishes*, 48, 1, 423–6, 1997.
94. FSSAI, *Report Milk Survey NMQS 18_10_2019* FSSAI, New Delhi, 2019. Available from: https://fssai.gov.in/upload/uploadfiles/files/Report_Milk_Survey_NMQS_18_10_2019.pdf.
95. Ortolani, C., Bruijnzeel-Koomen, C., Bengtsson, U., Bindslev-Jensen, C., Björkstén, B., Høst, A. *et al.*, Controversial aspects of adverse reactions to food. *Allergy*, 54, 1, 27–45, 1999.
96. Sarkar, S.N., Isolation from argemone oil of dihydrosanguinarine and sanguinarine: toxicity of sanguinarine. *Nature*, 162, 4111, 265–6, 1948.
97. Tyan, Y.-C., Yang, M.-H., Jong, S.-B., Wang, C.-K., Shiea, J., Melamine contamination. *Anal. Bioanal. Chem.*, 395, 3, 729–35, 2009.
98. Srivastava, S., Food adulteration affecting the nutrition and health of human beings. *J. Biol. Sci. Med.*, 1, 1, 65–70, 2015.
99. Manning, L. and Soon, J.M., Food safety, food fraud, and food defense: a fast evolving literature. *J. Food Sci.*, 81, 4, R823–34, 2016.

8

Natural Food Toxins as Anti-Nutritional Factors in Plants and Their Reduction Strategies

Naman Kaur, Aparna Agarwal*, Manisha Sabharwal and Nidhi Jaiswal

Department of Food and Nutrition, Lady Irwin College, University of Delhi, Sikandra Road, New Delhi, India

Abstract

Every year, a major portion of food is wasted due to food contamination which leads to food toxicity. On entering the human body through food or water, food contaminants result in several foodborne diseases (FBDs) and cause societal, and economic disturbances. The food contaminants are broadly categorized as microbial or environmental contaminants, naturally occurring toxic constituents, and those resulting from food additives or novel foods or ingredients. Some of these naturally existing plant-based food toxins are anti-nutritional factors (ANFs). Usually, ANFs are present in small quantities in fruits, vegetables, cereals, and pulses to protect them from insects, microbes, or pests, and may not be harmful when consumed. However, when consumed by people sensitive to an individual ingredient, they may induce hazardous health effects. ANFs inhibit the nutrient absorption and digestion in human body. A deeper knowledge of chemical structure of these ANFs may prove helpful in devising technological strategies to produce or process toxin-free plant-based foods, thus improving the quality of the food, without compromising the health of the consumers. In this chapter, we attempt to elucidate different anti-nutrients, their structures, and adverse effects as well as strategies to reduce their levels to improve the quality of food.

Keywords: Anti-nutritional factors, food toxins, natural food toxins, reduction strategies, adverse effects

---

*Corresponding author: aparnadt@gmail.com

Abbreviations

FBD	Foodborne diseases
ANF	Antinutrient factors
GA	Glycoalkaloids
PA	Pyrrolizidine alkaloids
HCN	Hydrogen cyanide
TAN	Tropical ataxic neuropathy
GSLs	Glucosinolates
TPO	Thyroid peroxidase
LAB	Lactic acid bacteria
HTST	High temperature short time

8.1 Introduction

Each bite of food can potentially lead to the risk of foodborne diseases (FBDs) caused due to food contamination [1]. WHO (2015) describes FBDs as "any disease of an infectious or toxic nature caused by consumption of food" [2]. Every year globally, large population falls ill and several die due to the consumption of unsafe food. In 2010, WHO estimated the worldwide prevalence of FBDs for 31 selected hazards at 33 million disability-adjusted life years (DALYs) with 40% of the total prevalence focused among children under 5 years of age [1].

The symptoms of FBDs may be mild and self-limiting such as vomiting, nausea, and diarrhoea and may be devitalizing and fatal like organ failure (kidney and liver), paralysis, brain, and neural disorders and cancers. These may lead to long duration of absenteeism and, subsequently, premature death. The risk of contracting and dying from FBDs is highest among infants, young children, pregnant women, the elderly, and individuals with a weak immune system. Infants and children with poor nutritional status are particularly exposed to foodborne hazards. They are potentially more susceptible to serious forms of FBDs which may further deteriorate their nutritional status and increase the risk of infant and child mortality. In cases of survival, a delay in cognitive as well as physical development maybe observed which may affect the productivity of an individual. Besides, FBDs also have an adverse effect to the socio-economic development of a country and it may pose several challenges to the tourism, agricultural, and food (export) sectors of the nation [1].

The FBDs may result from different types of food contaminants like biological or non-biological contaminants present in food. These contaminants can potentially cause food toxicity that can result in adverse health and life threats among all the age groups. Food contamination may occur at any step of the food supply chain starting from production to consumption. The chances of microbial infestation are higher in case of fresh produce; however, the likelihood of pathogenic growth and toxic chemicals both at pre- and post-storage and processing conditions [3, 4] is greater among grains and processed products like dairy and meat products. Therefore, both storage and processing conditions play a vital role in ensuring food safety.

Food contaminants can be broadly divided into three categories, namely, biological contaminants such as microbial or environmental, naturally occurring toxic compounds, and food additives or novel foods or ingredients that may be added intentionally or unintentionally at food processing stage [5]. The most hazardous contaminants are the ones produced by bacterial or mould infestations in food. The microbial infestation causes the release toxins that prevail in the food even after the destruction of the biological source. The environmental/chemical contaminants like pesticide residues or heavy metals are generally effectively controlled in modern food supplies; however, they can cause considerable hazards localities. Generally, naturally existing toxic compounds are mostly present in very small doses and, hence, do not have deleterious effects when such foods are consumed. However, in individuals who are sensitive to certain ingredients, consumption of even small doses of these compounds may have harmful effects. Of a variety of food contaminants, food additives or novel foods are the least hazardous as their toxicology is well researched, and hence, they are employed under scrutinized condition. Several plants comprise perilous levels of naturally present toxic components which may result in intoxications if they are misidentified as edible plants during harvesting [5]. There exist various plants such as cereals, pulses, legumes, fruits, and vegetables that possess higher concentrations of certain natural toxic constituents also known as anti-nutrient factors (ANFs). The plant-based foods mostly possess anti-nutrients like phytates, lectins, tannins, and oxalates. These ANFs play a significant part in the protection of the plant from animals, pests, insects, and birds. Nevertheless, the consumption of these

ANFs in large amount or their consumption by an individual sensitive to such ingredients can potentially result in hazardous effects. For example, oxalate is an anti-nutrient found in spinach and soya beans, which prevents the absorption of calcium in the body. Tannins are ANFs found in tea, chocolate, and wines that inactivate the absorption of protein in the body. Since sensitivity toward these ANFs varies from individual to individual and these compounds once ingested cannot be removed from the body, it is essential to be aware about different ANFs, their adverse effects on the health of human beings, and strategies for reducing their levels in food in order to produce healthy, nutritious, and safe food for consumption [6]. Table 8.1 enumerates common sources comprising certain ANFs.

This chapter deals with different types of ANFs, their structure, and their adverse effects on the health of human beings. Additionally, the chapter also elucidates various strategies to reduce their levels in the food to improve the food quality.

Table 8.1 Common sources of anti-nutrient factors.

Anti-nutrient factors	Common sources
Phytates	Legumes, peanuts, cereals, and oilseeds [7, 8]
Tannin	Pomegranate, berry fruits and cocoa bean, cereals (sorghum and barley) [9, 10], faba bean, lima bean, sunflower seed meal, and rapeseed [11]
Trypsin	Chickpeas, soybeans, red kidney beans, adzuki beans, mung beans and other representatives of the *Leguminoseae*, *Solanaceae*, and *Gramineae* families [12]
Saponin	Soybean, peanuts, chickpeas, broad beans, lentils, sunflower seeds, spinach leaves, tea leaves, quinoa seeds, sugar beet, allium species, oats, yucca, tomato seeds, fenugreek seeds, asparagus, aubergine, and yam [13]
Lectins and hemagglutinin	Cereals, beans, legumes, tubers like potatoes, and also in animals [11]

(*Continued*)

Table 8.1 Common sources of anti-nutrient factors. (*Continued*)

Anti-nutrient factors	Common sources
Alkaloids	Tobacco, leaves of coca plant, cinchona bark, dried latex of opium poppy and unripe potatoes and potato sprouts [11]
Oxalates	Cruciferous vegetables (kale, radishes, cauliflower, broccoli), chard, spinach, parsley, beets, rhubarb, black pepper, chocolate, nuts, berries (blueberries, blackberries) [14]
Cyanogenic glycosides	Cassava (*Manihot esculenta*), linseed (*Linum usitatissmium*), various sorghums (*Sorghum spp.*), and white clover (*Trifolium repens*). Lesser quantities are found in the kernels of such plants as almonds (*Amygdalus communis*), apricots (*Prunus armeniaca*), peaches (*Prunus persica*), and apples (*Malus sylvestris*) [15]
Goitrogens	Broccoli, cabbage, cauliflower, Brussels sprouts, and kale [16]

8.2 Anti-Nutritional Factor

8.2.1 Tannins

Tannins belong to a broad class of polyphenolic compounds and regarded as highly bioactive compounds. They have molecular weight ranging from 500 to 3,000 Daltons. Tannins are secondary compounds widely occur in the vascular tissues of leaf, bud, seed, root, and stem and are responsible for the astringent taste of commonly consumed plant foods and beverages. They are widely present in apple, grapes, pomegranate, berries, cocoa bean, legumes, whole grains, tea, etc. Various food crops and legumes have been reported to contain significant quantities of tannins as they are concentrated in the bran fractions. Cereals like sorghum and barley and pulses and legumes like pigeon pea, pea, faba bean, lima bean, and rapeseed are concentrated sources of tannins [7, 8].

8.2.1.1 Types

The chemical structure of tannin contains a hydroxyl group bonded to an aromatic ring. Tannins are chemically classified into two groups: hydrolyzable tannins and condensed tannins. Both the forms are widely distributed in plants and shows different nutritional and toxic effects. Hydrolyzable tannins are glycosylated gallic acid (including gallotannins and ellagitannins) and are selectively found in the diet. During the digestion process, these are broken down into various compounds or substances that can be toxic. They are readily broken down during the digestion process. Condensed tannins (proanthocyanidin), on the other hand, are oligomers and polymers of flavonoids [9]. Condensed tannins also known as catechin tannins, flavanols, or proanthocyanidins [include catechin, epicatechin (EC), epigallocatechin (EGC), epicatechin-3-gallate, and (-)-epigallocatechin-3-gallate (EGCG)] are the second-most abundant natural plant-derived polyphenols in the diet after lignin [10]. Leguminous forages and seeds are considered as concentrated sources of condensed tannins. It has been documented previously that adequate amounts of condensed tannins are present in peanut and millets [11, 12]. Condensed type of tannins is more resistant to hydrolysis, so it is neither hydrolyzed nor absorbed during digestion [13].

8.2.1.2 Adverse Effects

Although, tannins show antioxidant properties, due to their phenolic nature and high chemical reactivity as they have the capability to form intra- and inter-molecular hydrogen bonds with macromolecules like proteins and carbohydrates. Previous studies have demonstrated the various pharmacological activities performed by tannins which include anticarcinogenic, immunomodulatory, and cardioprotective activities. However, their anti-nutritional effects have also been reported because of their ability to act as chelators and thereby inhibiting the absorption of dietary minerals such as iron, copper, and zinc. Due to their interference with dietary iron absorption, tannins have been suggested as a contributor to iron-deficiency anemia [14, 15].

Tannins can also bind and precipitate proteins. Tannins impair the digestion of various nutrients by forming complexes with proteins. The complex form between the hydroxyl group of tannins and the carbonyl group of proteins prevents the body from absorbing important bioavailable substances [16]. These complexes involve both hydrogen-bonding and hydrophobic interactions and results in inactivation of digestive enzymes, making proteins partially unavailable and reduced protein digestibility [17, 18]. Decrease in amino acid availability and increased fecal nitrogen also occurs. Recent studies have shown that tannins are responsible for lower feed intake, growth rate, and protein digestibility in experimental animals. Tannins exhibit anti-amylase activity and also inhibits the activities of trypsin, chemotrypsin, and lipase, thus decreasing the protein quality of food and interfering in the digestion. They also have the ability to form complex with vitamin B [19]. The depression of intestinal digestion and microbial enzyme activity is observed with high concentration of tannin in the diet. Tannins-protein complexes resulted in low protein digestibility, decreased amino acid availability, and increased fecal nitrogen. These complexes may not be dissociated and may thus be excreted with the feces [20, 21].

8.2.2 Saponins

8.2.2.1 *Saponins*

Saponins are the secondary compounds found primarily in plants. They are water soluble, non-volatile constituents that can form soapy foam when mixed with water. They are either triterpene or steroidal glycosides that are widely occurring in commonly consumed plants. They contain a sugar moiety coupled with a non-sugar, non-polar aglycones called as sapogenin in their structure [22]. The soap-like behavior is due to the presence and combination of both polar and non-polar structural elements in them, which in turn explains their different properties like sweetness and bitterness, and foaming and emulsifying properties. Saponins are known to possess varied chemical and biological properties including hemolytic and medicinal properties, because of the complexity in their structure [23].

Saponins are present in varied number of plants, including legumes, oilseeds, potatoes, broccoli, eggplants, alfalfa, and ginseng root; however, the saponin content varies at different growth or maturity stage, different tissues, and different variety of plants [24]. Saponins are majorly present in legumes as compared to cereals having low saponin content [25]. It has been reported that macrosperma has significantly higher saponin content as compared to microsperma in lentils. Regarding different tissues, saponin content is higher in cortex as compared to flesh in yam [26]. Triterpenoid saponins are usually found in legumes such as chickpea, soybean, kidney bean, and groundnut. Apart from legumes, sunflower seeds, quinoa, spinach, and tea leaves are also the source of triterpenoid saponins. However, food plants like oats, fenugreek seeds, asparagus, and tomato seeds are sources of steroid saponins. The amount of saponin varies in a range of 0.05% to 0.23% among chickpea, mungbean, and pigeonpea [27]. While comparing different vegetable parts, saponin content of seeds is highest as compared to leaf and tuber.

8.2.2.2 Adverse Effects

Saponins possess various biological properties depending on their structure including hypocholesterolemic and anticarcinogenic properties. It has been reported that intake of saponin rich food helps in controlling plasma cholesterol level and thus reducing the risk of cardiovascular diseases. It is also known to inhibit platelet aggregation and can be used as adjuvants in application of vaccines.

Although saponins are known to have pharmacological properties, their toxic effects are also well documented. One of the major adverse effects of saponin is their strong hemolytic activity because of their interaction with cholesterol group of erythrocyte membranes and formation of irreversible pore like structures, thus disrupting the red blood cells. Previous studies have reported the lethal effect of saponins on fish and cold-blooded animals. The extreme toxicity of saponin can be attained through the reduction of surface tension of blood in cold-blooded animals. It has been observed that only few plants have saponins with strong hemolytic activity, especially in the seeds of Kochia scoparia as compared to other plants in which saponins have low hemolytic activity. It has also been documented that out of 10 amaranth leaves, only one leaf can be hemolytic because of the presence of saponin with oleanane type.

Besides hemolytic activity, previous studies have reported that high concentration of saponins can inhibit the metabolic and digestive enzymes and therefore affect the absorption of various nutrients. Gastrointestinal digestion is impaired as activities of various digestive enzymes like amylase, trypsin, chymotrypsin, and lipase are inhibited by saponins. It has been reported that enzymes such as protease, amylase, and cholinesterase are inhibited by soy saponin and enzymes like chymotrypsin and protease are inhibited by alfalfa saponins. Saponins have also shown to inhibit glucosidase activity, a carbohydrate hydrolyzing enzyme which is responsible for the breakdown of complex carbohydrates into simpler ones. Thus, saponins may lead to impaired digestion of carbohydrates, lipids, and protein and thus affecting protein digestibility and showing hypoglycemic and hypocholesterolemic effect. The hypocholesterolemic effects can be due to intralumenal physicochemical interaction of saponins in the gut. Saponins have the ability to form complexes with cholesterol which are generally insoluble and prevent the absorption of cholesterol from the intestine and thereby responsible in reducing the plasma cholesterol level. An increase in fecal excretion of bile acids is also reported due to intake of saponin from soybean and chickpea. These complexes not only alter the function of glycoproteins in the plasma membranes but can also change the functioning of ion channels, transporters, and receptors.

They interfere in the uptake of vitamins and minerals too as they can bind to various nutrients like iron, calcium, magnesium, or zinc, thereby affecting their absorption. This can be due to the increased permeability of intestinal mucosal cells by saponins and facilitation in the uptake of substances that are usually not absorbed and thereby resulting in the development of leaky gut. Again, this activity is strong in saponins present in alfalfa where iron absorption is strongly and significantly reduced as compared to soy saponin with low reduction in iron absorption when administered to rats.

Previous studies have reported that the binding of saponins to small intestinal cells leads to reduction in intestinal absorption of nutrients, where it effects the overall growth rate and responsible for bloating in animals, thereby affecting their metabolism and health. Saponins have adverse health effects on human, which includes reduction in bioavailability of iron and impairment of growth. It has also been reported that a diet high in triterpenoid saponins when fed to chicks results in the reduced absorption of vitamin A and vitamin E. Saponins when consumed in higher concentrations could be highly toxic and may lead to death as documented previously in a study where the sheep when fed higher saponin levels (equal to greater than 150 mg/kg body weight) were died.

8.2.3 Lectins and Hemagglutinin

The word "Lectin" has been derived from the Latin word "*legere*", which denotes "to select". Lectins are mostly present in plants, particularly cereals, beans, legumes, tubers like potatoes, and also in animals. Lectin activity has been detected in over 800 varieties of the legume family and 2%–10% of the total protein legume seeds are lectins [11]. Lectins usually have the tendency to bond with carbohydrate molecules; however, currently, proteins that agglutinate red blood cells with known sugar specificity are considered as "lectins". Proteins with unknown sugar specificity are known as "hemagglutinins" and those found in plants are referred to as "plant hemagglutinins" [11]. Fundamentally, both lectins and hemagglutinins are proteins or glycoproteins, with at least one non-catalytic domain that displays reversible binding to certain monosaccharides or oligosaccharides. They tend to bond with the carbohydrate molecules on the surface of erythrocytes and agglutinate the erythrocytes, without modifying the properties of the carbohydrate molecules [17]. Lectins bind with different types of carbohydrate moieties, and as a result, there may be variance in their bonding with intestinal wall based on the variety of carbohydrate molecule. One of the most vital properties of lectins is to avert absorption of digestive by-products in the small intestine. Additionally, lectins also interact with certain blood groups. Besides, they also facilitate several operations in mitotic division, annihilate carcinogenic cells, as well as exhibit deleterious effects in certain animals [11].

8.2.3.1 Adverse Effects

Various researches have been conducted to examine the adverse effects of plant lectins on the health of human beings. It has been reported that certain plant lectins hampered the process of exocytosis and that of repairing the plasma membrane, thereby becoming toxic to gut epithelial cells [18]. Certain legume lectins have been reported to obstruct several intestinal and brush border enzymes such as maltase, sucrase, dipeptidyl peptidase, and aminopeptidase. Additionally, a few legume lectins have also been reported to promote bacterial overgrowth with a likelihood of disturbing the equilibrium of the normal flora and thus, resulting in several symptoms of intestinal obstruction. Certain lectins have also been reported to injure the luminal membranes of the epithelium, inhibit digestion and absorption of nutrient, as well as alter the bacterial flora. Subsequently, they cause disturbance in the nutrient metabolism, elevate enlargement and/or deterioration of internal organs and tissues, and modify hormonal and immunological properties [19]. Studies report that presence of lectin from red kidney beans (*P. vulgaris*), named Phytohemagglutinins (PHA), if consumed in large doses in the form of uncooked or partially cooked kidney beans can result in food poisoning and poor epithelial resistance and, subsequently, acute gastroenteritis [19, 20]. The acute symptoms following ingestion of PHA include nausea, vomiting, and diarrhoea. It was also reported that intake of PHA over a long duration in rodent models was identified by weight loss, enhanced cell turnover, and gut hyperplasia.

Hence, an effective processing technique is essential to immobilize PHA prior to their consumption from PHA-containing legumes [21]. It was also observed that animals fed on lectins exhibited instabilities in the hormonal homeostasis. PHA was also found to be responsible in promoting pancreatic hypertrophy among rats in a dose-dependent manner. This was found to be associated with elevating plasma levels of a gut hormone and cholecystokinin, affecting pancreatic growth and gastrointestinal function [19].

8.2.4 Alkaloids

Alkaloids make a sizeable group of chemical compounds synthesized by plants from amino acids. They are small organic molecules found in almost 15% to 20 % of all vascular plants in the form of salts of plant acids (e.g., oxalic, citric, malic, or tartaric acid). The term "alkaloid" was coined by Wilhelm Meissner, a German pharmacist. The conventional definitions of alkaloids accentuate their plant origin, physiological actions, basicity, as well as bitter taste [22]. The alkaloids and other secondary metabolites in plants help in augmenting their reproductive rates. This is done either by enhancing the defence strategies against biotic and abiotic stresses or by influencing the pollinators and seed/fruit disperser visitation. The defence techniques involve repelling the predators by bitterness taste, toxicity, or damage repair by antioxidant system [23]. These compounds are mostly composed of numerous carbon rings with side chains, with at least one carbon atom substituted by a nitrogen atom, which is a common chemical feature among a variety of alkaloids [24]. Alkaloids can be classified as true alkaloids, proto-alkaloids, and pseudo-alkaloids. True alkaloids are the one obtained from amino acid and comprise a heterocyclic ring structure with nitrogen atom. The proto alkaloids are the ones with nitrogen atom derived from an amino acid. While, pseudo alkaloids are those that possess a basic carbon skeleton which is not obtained from amino [25]. Alkaloids are regarded as anti-nutrients as they influence the nervous system by impeding or by inaccurate enhancement of electrochemical transmission. Alkaloids are known to result in gastrointestinal as well as neurological disorders. Some of the well-known alkaloids are cocaine (coca plant leaves), morphine (dried latex of opium poppy), nicotine (tobacco), quinine (cinchona bark), and solanine (unripe potatoes and potato sprouts) [11].

8.2.4.1 Adverse Health Effects

Plants rich in alkaloid have extensive pharmacological properties which include neuroprotective [26, 27], antidepressant [28], antibacterial, analgesic [29], and anticancer However, ingesting even a small amount plant comprising toxic alkaloids can cause severe damage to the human body indicating the positive association between intoxication response and sensitivity of the target animal [13, 30]. Consumption of toxic alkaloids may hinder the normal functioning of human body metabolisms and cause symptoms ranging from mild ones such as vomiting, itching, nausea, and mild gastrointestinal perturbation to chronic ones like psychosis, arrhythmias, paralysis, teratogenicity, and sudden death [30–32] Various researches have evidently proved the physiological effects of toxic alkaloids on the health of humans. Potato plant (*Solanum tuberosum*) possesses two toxic and bitter glycoalkaloids (GA), namely, a-solanine and a-chaconine. Although, the level these GAs is generally low and does not result in hazardous health effects, their consumption at higher concentration (300–800 mg kg^{-1}) has been correlated with acute poisoning as well as gastro- intestinal and neurological disturbances, among humans [27]. Similarly, when tropane and tryptamine alkaloids are consumed at higher concentration, they result in rapid heartbeat, paralysis, and, in certain cases, may even cause death [11].

Plants belonging to the *Solanaceae family*, such as nightshades (*S. nigrum*), potato (*S. tuberosum*), tomato (*S. lycopersicum*), eggplant (*S. melongena*), pepper (*Capsicum annuum*), and petunia (*Petunia sp.*) possess GAs—solanine and chaconine. The ingestion of these GAs can lead to gastrointestinal and neurological disorders. This can be ascribed to obstruction in acetyl cholinesterase and calcium transport. The combination of the two GAs has been suggested to result in a synergic effect which is responsible for the increase in the toxicity [33]. Another group of plant family such as *Asteraceae, Boraginaceae*, and *Fabaceae* are known to possess pyrrolizidine alkaloids (PAs), also called ornithine-derived alkaloids which are present in over 6,000 plants. The PAs are considered to be highly efficacious to prevent predators, such as human beings as well as livestock [34]. Consumption of plants consisting PAs has been known to cause acute and chronic liver toxicity in humans and other animals. They are also known to cause nausea, vomiting, abdominal pain, diarrhoea, as well as oedema [35].

8.2.5 Oxalates

Certain organic acids have been studied to form ANFs. Oxalic acid is one such organic acid that forms strong bonds with minerals such as magnesium, calcium, potassium, and sodium. These bonds result in the formation of salts or esters known as oxalates. Oxalates are mostly found green leafy vegetables; however, they may also be produced in the human body [36]. Foods which possess high oxalate content are cruciferous vegetables like cauliflower, broccoli and radish, spinach, parsley, beetroot, chocolate, and nuts [16]. Oxalates can be of two types, namely, soluble and insoluble. Salts that are formed with minerals like potassium and sodium are known as soluble salts, while those formed with minerals such as calcium, iron, and magnesium are known as insoluble. Due to the tendency of the insoluble oxalates to precipitate in the kidneys or the urinary tract, when consumed in large amount, they form sharp-edged oxalate crystals that cannot be eliminated from the body via urinary tract. The crystals formed are known to play a vital role in the formation of kidney stones in the urinary tract during the excretion of acid in the urine [37]. Consumption of high concentration of oxalic acid results in the binding of the acid with the minerals in the diet and formation of the salts/oxalates. Oxalates have a direct impact on calcium and magnesium metabolism. They also interact with proteins and form compounds that induce an inhibitory effect during peptic digestion [11].

8.2.5.1 Adverse Effects

Consumption of oxalic acids leads to the binding of the acid and the minerals present in the diet reducing the absorption of the dietary minerals in the human body. This bonding results in the formation of sharp-edged stones or crystals. Oxalates are absorbed in the stomach, small intestines, as well as large intestines [38]. The soluble oxalates are mostly absorbed in the large intestines by passive diffusion [39]. Oxalates that are absorbed in the body are still capable of binding with the minerals present in the blood vessels but are subsequently removed from the body after 24 hours of oxalate ingestion via urine [40]. There exist certain oxalates which are broken down in the large intestine with the help of gut bacteria like *Oxalobacter formigenes*. These bacteria produce oxalate decarboxylase in order to break down oxalates into carbon dioxide and formate [37]. Insoluble oxalate such as calcium oxalate, however, cannot be absorbed in the body and hence eliminated from the body through feces [37, 41]. Studies have reported that the molar ratio of total oxalate: calcium higher than 9:4 can cause adverse effects among humans [37].

Although, majority of individuals can metabolize certain amount of oxalate, there are certain individuals with conditions like enteric and primary hyperoxaluria who require a lower intake of oxalates. Among individuals with sensitivity toward oxalates, consumption of even lower concentration of oxalates can lead to symptoms like burning sensation in the eyes, ears, mouth, and throat. Whereas, consumption of higher concentration of oxalates in the same group my result in nausea, abdominal pain, diarrhoea, and muscle weakness [42]. It has been reported that most of the calcium oxalates are not absorbed into the blood vessels; hence, its effect on toxicity is less. However, the sharp-edged crystals penetrate the tissues resulting in discomfort in the mouth and tongue [37]. It has also been found that ingesting higher concentration of oxalic acid leads to gastric hemorrhage, hematuria, as well as corrosion of the mouth and gastrointestinal tract [43].

8.2.6 Cyanogenic Glycosides

Cyanogenic glycosides are a group of amino-acid derived compounds present in plants that are produced as secondary metabolites. They contain nitrile group and produce hydrogen cyanide (HCN) following their breakdown by biocatalysts due to the process of hydrolysis. The release of HCN determines the toxicity of these compounds and their derivatives. Cyanogenic glycosides are present in almost 2,000 plant species, of which various species are used as food. Cyanogenic glycosides are commonly found in families like *Fabaceae, Linaceae, Rosaceae, Compositae*, as well as *Leguminosae*. There exist nearly 25 known cyanogenic glycosides which are mostly found in the edible parts of plants (especially their seeds) like apples, apricots, cassava, cherries, peaches, pome fruit, plums, almonds, bamboo shoots, linseed, chick peas, and cashews [44]. Different foods possess different types of cyanogenic glycosides, for example, bamboo shoots contain taxiphyllin and cassava contains linamarin.

Cyanogenic glycosides play a vital role in the chemical defence mechanism in plants as well as in plant-insect interactions [45]. Although, these compounds are not toxic in themselves, on the disruption of the cell structures of the plant they come in contact with β-glucosidase enzyme. The disintegration of plant cell structure may occur as a result of chewing or grinding, pounding, soaking, or fermentation in the presence of water which further leads to hydrolysis [46]. β-glucosidase plays a role in hydrolysis by making sugars and a cyanohydrin that immediately breaks down to HCN and a ketone or aldehyde [46, 47]. Cyanide toxicity can ensue among human beings as well as other animals if the dosage ranges between 0.5 and 3.5 mg HCN per kilogram body weight. The risk of cyanide toxicity is higher among children due to their small body size. The toxicity caused due to cyanogenic glycosides has been correlated with their potential to be hydrolyzed either spontaneously or in the presence of a biocatalyst to yield cyanide as a final product. Therefore, toxic levels of cyanogenic glycosides are evaluated by the amount of free cyanide produced after hydrolysis.

As a result, estimation of total cyanogenic glycosides in human diet is a cumbersome task [48].

8.2.6.1 Adverse Effects

The release of HCN determines the toxicity level of cyanogenic glycosides and their derivatives. Dietary exposure to high levels of cyanogenic glycosides results in toxicity which may cause acute cyanide poisoning, irreversible neurological disorders, and a variety of chronic diseases [49]. Cyanide toxicity in humans has been reported to result in symptoms like vomiting, stomach-ache, diarrhoea, convulsion, and, in severe cases, death. A study reported that an elevated and persistent intake of cyanogens at sublethal concentrations from cassava or cassava flour together with a reduced consumption of sulphur amino acids resulted in Konzo among women and children [50]. Konzo, an upper motor neuron disease, generally affects children and women of child-bearing age and is identified by irreparable but nonprogressive symmetric spastic paraparesis that has an abrupt onset [51–53]. Another health concern related to sustained consumption of inappropriately processed cassava products is Tropical ataxic neuropathy (TAN). TAN is a term used to represent different neurological syndromes resulting from persistent intake of improperly processed cassava products for a long duration. It has chiefly occurred in Africa, especially Nigeria [54] and is very common in people above the age of 40 years [50]. TAN has been known to result in symptoms like neuro sensory deafness, sore tongue, optical atrophy, and sensory gait ataxia [54].

8.2.7 Goitrogens

Goitrogens are characterized as substances present in nature which interfere with the normal functioning of the thyroid gland. The name "goitrogens" was derived from the word "goiter", which means the enlargement of the thyroid gland. They may either produce a direct effect on the gland or obliquely modify the regulatory mechanisms and peripheral metabolism of the thyroid glands. Goitrogens have been suggested to cause hinderance in the production and release of thyroid hormones. The difficulty to synthesize thyroid hormone by the thyroid gland results in the enlargement of the gland in order to indemnify insufficient production of hormones. Thyroid hormone plays a significant role in managing body metabolism; hence, their deficiency causes diminished growth and reduced reproducibility [24]. Goitrogens are mostly found in soy-based products as well as cruciferous vegetables like cauliflower, broccoli, Brussels sprouts, cabbage, mustard, and turnips in the form of glycosides [11, 55]. Besides, goitrogens are also found in the *Brassica* family and other plants in the form of glucosinolates (GSLs), a class of more than 120 compounds [56]. During the process of chewing and ingestion, GSLs with the help of myrosinase, an enzyme that is activated in damaged plant tissue and is secreted by human microflora, get converted into different types of compounds such as thiocyanates, isothiocyanates, nitriles and sulforaphane [56, 57] Based on various *in vivo* and *in vitro* animal model researches, compounds like resveratrol, isoflavones, and flavonoids have also been known to possess goitrogenic effects [58–61]. Isoflavones are compounds that are present naturally among the plants belonging to the flavonoid family. Genistein and daidzein are two isoflavones that are exclusively found in soy and soy-based products like tofu and tempeh; while, other flavonoids and resveratrol are found across the plant kingdom [55, 61, 62]. Goitrogenic compounds known as C-glycosyl flavones are found in millets which have been reported to hinder thyroid peroxidase (TPO) [63, 64].

8.2.7.1 Adverse Effects

Goitrogens are compounds that interfere with iodine absorption in the human body. They may either obstruct the transporters which manage the movement of iodine into the cell (e.g., the NIS) or they may inhibit the process of production of the thyroid hormones (e.g., T3 and T4). Occasionally, they may also immobilize TPO activity. TPO plays a crucial role in the bonding of iodine to the tyrosine to produce T4 and T3. Other substances can either restrict the deiodinization of the thyroid hormones or can obstruct the receptor sites for the thyroid hormones. Besides, high consumption of soy-based foods can result in increased fecal bulk which results in excessive excretion of T4. This further results in difficulty for the thyroid gland to compensate for the insufficient iodine stores [55].

Studies have also reported that exclusive or elevated consumption of diet rich in vegetables and/or seeds belonging to the *Brassicaceae* family can result in toxic effects on the human body. Presence of goitrin and isothiocyanates in the vegetable from the *Brassicaceae* family leads to the restriction of iodine uptake by the thyroid glands which causes depletion of iodine stores, which further inhibits T4 hormone. It has also been found that, consumption of high levels of GSLs can lead to certain adverse effects like enlargement of the thyroid gland, reduced levels of plasma thyroid hormone, liver and kidney abnormalities, diminished growth, reduced reproducibility, as well as mortality [65]. Thyroid enlargement has been associated with prolonged ingestion of vegetables from Brassica species containing anti-thyroid effects of GSLs can result in subclinical signs, such as decreased reproductive performance and growth or, in more severe cases, clinically evident goiter [66] Similarly, isothiocyanates present in the cruciferous vegetables have also been reported to diminish the functioning of thyroid gland by immobilizing TPO enzyme and by interrupting the messages transmitted across the membranes of thyroid cells [55]. Table 8.2 provides the range of anti-nutritional effects caused by different ANFs.

Table 8.2 Adverse health effects of anti-nutrient factors.

Anti-nutrient factors	Adverse health effects
Phytate	Reduced calcium and iron absorption [17]
Tannin	Impaired digestion, reduced protein digestibility, and poor absorption of nutrients [16]
Trypsin inhibitor	Reduced protein digestibility [11]
Saponin	Hypocholesterolemic effect [67], hypoglycemia [68] or reduced protein digestibility, impaired nutrient absorption, leaky gut [69]
Lectins (hemagglutinin)	Obstruction of several intestinal and brush border enzymes. Injured luminal membranes of the epithelium, inhibited digestion and absorption of nutrient, altered bacterial flora, elevated enlargement and/or deterioration of internal organs and tissues, and modified hormonal and immunological properties [19]
Alkaloids	Vomiting, itching, nausea, mild gastrointestinal perturbation, psychosis, arrhythmias, paralysis, teratogenicity, and sudden death [30, 32]
Oxalates	Reduced mineral absorption, kidney stone formation [70], gastric discomfort [16]
Cyanogenic glycosides	Acute cyanide poisoning, irreversible neurological disorders, gastric discomfort, convulsion, death [48]
Goitrogens	Impaired iodine absorption, immobilized thyroid peroxidase enzyme [55], enlargement of the thyroid gland, reduced levels of plasma thyroid hormone, liver and kidney abnormalities, diminished growth, reduced reproducibility and mortality [65]

8.3 Methods to Reduce Levels of Anti-Nutritional Factors in Foods

Edible plants like cereals, legumes, as well as nuts and oilseeds are composed of various essential nutrients including carbohydrates, proteins, dietary fiber, minerals, and vitamins. The bioavailability of these vital nutrients among humans is comparatively low, especially when consumed raw or in unprocessed form. The chief factor to reduce the nutrient bioavailability is the presence of naturally occurring ANFs like phytate, tannins, saponins, and oxalates [71]. Several researches conducted in the past have evidently proven the adverse effects of anti-nutrients on nutritional value of the food consumed. The ability of the anti-nutrients to reduce the mineral and vitamin content of the foods and to increase the risk of toxicity makes it necessary to lessen their concentration in the food before consumption. In the past, various traditional techniques and technological processing methods including milling, de-branning, soaking, roasting, irradiation, high pressure processing, microwave heating, extrusion, cooking, germination, and fermentation have been utilized for the reduction of anti-nutritional content in foods.

8.3.1 Soaking

Soaking is one of the traditional methods used for the removal of anti-nutrients present in foods. Besides reducing the levels of enzyme inhibitors and other anti-nutrients, soaking also helps in the reduction of cooking time and augments the release of enzymes such as endogenous phytases, present in grains, nuts, and oilseeds. It helps in creating moist conditions in grains, nuts, and oilseeds, which is essential for their germination which is known to improve the digestibility and nutritional value of food [73]. Additionally, soaking is also an essential step for fermenting foods for further reduction of levels of several anti-nutrients present in foods [74]. There exist various anti-nutrients that are water soluble in nature. This property helps in removing such anti-nutrients from foods by leaching. Generally, soaking escalates the hydration level of grains like legumes and cereals which helps in softening of the grains as well as in activating endogenous enzymes such as phytase that helps in enhancing the ease of further processing such as cooking or heating. Besides reducing the phytic acid content, soaking has also been reported to effectively improve the mineral concentration and protein availability of food grains [75]. It has also been reported that during soaking, endogenous or exogenous phytases can potentially improve the *in vitro* solubility of minerals like zinc and iron by 2% to 23% [76].

A study reported 27.9% and 36.0% phytic acid reduction on soaking *Mucana flagellipes* at room temperature for 6 and 24 hours, respectively [77]. The process of soaking increased the phytase activity, thereby reducing the phytate content in the grains. As a result of soaking and fermentation, phytochemicals are reduced due to leaching of water-soluble vitamins and minerals in grains and legumes [78, 79]. Similarly, other studies have recommended consumption of wheat and barley after soaking for a period [74], particularly for 12 to 24 hours [80–82]. It has also been found that soaking helped in the significant reduction of phytate content at 45°C and 65°C [83]. Likewise, in another study, it was found that phytic acid concentration in chickpea was reduced by 47.45% to 55.71% when the soaking time was increased from 2 to 12 hours [80].

8.3.2 Fermentation

Another potentially useful technique to reduce bacterial contamination of foods is fermentation [84, 85]. Fermentation is a metabolic process which involves oxidation of sugars for the production of energy. It significantly reduces the content of anti-nutrients present in foods including phytic acid, tannins, and polyphenols of cereals [86]. It also helps in enhancing mineral absorption from plant-based foods [87]. Additionally, fermentation has been reported to enhance the nutritive value of food grains by improving the levels of essential amino acids like lysine, methionine, and tryptophan [88]. Generally, phytic acid present in cereals tends to form complex compounds with metal cations such as iron, calcium, and zinc, which are degenerated by enzymes which require an optimum pH facilitated by the process of fermentation. This degeneration helps in the reduction of phytic acid and liberation of soluble iron, zinc, and calcium, which further aids in enhancing the nutritional level of food grains [89]. Researches have also reported reduction in various anti-nutrients such as protease inhibitors, phytic acids, and tannins as a result of millet fermentation for 12 and 24 hours [75].

Researches conducted in the past have recommended the use of fermented millet-based products as probiotics to treat diarrhoea among young children [84, 85]. It has also been reported that fermenting cereals with the help of lactic acid bacteria (LAB) augments free amino acids and their derivatives by the process of proteolysis and metabolic synthesis [88]. Lactic acid fermentation has also been reported to result in the reduction of tannin levels which helps in enhanced iron absorption, except in certain cereals with high tannin content, where slight or no enhancement in availability of iron was reported. *Lactobacillus spp.* plays a vital role in fermentation among several cereals. However, *Lactobacillus spp.* and *Streptococcus spp.* have been found to be not much suitable to ferment rice as they lack amylase, which is essential for saccharification of starch [90]. When cereals are employed as natural medium for lactic acid fermentation, either amylase is added before or during fermentation, or amylolytic *bifido* bacteria is utilized as they comprise enough amylase required for saccharification of the grain starch. It was reported in a study that fermenting germinated millets at 30°C with combinations of probiotics culture comprising *Lactobacillus brevis, Lactobacillus fermentum, Saccharomyces cerevisiae,* and *Saccharomyces diasticus,* for 72 hours, lead to a reduction (about 88.3%) of phytic acid content [91]. A recent study fermented maize flour with a consortium of LAB using a standard method with time interval of 12 hour to check the consequence of fermentation on ANFs [92]. The results exhibited significant ($p < 0.05$) reductions in anti-nutrients, like tannin, polyphenol, phytate, and trypsin inhibitor activity with increasing fermentation period [92]. Another study conducted to study the influence of microbial fermentation on anti-nutritional content of local cassava products reported substantially reduced ($P < 0.05$) levels of cyanide, tannins, phytate, oxalate, and saponins by 86%, 73%, 72%, 61%, and 92%, respectively [93].

8.3.3 Germination

Germination is another potentially suitable strategy to reduce the anti-nutritional elements present in plant-based foods [94]. Seed germination helps in activating phytase, which is responsible for the degradation of phytate and subsequently results in the reduction of phytic acid level in food. In addition, it is also responsible for modifying the nutrient content, biochemical and physical properties of the foods. This technique is used commonly for reducing anti-nutritional content in cereals [95–97]. Reduction of ANFs such as phytic acid and tannins in germinated cereals helps in augmenting the bioavailability of various minerals, which further leads to enhanced nutritional value of the food products [78, 96]. It has been observed that germination of cereal grains results in improved phytase activity [76]. It has also been reported that following the process of malting of millet samples for 72 and 96 hours, a reduction in the phytic acid content (23.95% and 45.3%, respectively) was observed [75, 98]. A previous study reported a significant reduction in the phytate content of cereal grains when determined after 10 days of germination [99]. Another study reported an increase in anti-nutrient components like total phenolic, flavonoid, and tannin in germinated buckwheat samples [100]. Recent studies have also reported modifications in isoflavone profile of soybean resulting due to germination which helps in activating β-glucosidases, thus, enhancing the nutritional value due to the chelating properties of isoflavones [101, 102]. Another recent research reported maximum reduction in the levels of polyphenols (up to 75%) in germinated millets in comparison to soaked, microwave treated, and fermented millets [72].

8.3.4 Milling

One of the most traditional strategies to reduce the ANFs in grains is milling, which is used to separate the bran from the grains by grinding the grains into flour. Although, milling helps in the removal of anti-nutrients like phytic acid, lectins, and tannins, which are mostly present in the bran of grains, it also eliminates certain vital minerals [74]. Various researches have reported a change in the chemical composition of millets resulting from milling. A research reported reduction in the levels of phytic acid and polyphenol in *chapatti* as well as substantially improved starch and protein digestion resulting due to milling and heating during making of *chapatti* [103]. A study was conducted on two varieties of pearl millets to evaluate their nutrient and anti-nutrient content as well as mineral bioavailability post-milling into whole flour, bran rich segment, and semi-refined flour. It was reported that no considerable difference in the nutrient composition of semi-refined flour and whole flour apart from fat content, which was 1.3% was observed. Although, reduced levels of phytate and oxalate were found in semi-refined flour in comparison to whole flour, due to removal of the bran [104].

8.3.5 Extrusion

Another technique used to reduce the levels of ANFs in foods is extrusion, which is a multi-step, multi-functional, and a combination of thermal-mechanical processes. Extrusion is a technique which involves high-temperature short-time (HTST) processing in combination with various other processes like mixing, shearing, particle size reduction, heat and mass transfer, melting, caramelizing, texturizing, and shaping. The process involves dry mixing of raw material (including flour or blended meal prepared from ground edible seeds, food additives, and other ingredients) which enters the extruder barrel through a hopper, mixed well, and then exposed to adequate preconditioning with the help of water or steam and at times low temperature heating. The preconditioned mix also known as extrudate is then permeated through the extruder while being enforced through an orifice or a die by a piston or a large, tight-fitting rotating screw inside the stationary barrel. The screw along the barrel helps in compressing and kneading the extrudate into plasticized mass along the barrel. The extrudate is then subjected to HTST cooking (>100°C) while increasing the pressure and shearing force provided by the screw in the barrel. At the end of the barrel, cooked extrudate is forced through a small die with a certain design/shape and is cut with a sharp blade to achieve a desirable size/length of the final product [105]. Commonly used extruders in the food industry include single- and twin-screw extruder [106]. Factors that play a vital role in extrusion in terms of the reduction of ANFs are moisture content and composition of raw material, barrel temperature and the feed rate of extruder [107].

HTST processing has been recognized as an efficacious technique as it retains the nutrients present in the food and eliminates the inhibitor as well as microorganisms causing food contamination. This technique has also been utilized to immobilize or to completely eliminate aflatoxins which generally need extreme conditions like high shear and high temperature [108]. Besides reducing ANFs, extrusion also helps in gelatinizing starch, increasing soluble dietary fiber, and reducing lipid oxidation [71]. It is also responsible for inactivating α-amylase inhibitors, trypsin, and chymotrypsin and hemagglutinin activity without altering protein content in food products [6].

8.3.6 Heating-Autoclaving (Wet Heating) and Roasting (Dry Heating)

Autoclave or wet heating is yet another technique which helps in the reduction of ANFs. It is heat treatment, when applied on cereals and other plant-based foods helps in activating phytase enzyme and increasing acidity [109]. Most foods have exhibited improved health benefits when autoclaved, for example, reduced ANFs in boiled food grains [110]. Generally, legumes are cooked by either the process of boiling or by the use of a pressure cooker before consumption. Studies conducted in the past have reported that boiling and cooking improved nutritive value of foods due to reduced anti-nutrients such as tannins and trypsin inhibitors [111]. Similarly, another study observed drastic reduction in the phytic acid content in legumes that were cooked and soaked [112]. A study conducted on whole wheat bread treated through autoclave and microwave application reported a decrease in the phytic acid content and an increase in the unbound mineral content. This can be attributed to the mineral-chelating property of phytic acid which reduces phytates and further decreases the level of bound minerals and increases the free mineral contents [113]. It was also reported that boiling taro leaves in water for 40 minutes resulted in reduction of oxalate content by 47%, although, no significant reduction in oxalate content was noted after baking the taro leaves for 40 minutes at 180°C [114]. It has also been reported that the process of roasting resulted in substantially reduced activity of trypsin inhibitor in soybean meal [115]. Various recently conducted researches have reported significant reduction in level of many ANFs following autoclaving, soaking, and cooking of legumes. Hence, autoclaving has been considered to one of the best ANFs reduction strategy in comparison to other processing methods [112, 116, 117].

8.3.7 Gamma Radiation

Recently, gamma radiation has been recognized as another technique for the reduction of ANFs in plant-based foods. It can be employed as a safe post-harvest technique to reduce anti-nutrients present in millet grains [118]. A study reported that gamma radiations resulted in the reduction of trypsin inhibitor, phytic acid, and oligosaccharides content (between 5% and 10%) in broad bean. Another study reported that low doses of gamma irradiation (0.5 and 1.0 kGy) in faba bean seeds showed a significant decrease in ANFs like tannin and phytic acid [119]. Although, there are studies that have reported that a 2 kGy dose had no significant change in the tannin content of two maize cultivar [120].

8.3.8 Genomic Technology

Recent researches have reported genomic technology as another technique to reduce the levels of ANFs in foods. Genomic resources have been suggested to be used as pathways to RNA interference and to eliminate ANFs, However, this technique has yet to be tried *in vivo*. A study designed zinc-finger nucleases constructs to mutate the IPK1 gene in maze, one of the phytic acid biosynthesis genes because corn contains high levels of phosphorus stored in the form of phytic acid. Genome editing technology can increase crop quality but there is an ongoing argument about genetically modified organisms' safety [121].

8.4 Conclusion

Every year, a large portion of global population falls ill due to FBDs. The FBDs are caused due to a variety of food contaminants, especially microbial contaminants, which are the most hazardous. Naturally existing toxic compounds commonly known as ANFs are another category of food contaminant that are predominantly present in the plant-based foods. These compounds are present in small doses in plants to protect them from animals, insects, and pests and may not to cause any deleterious effects in humans when consumed. However, if consumed in large doses or if consumed by an individual sensitive to these compounds, then they may result in adverse health effects. This poses a challenge for consumers, especially those dependent on a plant-based diets. Therefore, a deeper knowledge about various ANFs, their sources, structures, mechanism of action, and adverse effects on human health is critical. Additionally, extensive research on strategies to reduce the levels of ANFs in the foods is essential to produce foods with least or no ANFs without altering their nutritional or organoleptic value.

References

1. World Health Organization, WHO estimates of the global burden of foodborne diseases: foodborne disease burden epidemiology reference group 2007-2015, WHO Library Cataloguing-in-Publication Data, 2015, [Online]. Available: https://apps.who.int/iris/handle/10665/199350.
2. World Health Organization, Fact Sheet No. 399 Food Safety, Fact Sheet, no. 16 February, 2016, 6, 2015, [Online]. Available: http://www.who.int/campaigns/world-health-day/2015/fact-sheet.pdf.
3. Sivapalasingam, S., Friedman, C.R., Cohen, L., Tauxe, R.V., Fresh produce: A growing cause of outbreaks of foodborne illness in the United States, 1973 through 1997. *J. Food Prot.*, 67, 2342, 2004.
4. Chulze, S.N., Strategies to reduce mycotoxin levels in maize during storage: A review. *Food Addit. Contam. - Part A Chem. Anal. Control Expo. Risk Assess.*, 27, 651, 2010.
5. Williams, P., Food toxicity and safety, in: *Essentials Hum. Nutr.*, p. 449, 2012.
6. Soetan, K.O. and Oyewole, O.E., The need for adequate processing to reduce the anti- nutritional factors in plants used as human foods and animal feeds: A review. *Afr. J. Food Sci.*, 3, 223, 2009.
7. Lolas, G.M., Palamidis, N., Markakis, P., Phytic acid total phosphorus relationship relationship in barley, oats, soybeans and wheat. *Cereal Chem.*, 53, 867, 1976.

8. GarcõÂa-Estepa, R.M., Guerra-HernaÂndez, E., GarcõÂa-Villanova, B., Phytic acid content in milled cereal products and breads. *Food Res. Int.*, 32, 217, 1999.
9. Serrano, J., Puupponen-Pimiä, R., Dauer, A., Aura, A.M., Saura-Calixto, F., Tannins: Current knowledge of food sources, intake, bioavailability and biological effects. *Mol. Nutr. Food Res.*, 53, 310, 2009.
10. Morzelle, M.C., Salgado, J.M., Massarioli, A.P., Bachiega, P., de Oliveira Rios, A., Alencar, S.M., Schwember, A.R., de Camargo, A.C., Potential benefits of phenolics from pomegranate pulp and peel in Alzheimer's disease: Antioxidant activity and inhibition of acetylcholinesterase. *J. Food Bioact.*, 5, 136, 2019.
11. Thakur, A., Sharma, V., Thakur, A., An overview of anti-nutritional factors in food. *Int. J. Chem.*, 7, 2472, 2019.
12. Tibe, O. and Amarteifio, J., Trypsin Inhibitor Activity and Condensed Tannin Content in Bambara Groundnut (*Vigna Subterranea* (*L.*) *Verdc*) Grown in Southern Africa. *J. Appl. Sci. Environ. Manage.*, 11, 159, 2010.
13. Moses, T., Papadopoulou, K.K., Osbourn, A., Metabolic and functional diversity of saponins, biosynthetic intermediates and semi-synthetic derivatives. *Crit. Rev. Biochem. Mol. Biol.*, 49, 439, 2014.
14. Mamboleo, T., Nutrients and antinutritional factors at different maturity stages of selected indigenous African green leafy vegetables, Sokoine university of agriculture, Morogoro, Tanzania, 2015.
15. Zagrobelny, M., Bak, S., Møller, B.L., Cyanogenesis in plants and arthropods. *Phytochemistry*, 69, 1457, 2008.
16. Popova, A. and Mihaylova, D., Antinutrients in Plant-based Foods: A Review. *Open Biotechnol. J.*, 13, 68, 2019.
17. Gemede, H.F., Antinutritional Factors in Plant Foods: Potential Health Benefits and Adverse Effects. *Int. J. Nutr. Food Sci.*, 3, 284, 2014.
18. Miyake, K., Tanaka, T., McNeil, P.L., Lectin-based food poisoning: A new mechanism of protein toxicity. *PLoS One*, 2, 1, 2007.
19. Vasconcelos, I.M. and Oliveira, J.T.A., Antinutritional properties of plant lectins. *Toxicon*, 44, 385, 2004.
20. Banwell, J.G., Howard, R., Kabir, I., Adrian, T.E., Diamond, R.H., Abramowsky, C., Small intestinal growth caused by feeding red kidney bean phytohemagglutinin lectin to rats. *Gastroenterology*, 104, 1669, 1993.
21. Shi, J., Xue, S.J., Kakuda, Y., Ilic, S., Kim, D., Isolation and characterization of lectins from kidney beans (*Phaseolus vulgaris*). *Process Biochem.*, 42, 1436, 2007.
22. Eguchi, R., Ono, N., Morita, A.H., Katsuragi, T., Nakamura, S., Huang, M., Altaf-Ul-Amin, M., Kanaya, S., Classification of alkaloids according to the starting substances of their biosynthetic pathways using graph convolutional neural networks. *BMC Bioinf.*, 20, 1, 2019.
23. Sparg, S.G., Light, M.E., Van Staden, J., Biological activities and distribution of plant saponins. *J. Ethnopharmacol.*, 94, 2–3, 219–243, 2004.

24. Akande, K.E., Doma, U.D., Agu, H.O., Adamu, H.M., Major Antinutrients Found in Plant Protein Sources: Their Effect on Nutrition. *Pakistan J. Nutr.*, 9, 827, 2010.
25. Amirkia, V. and Heinrich, M., Alkaloids as drug leads - A predictive structural and biodiversity-based analysis. *Phytochem. Lett.*, 10, 48, 2014.
26. Deshpande, S.S. and Cheryan, M., Effects of Phytic Acid, Divalent Cations, and Their Interactions on α-Amylase Activity. *J. Food Sci.*, 49, 516, 1984.
27. Jadhav, S.J., Sharma, R.P., Salunkhe, D.K., Naturally occurring toxic alkaloids in foods. *Crit. Rev. Toxicol.*, 9, 21, 1981.
28. Helsper, J.P.F.G., Hoogendijk, J.M., van Norel, A., Burger-Meyer, K., Antinutritional Factors in Faba Beans (*Vicia faba L.*) As Affected by Breeding toward the Absence of Condensed Tannins. *J. Agric. Food Chem.*, 41, 1058, 1993.
29. Giri, A.P. and Kachole, M.S., Amylase Inhibitors Of Pigeon pea (Ca Janus Ca Jan) Seeds. *Phytochemistry*, 47, 197, 1998.
30. Shi, J., Arunasalam, K., Yeung, D., Kakuda, Y., Mittal, G., Jiang, Y., Saponins from Edible Legumes: Chemistry, Processing, and Health Benefits. *J. Med. Food*, 7, 67, 2004.
31. Rao, P.U. and Deosthale, Y.G., Tannin content of pulses: Varietal differences and effects of germination and cooking. *J. Sci. Food Agric.*, 33, 1013, 1982.
32. Osuntokun, B.O., Cassava diet and cyanide metabolism in Wistar rats. *Br. J. Nutr.*, 24, 797, 1970.
33. Yamashoji, S. and Matsuda, T., Synergistic cytotoxicity induced by α-solanine and α-chaconine. *Food Chem.*, 141, 669, 2013.
34. Shimshoni, J.A., Mulder, P.P.J., Bouznach, A., Edery, N., Pasval, I., Barel, S., Abd-El Khaliq, M., Perl, S., Heliotropium europaeum poisoning in cattle and analysis of its pyrrolizidine alkaloid profile. *J. Agric. Food Chem.*, 63, 1664, 2015.
35. Koleva, I.I., van Beek, T.A., Soffers, A.E.M.F., Dusemund, B., Rietjens, I.M.C.M., Alkaloids in the human food chain - natural occurrence and possible adverse effects. *Mol. Nutr. Food Res.*, 56, 30, 2012.
36. Akwaowo, E.U., Ndon, B.A., Etuk, E.U., Minerals and antinutrients in fluted pumpkin (*Telfairia occidentalis Hook f.*). *Food Chem.*, 70, 235, 2000.
37. Noonan, S.C., Oxalate content of foods and its effect on humans. *Asia Pac. J. Clin. Nutr.*, 8, 64, 1999.
38. Hatch, M. and Freel, R.W., Intestinal transport of an obdurate anion: Oxalate. *Urol. Res.*, 33, 1, 2005.
39. Hughes, J. and Norman, R.W., Diet and calcium stones. *Cmaj*, 146, 137, 1992.
40. Chai, W., Liebman, M., Kynast-Gales, S., Massey, L., Oxalate absorption and endogenous oxalate synthesis from ascorbate in calcium oxalate stone formers and non-stone formers. *Am. J. Kidney Dis.*, 44, 1060, 2004.
41. Hanes, D.A., Weaver, C.M., Heaney, R.P., Wastney, M., Absorption of calcium oxalate does not require dissociation in rats. *J. Nutr.*, 129, 170, 1999.

42. Natesh, N.H., Abbey, L., Asiedu, S.K., An overview of nutritional and anti nutritional factors in green leafy vegetables. *Hortic. Int. J.*, 1, 58, 2017.
43. Concon, J.M., Food toxicology. Part A: Principles and concepts; Part B: Contaminants and additives, 1371 Seiten, zahlr. Tab. Marcel Dekker, Inc., New York, Basel, 1988.
44. Haque, M.R. and Bradbury, J.H., Total cyanide determination of plants and foods using the picrate and acid hydrolysis methods. *Food Chem.*, 77, 107, 2002.
45. Ganjewala, D., Kumar, S., Devi, S.A., Ambika, K., Advances in cyanogenic glycosides biosynthesis and analyses in plants: A review. *Acta Biol. Szeged.*, 54, 1, 2010.
46. Taylor, S.L., Natural Toxins in Food, in: *Encycl. Lifestyle Med. Heal*, 2012.
47. Gruhnert, C., Biehl, B., Selmar, D., Compartmentation of cyanogenic glucosides and their degrading enzymes. *Planta*, 195, 36, 1994.
48. Bolarinwa, I.F., Oke, M.O., Olaniyan, S.A., Ajala, A.S., A Review of Cyanogenic Glycosides in Edible Plants, in: *Toxicol. - New Asp. to This Sci. Conundrum*, 2016.
49. World health Organization, Food, Book Review: Safety Evaluation of Certain Food Additives and Contaminants. *Nutr. Health*, 15, 74, 2001.
50. Food Standards Australia New Zealand (FSANZ), Final assessment report proposal P257, Advice on the preparation of cassava and bamboo shoots, March 2004, [Online]. Available: https://www.foodstandards.gov.au/code/proposals/documents/P257_Initial_Cassava.pdf
51. Casadei, E., Jansen, P., Rodrigues, A., Mantakassa: An epidemic of spastic paraparesis associated with chronic cyanide intoxication in a cassava staple area of Mozambique. 2. Nutritional factors and hydrocyanic acid content of cassava products. *Bull. World Health Organ.*, 62, 485, 1984.
52. Tylleskar, T., Banea, M., Bikangi, Cooke, R.D., Poulter, N.H., Rosling, H., Cassava cyanogens and konzo, an upper mtoneuron disease found in Africa. *Lancet*, 339, 208, 1992.
53. Ernesto, M., Cardoso, A.P., Nicala, D., Mirione, E., Massaza, F., Cliff, J., Haque, M.R., Bradbury, J.H., Persistent konzo and cyanogen toxicity from cassava in northern Mozambique. *Acta Trop.*, 82, 357, 2002.
54. Oluwole, O.S.A., Onabolu, A.O., Link, H., Rosling, H., Persistence of tropical ataxic neuropathy in a Nigerian community. *J. Neurol. Neurosurg. Psychiatry*, 69, 96, 2000.
55. Sinha, K. and Khare, V., Review on : Antinutritional factors in vegetable crops. *Pharm. Innov. J.*, 6, 353, 2017.
56. Fahey, J.W., Zalcmann, A.T., Talalay, P., The chemical diversity and distribution of glucosinolates and isothiocyanates among plants. *Phytochemistry*, 56, 5, 2001.

57. Felker, P., Bunch, R., Leung, A.M., Concentrations of thiocyanate and goitrin in human plasma, their precursor concentrations in brassica vegetables, and associated potential risk for hypothyroidism. *Nutr. Rev.*, 74, 248, 2016.
58. Egert, S. and Rimbach, G., Which sources of flavonoids: Complex diets or dietary supplements? *Adv. Nutr.*, 2, 8, 2011.
59. Giuliani, C., Iezzi, M., Ciolli, L., Hysi, A., Bucci, I., Santo., S.D., Rossi, C., Zucchelli, M., Napolitano, G., Resveratrol has anti-thyroid effects both *in vitro* and *in vivo*. *Food Chem. Toxicol.*, 107, 237, 2017.
60. Messina, M. and Redmond, G., Effects of soy protein and soybean isoflavones on thyroid function in healthy adults and hypothyroid patients: A review of the relevant literature. *Thyroid*, 16, 249, 2006.
61. Kozłowska, A. and Szostak-Wegierek, D., Flavonoids–food sources and health benefits. *Rocz. Państw. Zakł. Hig.*, 65, 79, 2014.
62. Dybkowska, E., Sadowska, A., Świderski, F., Rakowska, R., Wysocka, K., The occurrence of resveratrol in foodstuffs and its potential for supporting cancer prevention and treatment. A review. *Rocz. Panstw. Zakl. Hig.*, 69, 5, 2018.
63. Gaitan, E., Lindsay, R.H., Reichert, R.D., Ingbarf, S.H., Cooksey, R.C., Legan, J., Meydrech, E.F., Hill, J., Kubotae, K., Antithyroid and goitrogenic effects of millet: Role of C-Glycosylflavones. *J. Clin. Endocrinol. Metab.*, 68, 707, 1989.
64. Boncompagni, E., Orozco-Arroyo, G., Cominelli, E., Gangashetty, P.I., Grando, S., Zu, T.T.K., Daminati, M.G., Nielsen, E., Sparvoli, F., Antinutritional factors in pearl millet grains: Phytate and goitrogens content variability and molecular characterization of genes involved in their pathways. *PLoS One*, 13, 1, 2018.
65. Tripathi, M.K. and Mishra, A.S., Glucosinolates in animal nutrition: A review. *Anim. Feed Sci. Technol.*, 132, 1, 2007.
66. Prieto, M.A., López, C.J., Simal-Gandara, J., Glucosinolates: Molecular structure, breakdown, genetic, bioavailability, properties and healthy and adverse effects. *Adv. Food Nutr. Res.*, 90, 305, 2019.
67. Ikewuchi, C.C., Hypocholesterolemic effect of an aqueous extract of the leaves of *Sansevieria senegambica* baker on plasma lipid profile and atherogenic indices of rats fed egg yolk supplemented diet. *Excli J.*, 11, 318, 2012.
68. El Barky, A. and Hussein, S.A., Saponins and Their Potential Role in Diabetes Mellitus. *Diabetes Manag.*, 7, 148, 2017.
69. Johnson, I.T., Gee, J.M., Price, K., Curl, C., Fenwick, G.R., Influence of saponins on gut permeability and active nutrient transport *in vitro*. *J. Nutr.*, 116, 2270, 1986.
70. Lo, D., Wang, H.I., Wu, W.J., Yang, R.Y., Anti-nutrient components and their concentrations in edible parts in vegetable families. *CAB Rev. Perspect. Agric. Vet. Sci. Nutr. Nat. Resour.*, 13, 1, 2018.

71. Nikmaram, N., Leong, S.Y., Koubaa, M., Zhu, Z., Barba, F.J., Greiner, R., Oey, I., Roohinejad, S., Effect of extrusion on the anti-nutritional factors of food products: An overview. *Food Control*, 79, 62, 2017.
72. Samtiya, M., Aluko, R.E., Dhewa, T., Plant food anti-nutritional factors and their reduction strategies: an overview. *Food Prod. Process. Nutr.*, 2, 1, 2020.
73. Kumari, S.V., The Effect of Soaking Almonds and Hazelnuts on Phytate and Mineral Concentrations, University of Otago, Dunedin, New Zealand, 2017.
74. Gupta, R.K., Gangoliya, S.S., Singh, N.K., Reduction of phytic acid and enhancement of bioavailable micronutrients in food grains. *J. Food Sci. Technol.*, 52, 676, 2013.
75. Coulibaly, A., Kouakou, B., Chen, J., Phytic Acid in Cereal Grains: Structure, Healthy or Harmful Ways to Reduce Phytic Acid in Cereal Grains and Their Effects on Nutritional Quality. *Amer. J. Plant Nutr. Fertil. Technol.*, 1, 1, 2010.
76. Vashishth, A., Ram, S., Beniwal, V., Cereal phytases and their importance in improvement of micronutrients bioavailability. *3 Biotech.*, 7, 1, 2017.
77. Udensi, E., Arisa, N., Maduka, M., Effects of processing methods on the levels of some antinutritional factors in Mucuna flagellipes. *Niger. Food J.*, 26, 47, 2009.
78. Ogbonna, A.C., Abuajah, C.I., Ide, E.O., Udofia, U.S., Effect of malting conditions on the nutritional and anti-nutritional factors of sorghum grist. *Ann. Univ. Dunarea Jos Galati, Fasc. VI Food Technol.*, 36, 64, 2012.
79. Kruger, J., Oelofse, A., Taylor, J.R.N., Effects of aqueous soaking on the phytate and mineral contents and phytate:mineral ratios of wholegrain normal sorghum and maize and low phytate sorghum. *Int. J. Food Sci. Nutr.*, 65, 539, 2014.
80. Ertaş, N. and Türker, S., Bulgur processes increase nutrition value: Possible role in *in-vitro* protein digestability, phytic acid, trypsin inhibitor activity and mineral bioavailability. *J. Food Sci. Technol.*, 51, 1401, 2014.
81. Mahgoub, S.E.O. and Elhag, S.A., Effect of milling, soaking, malting, heat-treatment and fermentation on phytate level of four Sudanese sorghum cultivars. *Food Chem.*, 61, 77, 1998.
82. Onwuka, G.I., Soaking, boiling and antinutritional factors in Pigeon peas (*Cajanus cajan*) and cowpeas (*Vigna unguiculata*). *J. Food Process. Preserv.*, 30, 616, 2006.
83. Greiner, R. and Konietzny, U., Phytase for food application. *Food Technol. Biotechnol.*, 44, 2, 125–140, 2006.
84. Manisseri, C. and Gudipati, M., Prebiotic Activity of Purified Xylobiose Obtained from Ragi (*Eleusine coracana, Indaf-15*) Bran. *Indian J. Microbiol.*, 52, 251, 2012.
85. Nduti, N., McMillan, A., Seney, S., Sumarah, M., Njeru, P., Mwaniki, M., Reid, G., Investigating probiotic yoghurt to reduce an aflatoxin B1 biomarker among school children in eastern Kenya: Preliminary study. *Int. Dairy J.*, 63, 124, 2016.

86. Simwaka, J.E., Chamba, M.V.M., Huiming, Z., Masamba, K.G., Luo, Y., Effect of fermentation on physicochemical and antinutritional factors of complementary foods from millet, sorghum, pumpkin and amaranth seed flours. *Int. Food Res. J.*, 24, 1869, 2017.
87. Galati, A., Oguntoyinbo, F.A., Moschetti, G., Crescimanno, M., Settanni, L., The Cereal Market and the Role of Fermentation in Cereal-Based Food Production in Africa. *Food Rev. Int.*, 30, 317, 2014.
88. Mohapatra, D., Patel, A.S., Kar, A., Deshpande, S.S., Tripathi, M.K., Effect of different processing conditions on proximate composition, anti-oxidants, anti-nutrients and amino acid profile of grain sorghum. *Food Chem.*, 271, 129, 2019.
89. Gibson, R.S., Bailey, K.B., Gibbs, M., Ferguson, E.L., A review of phytate, iron, zinc, and calcium concentrations in plant-based complementary foods used in low-income countries and implications for bioavailability. *Food Nutr. Bull.*, 31, 134, 2010.
90. Ray, M., Ghosh, K., Singh, S., Mondal, K.C., Folk to functional: An explorative overview of rice-based fermented foods and beverages in India. *J. Ethn. Foods*, 3, 5, 2016.
91. Khetarpaul, N. and Chauhan, B.M., Fermentation of pearl millet flour with yeasts and lactobacilli: *in vitro* digestibility and utilisation of fermented flour for weaning mixtures. *Plant Foods Hum. Nutr.*, 40, 167, 1990.
92. Ogodo, A.C., Agwaranze, D.I., Aliba, N.V., Kalu, A.C., Nwaneri, C.B., Fermentation by Lactic Acid Bacteria Consortium and its Effect on Antinutritional Factors in Maize Flour. *J. Biol. Sci.*, 19, 17, 2019.
93. Etsuyankpa, M.B., Gimba, C.E., Agbaji, E.B., Omoniyi, K.I., Ndamitso, M.M., Mathew, J.T., Assessment of the Effects of Microbial Fermentation on Selected Anti-Nutrients in the Products of Four Local Cassava Varieties from Niger State, Nigeria. *Am. J. Food Sci. Technol.*, 3, 89, 2015.
94. Nkhata, S.G., Ayua, E., Kamau, E.H., Shingiro, J.B., Fermentation and germination improve nutritional value of cereals and legumes through activation of endogenous enzymes. *Food Sci. Nutr.*, 6, 2446, 2018.
95. G.L., N.C., S.R., The Impact of Malting on Nutritional Composition of Foxtail Millet, Wheat and Chickpea. *J. Nutr. Food Sci.*, 05, 5, 2015.
96. Oghbaei, M. and Prakash, J., Effect of primary processing of cereals and legumes on its nutritional quality: A comprehensive review. *Cogent Food Agric.*, 2, 1, 2016.
97. Onyango, C.A., Ochanda, S.O., Mwasaru, M.A., Ochieng, J.K., Mathooko, F.M., Kinyuru, J.N., Effects of Malting and Fermentation on Anti-Nutrient Reduction and Protein Digestibility of Red Sorghum, White Sorghum and Pearl Millet. *J. Food Res.*, 2, 41, 2013.
98. Makokha, A.O., Oniang'o, R.K., Njoroge, S.M., Kamar, O.K., Effect of traditional fermentation and malting on phytic acid and mineral availability from sorghum (*Sorghum bicolor*) and finger millet (*Eleusine coracana*) grain varieties grown in Kenya. *Food Nutr. Bull.*, 23, 241, 2002.

99. Azeke, M.A., Egielewa, S.J., Eigbogbo, M.U., Ihimire, I.G., Effect of germination on the phytase activity, phytate and total phosphorus contents of rice (*Oryza sativa*), maize (*Zea mays*), millet (*Panicum miliaceum*), sorghum (*Sorghum bicolor*) and wheat (*Triticum aestivum*). *J. Food Sci. Technol.*, 48, 724, 2011.
100. Zhang, G., Xu, Z., Gao, Y., Huang, X., Zou, Y., Yang, T., Effects of germination on the nutritional properties, phenolic profiles, and antioxidant activities of buckwheat. *J. Food Sci.*, 80, 1111, 2015.
101. Yoshiara, L.Y., Mandarino, J.M.G., Carrão-Panizzi, M.C., Madeira, T.B., da Silva, J.B., de Camargo, A.C., Shahidi, F., Ida, E.I., Germination changes the isoflavone profile and increases the antioxidant potential of soybean. *J. Food Bioact.*, 3, 144, 2018.
102. de Camargo., A.C., Favero, B.T., Morzelle, M.C., Franchin, M., Alvarez-Parrilla, E., de la Rosa, L.A., Geraldi, M.V., Junior, M.R.M., Shahidi, F., Schwember, A.R., Is chickpea a potential substitute for soybean? Phenolic Bioactives and potential health benefits. *Int. J. Mol. Sci.*, 20, 1, 2019.
103. Chowdhury, S. and Punia, D., Nutrient and antinutrient composition of pearl millet grains as affected by milling and baking. *Nahrung - Food*, 41, 105, 1997.
104. Suma, P.F. and Urooj, A., Nutrients, antinutrients & bioaccessible mineral content (invitro) of pearl millet as influenced by milling. *J. Food Sci. Technol.*, 51, 756, 2014.
105. Kaur, G.J., Rehal, J., Singh, A.K., Singh, B., Kaur, A., Optimization of extrusion parameters for development of ready-to-eat breakfast cereal using RSM. *Asian J. Dairy Food Res.*, 33, 77, 2014.
106. Singh, N., *Extrusion-Cooking Technology: Applications, Theory and Sustainability*, L. Moscicki (Ed.), WILEY-VCH, Weinheim, Germany, 2011.
107. Lević, J., Olivera, Đ., Sredanović, S., Heat treatments in animal feed processing, pp. 1–191, Institute for Food Technology, 2010.
108. Saalia, F.K. and Phillips, R.D., Degradation of aflatoxins by extrusion cooking: Effects on nutritional quality of extrudates. *LWT - Food Sci. Technol.*, 44, 1496, 2011.
109. Ertop, M.H. and Bektaş, M., Enhancement of Bioavailable Micronutrients and Reduction of Antinutrients in Foods With Some Processes. *Food Health*, 4, 159, 2018.
110. Rehman, Z.U. and Shah, W.H., Thermal heat processing effects on antinutrients, protein and starch digestibility of food legumes. *Food Chem.*, 91, 327, 2005.
111. Patterson, C.A., Curran, J., Der, T., Effect of processing on antinutrient compounds in pulses. *Cereal Chem.*, 94, 2, 2017.
112. Vadivel, V. and Biesalski, H.K., Effect of certain indigenous processing methods on the bioactive compounds of ten different wild type legume grains. *J. Food Sci. Technol.*, 49, 673, 2012.

113. Demir, M.K. and Elgün, A., Comparison of autoclave, microwave, IR and UV-C stabilization of whole wheat flour branny fractions upon the nutritional properties of whole wheat bread. *J. Food Sci. Technol.*, 51, 59, 2014.
114. Savage, G.P. and Mårtensson, L., Comparison of the estimates of the oxalate content of taro leaves and corms and a selection of Indian vegetables following hot water, hot acid and *in vitro* extraction methods. *J. Food Compos. Anal.*, 23, 113, 2010.
115. Vagadia, B.H., Vanga, S.K., Raghavan, V., Inactivation methods of soybean trypsin inhibitor – A review. *Trends Food Sci. Technol.*, 64, 115, 2017.
116. Shimelis, E.A. and Rakshit, S.K., Effect of processing on antinutrients and *in vitro* protein digestibility of kidney bean (*Phaseolus vulgaris L.*) varieties grown in East Africa. *Food Chem.*, 103, 161, 2007.
117. Torres, J., Rutherfurd, S.M., Muñoz, L.S., Peters, M., Montoya, C.A., The impact of heating and soaking on the *in vitro* enzymatic hydrolysis of protein varies in different species of tropical legumes. *Food Chem.*, 194, 377, 2016.
118. Mahmoud, N.S., Awad, S.H., Madani, R.M.A., Osman, F.A., Elmamoun, K., Hassan, A.B., Effect of γ radiation processing on fungal growth and quality characteristcs of millet grains. *Food Sci. Nutr.*, 4, 342, 2016.
119. Osman, A.M.A., Hassan, A.B., Osman, G.A.M., Mohammed, N., Rushdi, M.A.H., Diab, E.E., Babiker, E.E., Effects of gamma irradiation and/or cooking on nutritional quality of faba bean (*Vicia faba L.*) cultivars seeds. *J. Food Sci. Technol.*, 51, 1554, 2014.
120. Fombang, E.N., Taylor, J.R.N., Mbofung, C.M.F., Minnaar, A., Use of γ-irradiation to alleviate the poor protein digestibility of sorghum porridge. *Food Chem.*, 91, 695, 2005.
121. Kim, H., Kim, S.T., Kim, S.G., Kim, J.S., Targeted Genome Editing for Crop Improvement. *Plant Breed. Biotechnol.*, 3, 283, 2015.

9

Feeding the Future—Challenges and Limitations

Baishakhi De and Tridib Kumar Goswami*

Agricultural and Food Engineering Department, IIT Kharagpur, India

Abstract

The massive surge in global population is expected to surpass 9 billion by 2050. Though hi-tech agri practices and modern food processing techniques are producing enough food for the entire population, still incidences of malnutrition, poverty, hunger question the food supply and security. There is enormous food wastage every year in the food supply chain, in different stages of post harvest manufacturing and spoilage during storage. Global population escalation on the other hand will contribute to hike in food prices. By mid century it is necessary to design the 'food for future' on a sustainable basis in a world that will be facing food sourcing and sustainability concerns, biodiversity problems, environmental and climatic hazards. Health disorders relating to under nutrition and over nutrition, consumer's attitudes need to be addressed for maintaining future food security and sustainability. This book chapter will discuss in details the challenges and constraints in future food security; green technologies and associated research trends in food processing with major focus on topics like search for alternative protein sources, novel technologies in food processing, nutrigenomics, biomimicry, DNA bar-coding, 'lab-on-chip' and nanosensors in food safety technology as remedial solutions and associated regulations in attaining a future sustainable food supply.

Keywords: Hi-tech agri practices, sustainable food supply, malnutrition, food engineering, green technologies, food security, sustainable diets, food processing

---

*Corresponding author: tkg@agfe.iitkgp.ernet.in

9.1 Introduction

The food habits and patterns of the global population vary from region to region and from country to country and greatly depend on the consumer preferences. In addition, 7.7 billion of people around the world consume about 14.5 million tons of food daily. With constant escalation in global population that is likely to be projected to 9.8 million by 2050, the demand for food supply is supposed to increase by 50% and animal-based foods by 70%. Obviously, the question that haunts the mind: Will the global food supply assure enough "food security" to feed the global population by 2050? [1–4]. Global food industry needs to address the basic issues of "sustainability", "food wastage", and "nutritional issues", e.g., malnutrition, overnutrition and micronutrient deficiencies, etc. Though half of the world's cultivatable land is employed for the purpose of agriculture, still the incidence of extreme poverty and hunger persists. About one-third of the total amount of food produced every year is wasted either in pre- or post-harvesting stage, during the stages of food processing, and spoilage during storage often leading to insufficiency in food supply and scarcity under severe conditions. Despite rapid technological advancements and industrializations, the burning issues of "poverty" and "hunger" are yet to be addressed. Malnutrition and under nourishment on one hand and overnutrition and obesity on other hand is the root cause of different ailments along with micro nutrient deficiency disorders. Such "nutritional issues" are directly posing a threat to the public health and thence the nation [1, 5]. In order to achieve sustainability in "food for future", it is essential to increase food production but without further more exploitation of agricultural lands, simultaneous increase in animal-based food supply and fishes with improved wild fisheries and aquaculture, and adopt eco-friendly procedures in agricultural production, e.g., aim to reduce the emissions of greenhouse gases [5]. Procurement of food is not always simple and straight forward. Rather, the transition of food from "farm to fork" involves a lot many people. The farmers who grow the crops, food processing at industrial scale, i.e., involvement of manufacturers, sold to public by retailers and grocers, which are finally procured by end users, i.e., consumers. Thus, the food supply chain is highly interdependent [4]. In order to feed the expected global population of 9–10 billion by 2050, there will be an increase in food demand by 60%–70%. To meet this escalated demand, attain food sustainability, reduce food wastage, end chronic hunger and poverty, and attain nutrition for all is really a challenging task. Continuance of the existing trend of nutrition will further increase in the incidences of nutrition related health disorders. Further increase in land utilizations for agricultural purposes and increased emissions of greenhouse gases will increase pollution and affect biodiversity. The world is passing through a phase of "nutritional transition" from agriculture-based economy to technology-based economy. In order to develop a "sustainable food system" for future, it is necessary to give equal priority to "food security" i.e., "food for all" and also ensure "nutrition security" for all. Overnutrition and undernutrition is directly related to the economic status. Undernourished individuals are represented by underweight, wasting (too thin for height), and stunted growth. Hidden hunger is another form of undernourishment which mostly refers to the deficiencies of vitamins and minerals in the body. Overweight and obesity represent overnourished individuals. However, it is a matter of concern that malnutrition has a great repercussion on the economic development and health of the nation. Policy interventions should be such that it should keep a focus both on food sustainability and simultaneous fulfillment of adequate nutritional requirements [7–10]. Novel technological and scientific innovations play a crucial role in this regard. Precision farming, vertical farming, hydro phonics, utilization of natural renewable resources, genome editing and search for alternative protein sources, maintaining genetic diversity of seeds, cultivated plants, and domestic animals are some of the necessary steps. Achievement of such goals necessitates proper access to finance, easy and accessible innovative technology, and enough data storage, eliminate distortions in trade, and smoothen the food supply chain. Here, it is to be mentioned that the current food supply system is highly unsustainable as a huge amount of food is getting wasted before reaching the end users. Great inputs are required in the field of agriculture, aquaculture, fisheries, livestock, and animal husbandry in order to cope with the increased demand for food supply. Proper implementations of agricultural biotechnology play an important role in this regard. Some food items are procured "raw", whereas others as "processed items". Initial food processing is followed by final processing steps often to achieve a brand value of the product. A partnering relationship is to be maintained among producers or manufacturers, retailers and traders, and distributers so that a finished packed product reaches the end users [1, 5–7]. Experts have opined that a versatile application of robotics, genetics, information technology, and nano-processes combinedly known as GRIN technologies are expected to revolutionize human life. Genetically modified organisms (GMOs) in agriculture are already a debatable issue. Nanotechnology is going to provide a significant contribution in food engineering and animal feed industry [8]. This book chapter will discuss in details on future food sustainability, nutritional aspects, the associated challenges, green technological applications, and regulatory issues.

9.2 Early Life Nutrition and Healthy Future

Food, the fundamental right of every human being, is essential to provide nutrition and energy needed for work and survival. But still, the incidences of poverty, hunger, and "hidden hunger" of malnutrition are the unsolved burning issues of the day. The food system operates over a wide sector including agriculture, health and nutrition, business, environment, and culture. Food production, processing, and supply system involve people and professionals from different domains, e.g., farmers, manufacturers, engineers, scientists, researchers, and designers to complete the pathway from "farm to fork". The food supply system influences the international relationship in the form of trade. The environmental sustainability affects the food production and quality which further influences the survival and health status of a nation. There is a dire need to develop a global food system that is nutritious, affordable, and accessible to every person in the world anytime and anyplace so as to cope with the severity of hunger and poverty [10]. In order to achieve a sound health, a proper well-balanced diet with requisite calories is essential. Inappropriate nutrition in the early life of a child or during the pregnant condition of a woman can have fatal consequences in later life, creating irreparable health damage. Public health status has a societal impact that obviously affects the development of a nation. Proper nutrition in the first 1,000 days of a child's life affects the future potentiality and productivity. Just as extensive breastfeeding and proper hygiene is necessary for a child, proper education and diversified nutritional diet is necessary for the mother. Proper early life nutrition helps to build a child's immune system and combat with future allergic diseases. Here, it is to be mentioned that "nutrition" is often a double-edged sword. Both undernutrition and overnutrition (collectively known as malnutrition) affect the health of individual and community. Child stunting, wasting, overweight, and lifestyle ailments like Type 2 diabetes, obesity, and other co-morbidities is a long list awaited to be solved. Early life undernutrition is responsible for one-third of child's death [10–17]. Fetal development and pregnancy are the prime life stages during which rapid growth and development of organs and body system. The first 1,000 days of life starting from conception till 2 years is essential for brain and body development. Nutrition plays a crucial role in this stage; along with hormonal regulation, genetic and epigenetic factors. Definitive "nutritional programming" in early life will influence the future metabolic and physiological functioning of the body. Nutritional effect during the "developmental plasticity" will affect the future health and well-being of the off springs. When an embryo or a fetus develops, it is subjected to different external challenges and building of its adaptive capacity starts then [11, 12, 15]. Here, the term "programming" coined by Lucas (1991) [18] actually refers to the plasticity of cells and tissues during the development of embryo or fetus. Plasticity is a short lived characteristic of fetal or embryonic cells or tissues for developing adaptability to the environment. During this critical developmental phase of early life, "nutrition" plays a role in future health and disease. Nutritional deficiency of a mother during pregnancy is associated with greater risks of Type 2 diabetes and cardiovascular mortality. The period and extent of breastfeeding and quality of food used during weaning has a link with the future disease conditions. Breast feeding is found to improve cognitive function and provide protection against immune related disorders of Type 1 diabetes or inflammatory bowel disease (IBD). Research evidences have shown that exposure of a mother to famine during first trimester gave birth to higher weight babies than normal. Famine-exposed babies are at increased risk of coronary heart disease, raised concentration of blood clotting factors, and increased obesity due to higher level of circulating lipids. Thinness at birth is inversely related to Type 2 diabetes and glucose intolerance but is associated with coronary heart disease death and metabolic syndrome if body mass index increases in later childhood. Research evidences have shown that maternal nutritional intake during pregnancy is found to affect the health status of the child in later adulthood. Food supports life but one must be cautious about the choice of food. Some individuals are allergic to certain foods ingestion of which causes severe anaphylactic reactions in the body [11, 12, 15].

9.2.1 Choice of Food and "Nutrition Transition"

The choice of a food depends on personal preference, influenced by culture, habits, habitats, and health needs. Along with the transformation from agri-based economies to modernized economies, the world is passing through a phase of "nutrition transition". Human beings are having "omnivorous" food habits. Choice of a food is guided by ethnological or scientific perspectives. Ethnological perspectives on the role of food are focused mainly on culture. The scientific approach takes a more reductionist approach to food and health. The development of subject on nutritional genomics provides an understanding on the interaction between nutrition and genomics. The subject helps to understand the scientific basis on the role of nutrition in improving public health. The bioactive components in foods interact with genome at molecular, cellular, and system level and exert its effect on health. Nutritional genomics helps to understand the role of nutrients in gene expression and their effectivity in combating different ailments. Further, here comes the concept of "personalized nutrition" where designer or need-based tailor-made foods or customized diet are developed to satisfy the individualistic health needs [14].

Lack of diverse diets accounts to malnutrition. Now, incorporation of diverse foods in regular diet is related to income status. For diet diversity, agriculture-related interventions play an important role in this regard. Poor income status and non-availability of funds in backward and developing countries often accounts for poor food security and non-availability of diversified diets. Nutrition transition or alteration in food choices has an impact both at individual and community level. Change in the choice of food at mass scale obviously impacts on food supply, necessitates new food product development (formulation and reformulation of new food stuffs), and changes in legislative laws and need of a survey on health impacts. Nutrition transition should focus in tackling different forms of malnutrition prevalent in different parts of the countries. Undernutrition is correlated with low income status and extreme poverty, whereas overnutrition and obesity are prevalent among affluent and high-income zones. Thus, to achieve a smooth transition toward better public health, governmental policies should prioritize different intervention strategies where combating of undernutrition is to be focused in low income and lagging states and ailments like obesity as a consequence of overnutrition are to be handled in urbanized areas [7, 14].

Thus, to achieve food sustainability in 2050, it is necessary to eliminate poverty and hunger and develop a global food system. Such food systems will be safe, nutritious, affordable, utilize world's renewable natural resources, climate smart, more resilient, add to value chain, and easily accessible by all sections of people. Along with food sustainability, it is essential to develop strategic plans for eliminating poverty and hunger and elevate the living standard of marginalized population in all remote areas. This will help to eliminate the prevailing problems of under nutrition. Most of the poor populations dwell in rural areas and some may migrate to urban areas in search of livelihood. First, it is necessary to raise the income level of poor rural population. Exploring both the farm and nonfarm options for arrangements of employment for poor people in rural areas strengthens agricultural productivity, improves growth of agribusiness, generates rural non-farm income sources, and improves in livestock system, aquaculture, fisheries, and animal agriculture. Farmers in rural areas must be provided exposure and access to markets to generate income not only from the sale of agri goods but also from fisheries and livestock food products like eggs, milk, and non-food products like wool. A positive correlation exists between increased agricultural productivity and rural nonfarm employment. Since both skilled and unskilled manpower is required behind developing any industry, the food industry itself can provide a job opportunity. Effective policies and strategies are to be implemented to eliminate poverty and hunger that may differ from country to country. In extremely poor countries where poverty and hunger crossed the threshold, a much targeted direct approach is needed for starting initial agricultural and food production. In many places of the globe, animal rearing remains the only source of income and food. Laws and policies need to focus on the vulnerable population of the globe—the women and the children. It is to be remembered that women play a major part of the labor force globally in agricultural sector. Nutrition specific interventions and programs are vital to combat child malnutrition, e.g., wasting, stunting, and eliminating diseases like anemia, vitamin deficiency, and other micronutrient deficiency health disorders. It should be a major focus to ensure "food security" despite of geographical and climatic challenges; also global supply of good quality seeds, fertilizers, and proper access to finance are other strategic measures to end poverty [10].

9.3 Challenges and Opportunities in Developing the Future Food Systems

Food always played a fundamental role in the history of mankind. It is rightly stated by Niola that "humans eat to live" and "humans live to eat" [8, 19]. Intensification of global food production has given a direct challenge to future food security in 2050. Whether at national or at international level, there exists an interlink between food security, nutrition, and livelihood security. The demand and supply of food depends on the global eating trends and the necessity to feed the growing population. Owing to "nutrition transition" on one hand that demands for high fiber rich diet and inclination toward westernized diet with increased consumption of animal products put forth great environmental challenges further coupled with increased utilization of biomass and fossil fuels. In order to develop a future sustainable food system, it is necessary to overcome the constraints and challenges and utilize the opportunities [6, 20].

Rapid improvements in science, technology, and medicine have not only improved the living standard of the public but also has prompted steady rise in global population. Food and agriculture organization predicted rise in population mostly in urbanized areas of the developing countries. Experts have opined that food production is to be increased by approximately 70% to feed the escalated growing population by 2050. Often a major portion of food crops produced and harvested in underdeveloped and developing countries are not consumed due to lack of proper food processing, packaging, handling, and fair distribution, thus persists the burning problems of poverty and hunger [5, 6, 9, 20]. Though enough food is being produced globally, lack of proper fair supply networks, socioeconomic, geographical, and climatic challenges accounts for the inappropriate and inadequate food distribution in all countries and continents of the globe. Continuous environmental pollution, increased emissions of greenhouse gases, natural calamities like draught, flood, extremes of temperature like too much heat or too much cold in different parts of the globe, melting of glaciers, landslides, etc., pose a direct threat to global food security. Agricultural production must be expanded faster keeping in pace with the population growth. To develop a "sustainable food for future", it is necessary for more effective utilization of agricultural lands and water resources. However, rapid urbanization and population explosion necessitating development of infrastructure for habitation purposes has lead to loss of arable lands. Conversion of reserve lands to arable lands is not always feasible as, along with accommodation, there is an obvious need for industrial development. Water scarcity is also affecting food security as, along with fresh supply of drinking water, heavy use of water is necessary for agriculture, industry, and domestic usage. Natural water bodies and man-made reservoirs are needed to promote aquaculture and fisheries. To ensure food security, it is also essential to conserve our natural resources, energy conservation by reduction of exhaustion of fossil fuels, waste reduction, and finally maintenance of ecological balance, i.e., biodiversity. Increased production and consumption of bio-fuels will also affect food security adversely with a hike on food commodity prices. An approximate loss of 1.3 billion tons of food every year globally during post-harvest processing, storage, and distribution in supply chain is another challenge before food security. Other additional challenges include the tackling of rising incidences of non-communicable diseases, income inequality, fluctuations in consumer demand and psychology, geographical barriers, and less exposure of rural farmers to urban markets, etc. [5, 6, 9, 20].

Despite several challenges, financial support must be provided at the governmental level to boost agricultural production and agro-based economy. Proper linking of urban consumer food demands with rural prosperity, along with conserving environmental resources, will ensure both urban and rural food security [7].

9.4 Sustainable Diet for the Future

The food system across the world is highly diverse and depicts the historical, cultural, religious, social, and economic context of a continent, country, or a race. An *"ideal diet"* can be defined as one that is healthy and safe in quality, sufficient in quantity, available and affordable for the majority, and organoleptically and culturally acceptable by the mass population. The term *"sustainable diet"* is both complex and broad. In a nutshell, a sustainable diet can be defined as *"those diets that are nutritionally adequate, safe, healthy, culturally acceptable, and affordable but are also protective for the environment, biodiversity, and ecosystems and conserving for natural and human resources, thus providing food and nutrition security for the present and future generations"*. The purpose behind the concept of *"sustainable diet"* is to achieve a complete physical, mental, and social well-being of individuals, to support their optimal growth and development at all stages of life for present and future generations, to combat with all forms of malnutrition, e.g., overnutrition, undernutrition, hidden hunger, and micronutrient deficiencies, to reduce the incidences of non-communicable diseases, to conserve natural and energy resources and also be eco-friendly, and to preserve biodiversity [21–25]. Nutrition transition from *"omnivorous diet"* to *"sustainable diet"*, i.e., inclusion of more plant-based items rather than animal products or processed food, is found to reduce emission of greenhouse gases from food production by 29%–70% (compared with a reference scenario for 2050), to reduce non-communicable disease risk burden and global mortality rate by 6%–10%, and to reduce the incidences of Type 2 diabetes by 16%–41%, cancer by 7%–13%, and mortality due to coronary heart disease by 20%–26% [22]. A *"sustainable healthy diet"* starts in early life at the time of birth with the initiation of breast feeding and then proceeds with the intake of minimally processed foods. Along with the growth of age and development, whole grains, legumes, fruits, and vegetables are incorporated in diet followed by animal products (e.g., fishes, eggs, poultry, and little amounts of red meat) and dairy items in due course. A fresh and clean drinking water always forms an essential part of any diet. Energy or calorie intake and consumption of macro- and micronutrients should be optimal so as to remain healthy and active throughout the lifecycle. Sustainable diets must be either free or contain minimal toxins and pathogens, protect against foodborne diseases, and diet-related non-communicable diseases. Considering environmental impact, sustainable diet reduces atmospheric and chemical pollution and aims to maintain the levels within permissible limits. Maintenance of biodiversity, *viz.*, crops, livestock, plant-based products, and prevention of overfishing and overhunting, is another focus of sustainable diet from the environmental point of view. Sustainable diets are prepared keeping in mind the taste, consumption pattern, culinary pattern, and culture of the consumers. While in food supply chain, during production and processing, prefers minimal use of hormones and antibiotics, in packaging steps tries to minimize the use of plastics and finally aims in reduction of food loss and wastage in supply chain. The final crucial step is the fair distribution of food through all possible channels [21]. For implementations of sustainable diets for the public health, well-being and future food security definitive policies and guidelines are to be framed and followed. In order to make "sustainable diet" available to the most vulnerable, it is necessary to take into account the inequities and inequalities, the poor and deprived section must receive the attention on a priority basis. Government must promote capacity development including consumer empowerment and quantify and balance trade regulations; some food-based dietary guidelines are required to be developed that will encompass together society, culture, economy, environment, and ecology. Arrangements must be made for food and nutrition education so as to popularize "sustainable diets" among the common mass [21, 24].

In order to achieve food sustainability, a highly collaborative and integrative approach is necessary among food scientists, engineers, nutritionists, economists, policy makers, and governmental involvement. Development of renewable and sustainable food sources and meeting the global demands for proteins sustainably are the important upcoming challenges. Increased consumption of traditional animal proteins (e.g., fish, poultry, and meat) will put a pressure on plant-based items as it is necessary to arrange for the animal feed. Also, there is an increase in demand for plant-based proteins due to increase in the number of vegetarian consumers. There are ongoing researches for alternative proteins sources, e.g., insect-based animal protein sources or marine flora and fauna for plant-based proteins. In order to develop a sustainable food system, it is urgent to reduce and reuse waste generated. Utilizations of microorganisms to increase the protein content of the organic substrates or valorization of biomass (cladodes and sea food residues) treated as waste provide a sustainable basis. Technological innovations are essential in the steps of food processing, product diversification, value addition, restricting of food matrices, and to speed up passing of food products through the food supply chain [26, 27].

9.5 Research Trends and Green Food Technologies

Technological innovations and approaches are the need of the hour for future food sustainability and security. A global food technology revolution is needed for the same. "Green food technology" refers to develop an eco-friendly harmonized food system that will supply the planet with sustainable, secure, and safe food without overhauling the existing resources of fresh water, arable lands, plant and the animal kingdom, energy, etc. Food processing and technology play the crucial role in the transformation of food from "farm to fork". Successful implementation of a technology requires continuous academic and research initiatives, infrastructure, finance, enabled channels for commercialization, and proper framing of policies and regulations. Any new technology in food sector is often challenged by consumer's attitude and acceptance. Consumer psychology is also guided by social norms, risk perception, affordability, and other emotional and moral judgments. There is a high rejection level among consumers for a technology about which they lack proper knowledge. Debates on GMOs persisted for decades in different countries throughout the globe. The emerging technologies of fourth industrial revolution to attain sustainable development goals (SDGs) in the food sector necessitate great advancements in next-generation biotechnologies (GMO and animal cloning), nutrigenomics, rapid prototyping and customized diet, nanotechnology, advanced robotics, internet of things, artificial intelligence, machine learning, advanced biomaterials, and energy technologies. Some transformative changes are to be taken to achieve the desirable impact on food systems by 2030. The changes include eco-friendly technologies, conservation of natural resources, search for alternative sources for protein, reduce food wastage, development of food sensing technologies for close monitoring of food safety and quality, and advanced research in nutrigenetics so that personalized nutritional therapy helps to combat diseases like obesity and finally promote value chain linkages by mobile service delivery, real-time supply chain transparency and traceability, big data and advanced analytics for insurance, and block chain–enabled traceability [6, 27–30].

The historical development in food processing (high or low temperature), dated back to 1750, when William Cullen [27] introduced the first artificial cold process, that gradually developed to today's energy-efficient refrigeration and freezing systems was the onset of development. In early 1800s, Nicholas Appert used heat for preservation of food stuffs in response to Napoleon Bonaparte's requirements to feed the French army that later gave rise to modern energy-efficient thermal processing units. Alternatives to traditional thermal and chemical processing are being developed that are considered to be more sustainable. Application of high pressure and pulsed electric field in food processing is found to be energy-efficient and eco-friendly. Pulsed lights, ohmic heating, microwave irradiation, ultrasound, and cold atmospheric plasma are also other sustainable alternative processes [6, 27, 28].

In contrast to traditional approaches, the emerging trends are strongly focusing in going "green" the procedures and technologies associated with agriculture and food processing (Table 9.1). *Organic farming*, a typical cradle-to-cradle approach, fulfills the essence of "green technology". Organic farming utilizes a small area of land for cultivation of crops and, in many cases, shows a better performance than conventional farming. This farming procedure helps to maintain biodiversity and eco-friendly and to conserve natural resources. Application of "green" technologies in food processing and engineering has gained the research momentum.

9.5.1 Green Technologies in Food Processing

To start with the extraction procedure, "green technologies" recommend the reduced utilization of water, energy, and waste generation, and thus, the ongoing trends make use of techniques like solvents under negative pressure, pressure freezing-air drying, application of magnetic fields,

Table 9.1 Emerging "green" trends in food processing and technology [27].

Steps/procedures	Traditional methods	Emerging trends
Quality of raw materials	Physical inspection, visual quality	Non-destructive techniques of spectral imaging
Processing techniques	Chemical preservative (gentle processing); thermal methods (intensive processing)	Ozone processing, cold plasma technology (for gentle purpose), high pressure processing, pulsed electric field, cavitations technologies (intensive processing)
Food packaging	Plastics, glass, cardboards	Edible coatings and films; nanotechnology-based smart packaging
Freezing purpose	Liquid nitrogen and other refrigerants	Individual quick freezing (IQF), cells alive system
Food preparation	Use of ovens for cooking (as per items to be prepared)	Instant cooking items (2–3 minutes), ready to use packed foods have gained popularity
Storage and distribution	Metal silos and relevant containers for air, road, and water transport	Use of sensing devices, cold chain distribution have gained popularity
Food processing waste	Landfill, incineration	Focusing on three "R's": reduce, recycle, and reuse

high hydrostatic pressure–assisted freezing or thawing, and pulsed force applications. *Biopreservation* of foods recommends the use of bacteriocins, organic acids, and probiotics. The other emerging technologies for microbial control in food processing include electromagnetic wave heating (*microwave heating* and *radiofrequency heating*), electric and magnetic fields (*ohmic heating, inductive heating*, and *moderate electric field heating*), and non-thermal technologies (*pulsed electric field, high pressure processing, ionization radiation, ultraviolet radiation, high intensity pulsed light, cold plasma processing, ultrasound*, and *ozonization*). There is a need to move from carbon-based energy sources to air, water, and solar-based sources as they are natural, abundant, and environmentally benign. *Green drying* operation unit aims to reduce energy expenditure where heat is recovered from the exhaust gases and re-circulated in the drying operation. Exhaust gases after heat recovery can also be further scrubbed to remove greenhouse gases. Further, enzyme-assisted food production is in progress that aims at developing enzymes with superior activities to be used under mild processing conditions or is capable to resist extreme conditions of pH, temperature, and pressure [6, 27, 28].

9.5.2 Nanotechnology in Food Processing and Food Safety

Nano-technical approaches have great potential to sustainable food chains. It helps in the development of novel nutraceutical formulation. Noteworthy applications include nanoencapsulation and controlled delivery of bioactive, *viz.*, omega fatty acids, vitamins, plant secondary metabolites, flavoring agents, and coloring agents; sustained delivery of antimicrobials, smart sensors for improved food safety, superior nano-packaging for enhancing the shelf life of the products, etc. Nano-encapsulation of bioactive, nano-emulsions not only provides protection to sensitive ingredients but also improves bioavailability. Nanotechnology-based products play a vital role in food safety. Rapid detections of pathogens is essential for food safety and quality; a versatile miniaturized, automated device *"Lab-on-a chip"* helps in rapid detection of pathogens and mycotoxins; *nanosensors* help to detect pathogens like *Salmonella*. Adulteration and mislabeling of food stuffs pose a direct threat to public health. *DNA bar-coding technology* can help in determining the original identity of the source material by comparing short genetic mark ups in the product with a reference DNA sequence [6, 27].

9.5.3 CRISPR-Based Technologies

Clustered Regularly Interspaced Short Palindromic Repeats (CRISPR) and CRISPR-associated proteins have emerged as a novel technology that target DNA sequences using programmable RNAs, thus opening new horizons in genome editing. This technology plays a crucial role in prokaryotic adaptive immune system. CRISPR genome editing can benefit the entire agriculture industry. CRISPR is an emerging technique for creating targeted genetic diversity in a crop. Both CRISPR and Genetically Modified Organism technologies can ensure sustainable global future food security. Breeding crops of better quality and performance (sustainable through adverse climatic conditions, resistant to pathogens) account to future food security. CRISPR can revolutionize food science from farm to fork.

Its versatile applications include agricultural crop trait enhancement, livestock breeding, formulation of next-generation food products of improved quality and health potentiality, high resolution typing of food pathogens, and vaccination of starter cultures against phages [31, 32]. As a potential gene editing tool, CRISPR/CaS has revolutionized the field of agriculture. Global population explosion, natural disasters, and climatic effects are constantly challenging the future global food security. In order to ensure global food security, it is necessary to urgently introduce crop varieties highly resistant to abiotic and biotic stresses and with nearly doubled yield potentials. Hybridization and mutational breeding are the conventional techniques for improving crop varieties; transgenic techniques have been used for crop improvement and understanding the plant biology. Recently, sequence specific nucleases (SSNs) are finding extensive use in precise genome editing of the crop varieties. In genome editing technology, different nucleases, e.g., zinc finger nucleases (ZFNs), transcription activator like effector nucleases (TALENs), and CRISPR/CaS9 system, are used to directly cleave the target gene that is further repaired by either non-homologous end joining (NHE) or homology-directed recombination pathways (HDR). However, ZFNs and TALENs are limited due to tedious procedures, cost factor, long turn over time, and less reliability in comparison to the second-generation CRISPR/CaS9 system. This genome editing tool has great impact on plant biology and crop breeding. CRISPR/CaS9 system is successful in improving crop yield, developing resistance against abiotic stresses like draught and salinity and biotic stresses like development of host resistance against targeted pathogens [33–38]. Genome editing is successful in improving qualities of agricultural crops, incorporation of traits like fragrance, long-term storage capacity, improve content of omega fatty acids, modulate starch content, etc. The significant applications of CRISPR/CaS9 in agriculture (Table 9.2) not only improves crop quality or develop abiotic- and biotic-resistant varieties, but CRISPR/CaS-mediated gene knockout is successful in high quality hybrid variety breeding, domestication of wild varieties, and developing herbicide-resistant crop varieties by gene insertion and replacement. Other novel breakthroughs of the technology include base editing (essential for plant breeding and crop improvement) by the use of Cas9 nickase or dead Cas9 with base conversion ability, multiplex genome editing (targeting several genes with a single molecular construct), generation of high-throughput plant mutant libraries, and DNA-free genome editing that was possible by transfection of CRISPR/CaS9 ribonucleoproteins in the plant protoplast by particle bombardment methods [34, 39]. Other than agriculture, CRISPR has been implemented among animals in the cattle, chicken, and pigs. Use of nCas9 helped to

Table 9.2 CRISPR/CaS9 in improving traits in agricultural crops [34].

Name of the crops	Targeted genes	Targeted traits
Rice	OsPDS, OsMPK2, OsBADH2, OsMPK5, OsAOX1a, OsAOX1b, OsAOX1c, OsBEL	Various abiotic stress tolerance and disease resistance
	OsDERF1, OsPMS3, OsEPSPS, OsMSH1, OsMYB5	Drought tolerance
	OsMPK2, OsDEP1	Yield under stress
	ALS, EPSPS	Herbicide resistance
	SBEIIb	High amylose content
	OsERF922	Enhanced rice blast resistance
Wheat	TaMLOA1, TaMLOB1 and TaMLOD1	Resistance to powdery mildew
	TaGW2	Increase in seed size
	GW2	Increase in grain weight and protein content
Maize	Wx1	High amylopectin content
	TMSS	Thermosensitive male sterile
	ARGOS8	Drought stress tolerance
	ALS	Herbicide resistance
Mushroom	PPO	Anti-browning phenotype
Potato	ALS	Herbicide-resistant
	Wx1	High amylopectin content
Tomato	SP, SP5G, CLV3, WUS, GGP1	Tomato domestication
	SlAGL6	Parthenocarpy
	SlJAZ2	Bacterial speck resistance
Soybean	ALS	Herbicide-resistant
Flax	EPSPS	Herbicide-resistant

develop increased resistance to tuberculosis among cattle; genome editing in pigs to develop the "lean pigs" that showed lower risk of mortality, and removal of endogenous porcine retro viruses has a potential impact on human health; chicken eggs have been edited to remove a protein from egg white that are known to cause allergic reactions. Applications of CRISPR in microbiological domain are being highly explored. CRISPR's genetic engineering helps to distinguish among completely identical strains of bacteria, and CRISPR is being effectively used in bacterial strains, e.g., *Bacillus subtilis*, *Clostridium*, *E. coli*, and *Corynebacterium glutamicum* [34, 35, 37, 38].

CRISPR is a precise, impactful global technology. Monsanto has already acquired the commercial license for implementing the technology for agricultural use. DuPont in collaboration with Caribou Biosciences is growing CRISPR-edited corn and wheat, other products being non-browning mushrooms, drought-tolerant corn, virus-resistant pigs, etc. LifeEdit is using CRISPR to promote agribusiness, DuPont is using CRISPR for typing and phage protection in dairy strains, Novozymes is using CRISPR to edit fungi, and CRISPR is being repurposed as a antimicrobial agent with Type I exonuclease Cas3 against *Enterobacteriaceae*, *Pseudomonas aeruginosa*, and *Clostridium difficile*. Thus, some of the highlighted potentials of CRISPR in food industry include development of new products with consumer preferred traits, increasing shelf life of perishable foods, knock out antibiotic resistance among food pathogens and improve food safety [34, 35, 37, 38]. Though this technology is of immense potency, however, final success lies on the acceptance among the end users, i.e., consumer world. GMO products are results of transgenic modifications. It is likely that consumers may not distinguish CRISPR products from GMO; however, it is doubtful if CRISPR products can achieve full market potential in comparison to conventional food systems. Such consumer psychology may have a detrimental effect on future food security. Governmental involvement in developing appropriate regulatory status, fostering communications between manufacturers and consumers, and creating public awareness about the benefits of genome editing are essential for integrating transgene free crops into the society [40, 41].

9.5.4 Future Directives

The growing population demands greater protein consumption, and thus, there is a constant search for alternative protein sources. Excess consumption of animal protein will affect food security in long run. Utilizations of plant proteins can serve as an alternative source. Insect proteins can be a rich alternative source. Tissue engineering technology has potentially progressed to produce "cultured meat" and the products thereof. Nutridense, stable, portable, and multi-component food items, *viz.*, canned soups and beverages, can be developed, each serving of which can replace a meal. Such food systems can fight the emergency disasters, hunger, and nutritional imbalances. Nutrigenomics that studies the genetic variance that alters the response to food in an individual and how different types of foods affects the genes of individuals is helpful in treating different ailments by fulfilling a person's nutritional need. Here comes the role of "personalized nutrition" and "customized diet". These are "tailor made" or "designer foods" that are developed with a purpose to meet the individualistic needs depending on their taste and physical and health status. Mass customized food items are developed to meet the needs of a wider consumer community. Further, it necessitates in-depth understandings of human genome and subjects of nutrigenetics (*interactions between food and human body*) and nutrigenomics (*the effect of diet on gene activity*). It can be expected that in coming future designer or customized food will not only satisfy the health needs but also will serve the purpose of adjunct therapy in many chronic ailments [6, 42].

9.5.4.1 3D Food Printing and Mass Customization of Diet

The well-informed and health conscious consumer world demands for tasty, healthy, and easy to cook food items. This has put forth a challenge before food processing and food engineering industries. It has stirred the need to develop novel technologies. Emergence of personalized nutrition and developing customized food as per individual health needs and conditions, consumer categories, *viz.*, infants, growing child, pregnant ladies, lactating mothers, athletes, geriatric population, and great change in taste are some of the challenging issues in the domain of food science and engineering. Developing value added food items with successful incorporation of pharmacologically active molecules in vegetal and fruits and altering food matrix design by structural and chemical modifications for achieving enhanced shelf life and retention of thermo-labile and sensitive yet valuable food components have become an important topic of research and also important issues to be addressed [43]. Three-dimensional printing (3DP), which consists of a raft of technologies with wide applicability in engineering and biomedical domain, has also gained successful implementation in food engineering [44, 45]. 3D food printing has a significant role in personalized nutrition. This innovative technology which is also considered as "green" has a huge market potential. With a customized approach, the technology can tailor food items as per individualistic needs, deal with geometrically complex food architecture and thus modulate food functionality, quality, appearance with enhanced organoleptic acceptability, and nutrotherapeutic potentials, extend shelf life of the 3D printed food items, and develop mass customized diet [44–48]. This technology also known as additive manufacturing (AM) develops 3D physical objects based on a computer model created in the program for graphic engineering in the form of .STL files. In other cases of 3DP, materials like plastics, resins, polymers, metals, and composites are used; but in 3D food printing, the printable food materials or the dough consists of edible food materials, e.g., chocolate, vegetable smash, fruit blend, and essential food ingredients, and falls under three categories. Chocolate, pasta dough, butter jelly, hydrogel, powdered starch, or sugar form the first category of food grade printable materials. Rice, meat, fruits, vegetables, etc., are traditional non-printable materials that can be made printable or suitable for extrusion by adding hydrocolloids like xanthum gum or gelatin. Established sources of major food components that are replaced by alternative sources, e.g., insect powder, can serve as an immense protein source and thus replace traditional meat that are among the 3D printable food components, which comes under the third category [47, 49, 50]. Often, the food dough to be 3D printed may be a synergistic value additive combination of different food components or food-derived bioactive principles. While making such combinations, due considerations are to be given, keeping in mind their intrinsic properties and binding mechanisms during deposition of layers [43, 46]. Designing our food items with the state of culinary arts, mass customized diets with customized organoleptic profile and personalized nutrition are the main focus in 3D food printing [46, 48]. The versatility of 3D food printing helps to achieve aesthetic and functional customization simultaneously. Firstly, it is a green technology, being automized reduces human labor, with a personalized nutritional approach can treat several ailments and thus has contribution on human health and nutrition. The customized approach helps to achieve personalized taste, meal composition as per health needs, and individualistic nutritional requirements, to develop novel food textures, to create culinary intricate state of art in our food items, to extend shelf life of 3D printed food products, and to ease transportation in the most remote parts of the world or into space along with the facility of customized mass scale production. 3D food printing does not require costly setup and hence is economical in small quantity production. It is a complementing technology in food processing. The use of food printers on domestic scale may not be possible at this present juncture; but this automated customized food fabrication technology has revolutionized the food industry and is sure to bring a versatile change in domestic kitchen in future. Above all, bridging the food industry with the digital era is the groundbreaking application of this technology [45–47].

Regarding technical aspects, the methods involved in 3DP food printing include fused deposition manufacturing or hot melt extrusion, room temperature extrusion, selective laser sintering or hot air sintering, binder jetting, and inkjet printing.

In *fused deposition manufacturing* or *hot melt extrusion*, the material is extruded from the nozzle layer by layer. The food material to be printed is heated to a certain temperature to obtain a semi-molten state followed by quick and easy cooling on combination with the previous layer. This technique is mostly used for 3D chocolate printing. In *extrusion* method of food printing, firstly, a virtual 3D model is designed, and then with the help of slicing software, this virtual 3D model is translated into individual layer patterns and the machine code for printing (G-code generation) is generated. After uploading the codes into the printer, the next step is to choose the preferred food recipe, and then, the food printing starts. According to the layer patterns generated from the 3D model, the extruded material is dispensed either by moving the nozzle above the motorized stage or by moving the stage underneath the nozzle to form a layer. Each layer then binds with the previous layer on the stage to form the layer-by-layer 3D structure. In layer-by-layer 3D food printing, it is important that each layer must have sufficient strength to maintain its own weight as well as the weight of the other layers without significant deformation or change of shapes. Layers are often fused by spraying binder solution, laser heating, or hot air treatment. Printed foods must undergo a post-deposition cooking process like baking and frying [46, 47–49]. In the extrusion process of 3D food printing, the extruder is the key component that mixes, heats, cools, and shapes highly viscous raw ingredients (both solid and liquid ingredients). Pre-ground and conditioned food ingredients are fed into extruder and, with the aid of mechanical or thermal energy, are pressurized into viscoelastic fluid, texturized, and shaped into die located at the end of the extruder and transited from high pressure to low pressure. This digitally controlled robotic process builds layer-by-layer 3D food objects starting from loading with food ingredients, well-controlled pushing of the ingredients out of the nozzle, moving the treated food material stream in a pre-defined path, and eventually bonding the deposited layers to get the desired 3D food items [49].

Selective sintering technology is based on melting together the particles of powders layer by layer. Powders rich in sugar content are very suitable for the procedure. This technology is suitable for sugar and fat-based materials with relatively low melting point. The CandyFab applies a selective low velocity stream of hot air to sinter and melt a bed of sugar. Here, computer controlled laser irradiation is used for the purpose of sintering. A very thin layer of powder is first applied evenly on the bed and exposed to laser irradiation or hot air that melts and sinters the particles together. The substrate is lowered and the same powder layer applied over, melted, and sintered together as done in previous layer. The process is repeated till the total 3D printed food material is prepared [46, 47].

In *powder bed jetting* or *binder jetting*, each powder layer is evenly distributed across the fabrication platform and a binder solution is sprayed to bind the consecutive powder layers. The powder material is stabilized through water mist to minimize disturbance caused by binder dispensing. The same procedure is repeated to form layer-by-layer complex food structures. The unbounded powder at each side is recovered for further reutilizations. Often, sugar starch mixtures with flavored binders are used for designing complex customized sculptural cakes, etc. Though it is a fast fabrication procedure, however, high machine costs and rough surface finishing are the associated limitations [46, 47, 49].

The *Inkjet food printing technology* dispenses stream or droplet from syringe-type print head in a drop-on demand. In this method, the ejected stream or droplets fall under gravity, impact on the substrate, and dry through solvent evaporation. The drops can form a two and half dimensional digital image as decoration or surface fill. This technology finds applications in creating multilayered printable 3D edible food products such as cookies, cakes, or pastries. The De Grood Innovations' FoodJet Printer uses pneumatic membrane nozzle-jets to deposit selected material drops onto pizza bases, biscuits, and cupcakes [47].

Other than the availability of designer confectionary items, 3D chocolates and 3D food printing has taken a pivotal role in manufacturing 3D fruit-based snacks with further value additives like dried mushrooms, beans, and citrus fruit juices containing requisite amount of vitamins, minerals, and amino acids beneficial for children, mothers, and athletes [48]; blends of chosen fresh vegetables with fruit mixtures and other added ingredients like milk or honey that are really considered as superfoods with high health benefits are used to develop 3D printed smoothies [50]. Rapid advancements in 3D food printing has lead to the development of customized fibrous meat with customized protein, fat, amino acid profile, and sensory attributes [51]. Modernized food printers are using lasers and robotic arms to develop intricate decorative designs in food architecture. The culinary intricate state of art in our food items depends on the process and planning rather than people's skill. In this era of personalized medicine, personalized nutrition, and nutrotherapeutic approach, food engineering and food processing technologies are also being revolutionized. With the help of domestic scale 3D food printers, customized diet or individualistic health need based food can be printed, thus bringing a versatile change in domestic kitchen in future.

9.6 Regulations and Trade

Availability, accessibility, proper utilizations, and stability are the different dimensions of food security. Due to increased globalization of world food supply, many countries are depending heavily on food imports. In order to minimize food related hazards, coordinated inspection and monitoring is essential to enforce the food safety regulatory systems. In case of imported food, often competent authorities have no direct control over the production process of their commercial partners. It is essential to develop trade relationships, increased interactions between the importing and exporting countries, implementation of certification mechanisms, and greater oversight of the importers that will help to strengthen the quality of imported food. Both international standards and domestic regulations are to be enforced to ensure food safety. Food laws and regulations vary from country to country. To bring a harmony among different laws, Codex Alimentarius Commission was developed by joint effort of WHO and FAO. Countries are required to adopt international standards set by this commission for the purpose of food safety. Significant and continuous investments in research and development is required that will enhance agriculture, livestock, and aquaculture. A continuous research boosting is needed for developing innovative production techniques, improving crop and livestock genetics. Farmers and small holders are to be promoted for increasing production with limited land and water resources. Proper investments are to be done for arranging education and training programs so as to acquaint the farmers with the modern agri tech procedures. Private sector investments and public-private partnerships are essential for the development and implementation of novel technologies [6, 52–54].

9.7 Conclusion

The challenge to feed the expected 9 billion populations by 2050 needs the conservation of agro-ecosystem. Simultaneously nutritious and quality healthy diets are essential to ensure overall food security. The SDGs target "food for all" by 2030 and end hunger among poor and vulnerable population. By 2030, SDGs target to eliminate all forms of malnutrition, double agricultural productivity, promote value addition and diet diversification, open up market and trade opportunities, nonfarm employment, modulate trade restrictions and distortions, focus on extensive agricultural and food engineering research, maintain genetic diversity of seeds, livestock gene banks, and finally regulate the proper functioning of the food supply chain.

References

1. Boyle, K., *Feeding the future.* How innovation and shifting consumer preferences can help feed a growing planet, pp. 1–152, Citi GPS: Global perspectives and solutions, 2018, https://www.citibank.com/commercialbank/insights/assets/docs/2018/feeding-the-future.pdf
2. McCouch, S., Bramel, P., Buckler, E., Burke, J.M., Feeding the future. *Nature*, 499, 23–24, 2013.
3. Floros, J.D., Feeding the World Today and Tomorrow— A Look into Our Future Food System. *64th Annual Reciprocal Meat Conference*, pp. 1–6, 2011.
4. Labine-Romain, A., Terrill, D., Mizrahi, J., Smith, X., Yetsenga, R., Koh, S.B., Fraser-Jones, J., Matthijssen, V., Khanvilkar, D., Brilliant, V., Future of food. How technology and global trends are transforming the food industry, Deloitte, 2019, https://www2.deloitte.com/au/en/pages/economics/articles/future-of-food-uber-eats.html.
5. Searchinger, T., Waite, R., Ranganathan, J., Creating a sustainable food future. A menu of solutions to sustainably feed more than 9 billion people by 2050. World resources report 2013-14 : interim findings. Edition: Interim Findings Publisher: World Resources Institute, 1–144, 2014. https://www.researchgate.net/publication/280755107_Creating_a_sustainable_food_future_A_menu_of_solutions_to_sustainably_feed_more_than_9_billion_people_by_2050_World_resources_report_2013-14_interim_findings
6. Tian, J.J., Bryksa, B.C., Yada, R.Y., Feeding the world into the future – food and nutrition security: the role of food science and technology. *Front. Life Sci.*, 9, 3, 155–166, 2016.
7. Pingali, P.L., Aiyar, A., Abraham, M., Indian Food Systems towards 2050: Challenges and Opportunities. In: *Transforming Food Systems for a Rising India.* Palgrave Studies in Agricultural Economics and Food Policy, Palgrave Macmillan, Cham, 1–15, 2019, https://doi.org/10.1007/978-3-030-14409-8_1
8. Eating in 2030: trends and perspectives. Barilla center for Food and nutrition, pp. 1–52, 2012, https://www.barillacfn.com/m/publications/eating-in-2030-trends-and-perspectives.pdf.
9. Pingali, P., Aiyar, A., Abraham, M., Rahman, A., *Transforming Food Systems for a Rising India.* Palgrave Studies in Agricultural Economics and Food

Policy, Palgrave Macmillan, pp. 1–382, 2019. https://www.researchgate.net/publication/331500711_Transforming_Food_Systems_for_a_Rising_India
10. Kim, J.Y., Ending poverty and hunger by 2030. An agenda for the global food system, pp. 1–32, 2015.
11. Langley-Evans, S.C., Nutrition in early life and the programming of adult disease: a review. *J. Hum. Nutr. Diet.*, 28, 1, 1–14, 2015.
12. Reynolds, C.M., Gray, C., Li, M., Segovia, S.A., Vickers, M.H., Early Life Nutrition and Energy Balance Disorders in Offspring in Later Life. *Nutr.*, 7, 8090–8111, 2015.
13. Davies, P. S. W., Funder, J., Palmer, D. J., Sinn, J., Vickers, M. H., and Wall, C. R., Early life nutrition and the opportunity to influence long-term health: an Australasian perspective. *J. Dev. Orig. Health Dis.*, 7, 5, 1–9, 2016. https://doi.org/10.1017/S2040174415007989
14. Nordström, K., Coff, C., Jönsson, H., Nordenfelt, L., Görman, U., Food and health: individual, cultural, or scientific matters? *Genes Nutr.*, 8, 357–363, 2013.
15. Pietrobell, A., Agosti, M., the MeNu Group, Nutrition in the First 1000 Days: Ten Practices to Minimize Obesity Emerging from Published Science. *Int. J. Environ. Res. Public Health*, 14, 1491, 2017.
16. Ohlhorst, S.D., Russell, R., Bier, D., Klurfeld, D.M., Li, Z., Mein, J.R., Milner, J., Ross, A.C., Stover, P., Konopka, E., Nutrition research to affect food and a healthy life span. *J. Nutr.*, 143, 1349–1354, 2013.
17. Irmak, S.J., The Importance of Nutrition for Health and Society. *Food. Nutr. Health*, 3, 1, 1, 2020.
18. Lucas, A., Programming by early nutrition in man. *Ciba Found. Symp.*, 156, 38–50, 1991.
19. Niola, M., *Non tutto fa brodo, il Mulino*, Intersezioni, Bologna, 2012.
20. Fróna, D., Szenderák, J., Harangi-Rákos, M., The Challenge of Feeding the World. *Sustainability*, 11, 5816, 2019.
21. FAO and WHO. *Sustainable healthy diets – Guiding principles*, Rome, 2019. http://www.fao.org/3/ca6640en/ca6640en.pdf
22. United Nations system standing committee on nutrition. Sustainable Diets for Healthy People and a Healthy Planet. pp. 1–36. 2017.
23. Green, H., Broun, P., Cook, D., Cooper, K., Drewnowski, A., Pollard, D., Sweeney, G., Roulin, A., Healthy and sustainable diets for future generations. *J. Sci. Food Agric.*, 98, 9, 3219–3224.
24. Fanzo, J., Healthy and Sustainable Diets and Food Systems: the Key to Achieving Sustainable Development Goal 2? *Food Ethics*, 4, 159–174, 2019.
25. Lake, A., Sustainable development starts and ends with safe, healthy and well-educated children, United Nations Children's Fund (UNICEF), 1–22, 2013.
26. Yakovleva, N., Flynn, A., Green, K., Foster, C., Dewick, P., A Sustainability Perspective: innovations in the food system. *Joint 4S/EASST Conference 2004 "Public proofs – sciences, technology and democracy"*, Paris, 2004.

27. Knorr, D., Augustin, M.A., Tiwari, B., Advancing the Role of Food Processing for Improved Integration in Sustainable Food Chains. *Front. Nutr.*, 7, 34, 2020.
28. Boye, J.I. and Arcand, Y., Current Trends in Green Technologies in Food Production and Processing. *Food Eng. Rev.*, 5, 1–17, 2013.
29. Riley, W.W. and Hussain, M.A., Green food technology. *Asia Pacific J. Food Saf. Secur.*, 2, 4, 1–2, 2016.
30. *Global food security: Emerging technologies to 2040*. National Intelligence council report, pp. 1–56, 2012.
31. Selle, K. and Barrangou, R., CRISPR-Based Technologies and the Future of Food Science. *J. Food Sci.*, 80, 11, R2367–R2372, 2015.
32. Ma, X., Mau, M., Sharbel, T.F., Genome Editing for Global Food Security. *Trends Biotechnol.*, 36, 2, 123–127, 2018.
33. Arefin, P., Application of Crispr-Cas 9 in Food and Agriculture Science: A Narrative Review, A Review Article. *Op. Acc. J. Bio Sci. Res.*, 2, 2, 2020.
34. Adhikari, P. and Poudel, M., CRISPR-Cas9 in agriculture: Approaches, applications, future perspectives, and associated challenges. *MJHR*, 3, 1, 1–11, 2020.
35. Brandt, K. and Barrangou, R., Applications of CRISPR technologies across the food supply chain. *Annu. Rev. Food Sci. Technol.*, 10, 133–50, 2019.
36. Zhang, Y., Pribil, M., Palmgren, M., Gao, C., A CRISPR way for accelerating improvement of food crops. *Nat. Food*, 1, 200–205, 2020, https://doi.org/10.1038/s43016-020-0051-8.
37. Donohoue, P.D., Barrangou, R., May, A.P., Advances in Industrial Biotechnology Using CRISPR-Cas Systems. *Trends Biotechnol.*, 36, 2, 134–146, 2018.
38. Ismail, E., Gavahian, M., Marti-Quijal, F.J., Lorenzo, J.M., Khaneghah, A.M., Tsatsanis, C., Kampranis, S.C., Barba, F.J., The application of the CRISPR-Cas9 genome editing machinery in food and agricultural science: Current status, future perspectives, and associated challenges. *Biotechnol. Adv.*, 37, 410–421, 2019.
39. Khang, D.T., Potential application and current achievements of CRISPR/Cas in rice. *Ann. Biotechnol.*, 1, 1003, 2018.
40. Shew, A.M., Nalley, L.L., Snell, H.A., Jr., Rodolfo, M.N., Dixon, B.L., CRISPR versus GMOs: Public acceptance and valuation. *Glob. Food Sec.*, 19, 71–80, 2018.
41. Tetsuya, I. and Araki, M., Consumer acceptance of food crops developed by genome editing. *Plant Cell Rep.*, 35, 7, 1507–1518, 2016.
42. De, B., Bhandari, K., Goswami, T.K., Innovative Technologies in Tailoring designer food and personalized nutrition. *Nov. Tech. Nutr. Food Sci.*, 4, 2, 1–5, 2019.
43. Godoi, F.C., Prakash, S., Bhandari, B.R., 3d Printing Technologies Applied for Food Design: Status and Prospects. *J. Food Eng.*, 179, 44–54, 2016.

44. Katakam, P., Dey, B., Assaleh, H.F., Hwisa, N.T., Adiki, K.S., Chandu, R.B., Mitra, A., Top-Down and Bottom-Up Approaches in 3D Printing Technologies for Drug Delivery Challenges. *Crit. Rev. Ther. Drug*, 32, 1, 61–88, 2015.
45. Singh, P. and Raghav, A., 3D Food Printing: A Revolution in Food Technology. *ASNH*, 2, 2, 11–12, 2018.
46. Izdebska, J. and Żołek-Tryznowska, Z., 3D food printing – facts and future. *Agro Food Ind. Hi. Tech*, 27, 2, 33–37, 2016.
47. Sun, J., Peng, Z., Zhou, W., Fuh, J.Y.H., Hong, G.S., Chiu, A., A Review on 3D Printing for Customized Food Fabrication. *Proc. Manuf.*, 1, 308–319, 2015.
48. Derossi, A., Caporizzi, R., Azzollini, D., Severini, C., Application of 3D printing for customized food. A case on the development of a fruit-based snack for children. *J. Food Eng.*, 220, 65–75, 2018.
49. Sun, J., Zhou, W., Yan, L., Huang, D., Lin, L., Extrusion-based food printing for digitalized food design and nutrition control. *J. Food Eng.*, 220, 1–11, 2018.
50. Severini, C., Derossi, A., Ricci, I., Caporizzi, R., Fiore, A., Printing a blend of fruit and vegetables. New advances on critical variables and shelf life of 3D edible objects. *J. Food Eng.*, 220, 89–100, 2018.
51. Liu, C., Ho, C., Wang, J., The development of 3D food printer for printing fibrous meat materials. *IOP Conf. Series. Materials Science and Engineering*, vol. 284, p. 012019, 2017.
52. Walls, H., Baker, P., Chirwa, E., Hawkins, B., Food security, food safety & healthy nutrition: are they compatible? *Glob. Food Sec.*, 21, 69–71, 2019.
53. Wichelns, D., Achieving Water and Food Security in 2050: Outlook, Policies, and Investments. *Agri.*, 5, 188–220, 2015.
54. The future of food safety. *First FAO/WHO/AU International Food Safety Conference*, Addis, pp. 1–24, 2019.

10
Alternate Food Preservation Technology

Pratik S. Gaikwad[1*], Chayanika Sarma[2], Aditi Negi[3] and Akash Pare[4†]

[1]Department of Food Packaging and System Development
[2]Department of Food Biotechnology
[3]Department of Primary Processing, Storage and Handling
[4]Department of Academics and Human Resource Development, Indian Institute of Food Processing Technology, Ministry of Food Processing Industries, Govt. of India, Thanjavur, Tamil Nadu, India

Abstract

Food preservation is practiced since ancient civilizations to prevent food spoilage. Various conventional preservation techniques are practiced globally for extending the shelf life of food products', but all these techniques are not viable as they affect the nutritional value, sensory attributes and increases the chances of contamination. Alternative food preservation techniques have emerged to overcome these shortcomings that utilizes less energy compared to conventional treatments. It also meets the food industry's challenges, consumers; for fresh, safe, and improved quality of foods to enhance the production, distribution, and storage condition. Investigations on non-thermal and novel thermal methods are carried out to understand their impact on food products' quality. All the non-thermal technologies are not suitable for processing every kind of food. However, utilizing a hurdle approach is essential for preserving foods. Novel food preservation technologies are energy efficient in maintaining the structure, function, and nutritional value of foods and beverages. These technologies can be applied separately or collectively to optimize quality, microbial load, processing period, food products' enzyme activity, and applications that hold a promising future for sustainable food product development. Thus, this chapter mainly discusses alternative food preservation techniques for improving quality and shelf life of food products.

Keywords: Non-thermal, novel thermal, nutritional quality, irradiation, hybrid-hurdle

---

*Corresponding author: gaikwad.pratik45@gmail.com
†Corresponding author: akashpare@iifpt.edu.in

Mousumi Sen (ed.) Food Chemistry: The Role of Additives, Preservatives and Adulteration, (275–340) © 2022 Scrivener Publishing LLC

10.1 Introduction

The spoilage of food products is mainly caused by microorganisms, insects, rodents, and oxidation. Globally, one-third of human food is lost every year owing to the spoilage of food. The conventional preservation techniques such as salting, sugaring, smoking, pickling, drying, fermentation, canning, boiling, and cold heating storage were utilised to extend the shelf life of food products. Alternative food preservation technologies are advantageous over conventional technologies for beverages and solid foods as they can retain nutritional qualities, sensory attributes, lower the chances of cross-contamination, and increase the shelf life. The alternate preservation technologies mainly focus on protocol design, functions of product and mechanism of microbial inhibition [1].

New processing and preservation methods are evaluated based on their efficiency against food-borne pathogens and microorganisms. Thermal processing is the most utilized technique for preserving food products by controlling pathogenic microorganisms' growth. However, this technology has some disadvantages such as overheating, which causes textural damage, lowering nutritional value, changes in color, flavour and sensorial attributes [2].

There is an increasing demand of consumers towards safe and minimally processed food products with improved quality. Therefore, emerging technologies such as non-thermal and novel-thermal technologies have been developed and evaluated as alternate food preservation technologies that produce safe and shelf-stable foods with few alterations in sensory and nutritional profiles [3].

Nowadays, the operations of heat treatments are eliminated from most of the processing techniques, as various new treatments like novel processing, cold pasteurization, non-thermal processing, hybrid-hurdle technology, etc., are introduced. Generally, non-thermal techniques have third-generation processing alternatives aimed to remove heat treatment from various food products [3]. The non-thermal and novel thermal preservation technologies have a precise application for various types of processed food.

The food process industries are often chooses new technology effective against microorganisms, less reactive to food, and consume less energy [1]. Various processing techniques are trending such as ohmic heating, microwave heating, infrared heating, radio frequency heating, high hydrostatic pressure, ultrasound, pulsed electric fields, irradiation, cold plasma, and membrane technology. The industrial application of new techniques relies on the properties of food products and process formulation. This chapter describes both non-thermal and novel-thermal preservation technologies that embrace food science and technology. The main principle behind each preservation technology, the application of technology in food, factors influencing the technology, their impact on food product quality, strengths, limitations along with future scope is also discussed in this chapter and focused on illustrating an alternate food preservation technologies over thermal preservation technologies.

10.2 Non-Thermal Preservation Technique

Non-thermal processing refers to indirect-thermal-energy-input processing, which enhances thermal energy within the food product [3]. Non-thermal processing technologies aim to develop efficient food processing techniques to eliminate microbial growth and improve end-product quality. These technologies are effective to operate at ambient temperature. Non-thermal processing utilized in food processing is summarized below.

10.2.1 Packaging Technology

The packaging is a preservation method that acts as a barrier to food from moisture, light, and external gases to control or extend packaged food products' shelf life. The role of packaging material in food packaging is to sustain or prevent the quality of packaged food products from external deteriorative influences and provide safety until the consumption of food products [4]. The primary function of packaging is to protect food, provide information, promote the product, and promote efficient distribution until end use [5]. There may be chances of food contamination at the time of distribution due to physical, chemical, and biological agents.

The packaging materials utilized for food products are flexible, semi-rigid, and rigid. Active packaging is a new approach to the preservation of food products. Active packaging includes scavenging substances, antimicrobial action, control atmosphere, and intelligent functions to enhance packaged food products' safety and shelf life until the end use [4]. An active packaging system can control the package's internal atmosphere for extending the shelf life of packaged food products. Modified atmospheric packaging (MAP) is another approach for the preservation of food products. MAP can modify or control the environment by changing gas composition in closed packages to extend shelf life and reduce oxidation and thus stops spoilage of food products (Figure 10.1).

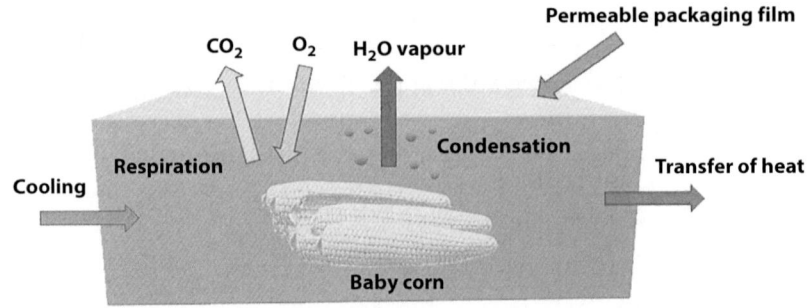

Figure 10.1 Overview of MAP packaged minimal processed fresh product.

The atmosphere of a package can be modified by active and passive MAP. Food Drug Administration (FDA) defines active MAP as "the movement of gases present in the package that can be replaced with the desired combination of gases". Active MAP uses scavengers such as oxygen (O_2), carbon dioxide (CO_2), ethylene, moisture absorbent, and ethanol emitter. On the other hand, passive MAP as "the product is stored in the selected package and the desired environment produces naturally as a significant source of product respiration and diffusion of gases through the package" [6, 7].

Nowadays, MAP is practically utilized commercially for many food products like fresh fruit and vegetables, meat and poultry, dairy, seafood, and baked foods [8]. The amount of different combinations of gasses for MAP depends on packaging material, different product categories, and storage temperature. The MAP package consists of a different combination of gases like O_2, CO_2, and nitrogen (N_2). The concentration of MAP gases with different food products is illustrated in Table 10.1. Absence or lower the amount of O_2 concentration in the package atmosphere helps to slow down the deteriorative reaction (bacterial growth, browning, and oxidation of lipid) and respiration rate; however, the higher concentration of CO_2 helps extend the shelf life of packaged food products [8, 10]. However, O_2 concentration will be higher in MAP stored red meat products to decrease myoglobin's oxidation, which retain the desirable red color of meat [11]. In MAP, most of the products eliminate or reduces the concentration of O_2 from atmosphere, except fresh or minimally processed fruit, vegetables and particular meat products [8]. CO_2 has a strong bacteriostatic impact on gram-negative bacteria and prevents the growth of some enzymes. CO_2 concentration helps to lower the pH, which increases the acidity of the product [12]. The solubility of CO_2 in water is 28 times higher than O_2 at 20°C [8].

Table 10.1 Gas composition of MAP with different food products.

Sr. no.	Food product	Gas combination			Type of package
		O$_2$ (%)	CO$_2$ (%)	N$_2$ (%)	
1.	Raw red meat	65–70	30	0	Tray sealed and thermoformed package
2.	Raw offal	80	20	0	
3.	Raw poultry with skin	0	30–100	70	
4.	Raw poultry without skin	70	20–30	0–10	
5.	Raw low fat white fish and seafood	30	40–70	30	
6.	Raw high fat oily fish and seafood	0	40–70	30–60	
7.	Shellfish, crustaceans, and molluscs	30	40–70	30	
8.	Ready meals	0	30–50	50–70	
9.	Cooked and dressed vegetable products	0	30–50	50–70	
10.	Cooked, cured, and processed meat products	0	30–50	50–70	Tray sealed, thermoformed, flow package or can

(*Continued*)

Table 10.1 Gas composition of MAP with different food products. (*Continued*)

Sr. no.	Food product	Gas combination			Type of package
		O_2 (%)	CO_2 (%)	N_2 (%)	
11.	Cooked, cured, and processed seafood products	0	30–70	30–70	
12.	Cooked, cured, and processed poultry products	0	30–70	30–70	
13.	Convenience food products	0	30–50	50–70	
14.	Fresh pasta products	0	50	50	
15.	Bakery Products	0	50–70	30–50	
16.	Hard cheese	0	30–100	0–70	
17.	Soft cheese	0	10–40	60–90	
18.	Sliced cheese	0	30–40	60–70	
19.	Cream cheese	0	100	0	
20.	Yoghurt	0	0–30	70–100	
21.	Fresh fruit and vegetables	5	5	90	
22.	Snacks and crispy products	0	0	100	

(*Continued*)

Table 10.1 Gas composition of MAP with different food products. (*Continued*)

Sr. no.	Food product	Gas combination			Type of package
		O_2 (%)	CO_2 (%)	N_2 (%)	
23.	Milk powder	0	0–20	80–100	Carbonated gable, glass or plastic bottles and aluminium or steel cans
24.	Dried food products	0	0	100	
25.	Liquid food and beverages	0	0	100	
26	Carbonated soft drink	0	100	0	

Source: [9].

N_2 is an inert and tasteless gas utilized in MAP to replace O_2 or as a filler to prevent the breakdown of the package from high concentration of CO2, and it also prevents packaged food from rancidity for extending the shelf life [8, 13]. For the replacement of N_2, MAP also utilizes noble gases including, Argon (Ar), Helium (He), Xenon (Xe), and Neon (Ne) for preserving and extending the shelf life of perishable food products. The recent studies related to MAP concentration with different gas compositions are illustrated in Table 10.2.

10.2.1.1 Challenges and Future Scope of MAP Processing

The significant challenges during MAP are to ensure the quality and safety of the packaged food products. The use of hurdle technology with the combination of MAP will provide high quality and safety of packaged food at a reasonable rate. Food preservation techniques including high hydrostatic pressure (HHP), irradiation, fermentation, ultrasound, pulsed electric light, microwave (MW) heating, and freezing may be used before or after MAP of food products to improve the quality and shelf life of the product. The effect of various MAP gases on biological, chemical, and physical properties needs to be understood for various food products.

10.2.2 Ozone (O_3) Treatment

The increasing expectations of consumers toward high-quality, safe, healthier, and less processed foods are the main challenges for food producers and industry. O_3 treatment is a promising technology that improves food products' quality and safety to feel the consumer's demand fully. Nowadays, O_3 finds various applications for disinfecting the surfaces, plant equipment, production area, fumigation, and sterilization. It is also utilized for preservation and shelf life extension of food products [35].

O_3 is a non-thermal food preservation process that inhibits microorganisms' growth from foodstuffs' surfaces (fruits, vegetables, meat, spices, herbs, beverages, etc.) in aqueous or gaseous phases to enhance the quality and safety of food products [36–38]. The required energy input for O_3 treatment is lower compared to another non-thermal processes, including MW, radiation, and thermal treatment [39]. For decades O_3 is used for purifying portable water. Now O_3 has been declared (US FDA, on 26th June 2001) as Generally Recognized as Safe (GRAS) substance for disinfecting food contact products and other purposes, including medical applications, spas, swimming pools, cooling towers, marine, municipal water and sewage [37, 40]. As a disinfectant, O_3 is about 50% stronger than chlorine over

ALTERNATE FOOD PRESERVATION TECHNOLOGY 283

Table 10.2 Application of non-thermal technologies for different food products.

Application	Purpose	Treatment	Results	References
MAP treatment				
Fig fruit	Evaluate the effect of the respiration rate of stored fig fruit at different temperatures (5–35°C)	O_2 of 0.21 cm³, CO_2 of 0.00035 cm³	Fig stored at 35°C, decreased O_2 consumption (43.3%) and CO_2 evolution (42.8%) rate with increase in storage duration. MAP helps to the extend shelf life of fig upto 14 h with equilibrium concentration of O_2 (0.0878) and CO_2 (0.0971).	[14]
Cantaloupe fruits	Determine the effect of MAP with different concentration on shelf life of cantaloupe fruits	80% N_2 + 10% O_2 + 10% CO_2 at 12°C	Cantaloupe fruits packed with different combination of MAP gases significantly affect the physico-chemical, microbial, and sensory attributes, and showed stable shelf life up to 14 days	[15]

(Continued)

Table 10.2 Application of non-thermal technologies for different food products. (*Continued*)

Application	Purpose	Treatment	Results	References
Kinnow (*Citrus nobilis* Lour x *C. deliciosa Tenora*) mandarin	Investigate the efficiency of different MAP films and quality of Kinnow mandarin stored in MAP for 60 days at refrigerated condition	PP-film (25 μ) with pinholes	MAP treatment help to maintain the quality parameters like ascorbic acid, soluble solid content, titratable acidity, sensory attributes of Kinnow mandarin. Also, slow down the activity of cellulase and pectin methylesterase enzyme	[16]
O_3 treatment				
Green bell pepper (*Capsicum annuum*)	Standardize the effect of O_3 concentration and application time for enhancing shelf life of minimally processed Indian green capsicum	1–3 mg/L for 1–5 min	O_3 treatment given for 5 min stored in polypropylene packages kept at 5 ± 0.5°C, 85 ± 5% RH showed prolonged shelf life upto 14 days as compared to control samples in which shelf life was upto 8 days and maintained the quality of minimally processed capsicum	[17]

(*Continued*)

Table 10.2 Application of non-thermal technologies for different food products. (*Continued*)

Application	Purpose	Treatment	Results	References
Whole maize flour	Evaluate the use of O_3 for reducing the ZEN contamination of whole maize flour (WMF)	51.5 mg L^{-1} for 60 min	O_3 is an effective for reducing ZEN contamination in WMF. However, it changes the fatty acid profile, pasting properties and peroxide value, affecting functional and technological aspects of WMF	[18]
Cantaloupe melon juice	Determine the impact of O_3 exposure on the quality and safety attributes of Cantaloupe melon juice	7.7 g/L for 10 min	O_3 exposure on Cantaloupe melon juice reduced the growth of *L. innocua*. Although, O_3 deducted intrinsic microflora loads and retained concentration of vitamin C upto 68%	[19]
HPP treatment				
Atemoya puree	Determine the impact of HPP on the glycaemic index (GI) of atemoya puree (AP) in rats	600 MPa for 15 min	HPP treatment delayed in increase of postprandial blood glucose levels (76.1%) and decreased in GI of AP to 49.8 compared to 65.4 in the control. HPP treatment increased dietary fibre content and viscosity of the puree but did not show any significant effect on glucose and pectin contents and decreased GI of AP	[20]

(*Continued*)

Table 10.2 Application of non-thermal technologies for different food products. (Continued)

Application	Purpose	Treatment	Results	References
Chicken meat	Determine the impact of HPP and *trans* cinnamaldehyde (tCinn, 0.016–0.084%, w/w) on reducing *Salmonella* and *Listeria monocytogenes*	266–434 MPa for 3.3–11.7 min	Chicken meat treated with tCinn did not show any impact but, combination of HPP (375 MPa, 8.0–8.5 min) and tCinn (0.05–0.07%) reduced five-log cycle of *Salmonella* and *L. monocytogenes*	[21]
Gooseberry juice	Determine the effect of thermal assisted HPP (THPP) on antioxidant capacity, bioactive functional compounds and physicochemical properties of Indian gooseberry juice	200–500 MPa, 1 s–20 min and 30–60°C	THPP did not affect the pH, titrable acidity and TSS of gooseberry juice but, increased L^*, antioxidant capacity and total phenolic content, and retain ascorbic acid upto 85%	[22]
Ultrasound treatment				
Bovine milk	Investigate the effect of ultrasonic treatment on degradation of parathion methyl (PM) in bovine milk	25 kHz, 900 W	Ultrasonic intensity showed significant effect on degradation of PM in bovine milk up to 97.10%	[23]

(Continued)

Table 10.2 Application of non-thermal technologies for different food products. (Continued)

Application	Purpose	Treatment	Results	References
Chickpea cooking water (aquafaba)	Investigate the effect of ultrasound on emulsifying properties and foaming capacity in chickpea cooking water	20 kHz, 50–100% power capacity for 10–30 min	Ultrasound treatment increased color, foaming expansion (259–548%), stability, texture, and viscosity and did not harm the density and protein solubility of chickpea cooking water	[24]
Tomato fruits	Evaluate the effect of ultrasound treatment on antioxidant capacity and secondary metabolites in mature tomato fruits	25 kHz, acoustic power density of 26 W/L	Ultrasound treatment increased ABTS (11.55%) and DPPH (22.69%) antioxidant activity. Also, increased accumulation of secondary metabolites, including ascorbic acid, carotenoids, lycopene, and total phenolic during storage of tomato fruits	[25]
PEF treatment				
Beef semitendinosus muscle (BSM)	Investigate the effects of PEF treatment on the properties of beef BSM	1–2 kV/cm	PEF treatment significantly decreased in chewiness, cutting force (35%), hardness, redness value (a*), and myoglobin content of BSM	[26]

(Continued)

Table 10.2 Application of non-thermal technologies for different food products. (*Continued*)

Application	Purpose	Treatment	Results	References
Strawberries and red bell peppers	Determine the effect of PEF pre-treated on physical properties of strawberries and red bell peppers	1 kV/cm, 0.3 and 6.0 kJ/kg for 2.0–28.6 ms	Pre-treated matrices with PEF significantly increased mechanical properties and rehydration capacity upto 50%, and reduced firmness upto 60% for both matrices	[27]
Bell pepper	Investigate the quality of PEF treated vegetable prepared bell pepper juice and powder obtained from this juice	3 kJ/kg	Bell pepper treated with PEF significantly increased vitamin C and lowered the total phenolic content of prepared juice. The powder obtained from the juice, retained vitamin C and carotenoids, and lower in total phenolic content	[28]
CP treatment				
Rice	Investigate the properties of plasma treated rice fortified with iron and ascorbic acid	20 kV for 10–15 min	Plasma treatment significantly improved cooking time, hydrophilicity, and surface energy of fortified rice. The iron and ascorbic acid for 100 g of fortified rice was 862.93 mg and 1,398.27 mg	[29]

(*Continued*)

Table 10.2 Application of non-thermal technologies for different food products. (Continued)

Application	Purpose	Treatment	Results	References
Fenugreek seed	Investigate the effect of high voltage atmospheric cold plasma (HVACP) on galactomannan extraction from dry fenugreek seeds	80 kV for 30 min	HVACP treatment significantly increased extraction yield of galactomannan (122%) and soaked and dry fenugreek seeds (67%). Galactomannan raised viscosity, swelling index and water-binding capacity, as well as lower melting enthalpy	[30]
OMF treatment				
Beef	Determine the effect of OMF on ice nucleation and iron-oxide nanoparticle diffusion (IND) through supercooling of beef samples	-	OMF treatment at −10°C significantly increased in IND (36%) and stability of distilled water (32%). OMF treatment with supercooled meat prolonged shelf life for a week at −4°C	[31]
Deionized water	Determine the effect of OMF on freezing rate, supercooling, and phase transition time of deionized water during freezing	0 to 12 mT (50 Hz)	OMF treatment with different strength did not showed any significant difference between freezing rate and phase transition time, as well as increased occurrence of supercooling (33%) compared to without OMF	[32]

(Continued)

Table 10.2 Application of non-thermal technologies for different food products. (*Continued*)

Application	Purpose	Treatment	Results	References
Membrane filtration technology				
Fresh milk	To prepare whey from bovine and buffalo fresh milk with the help of ultrafiltration membrane technology	–	Ultrafiltration process showed separation, concentration and purification of whey protein. Ultrafiltered milk showed low ratio moisture content, lactose, and salt	[33]
Yerba mate extract	To increase the chemical and physical stabilities of the yerba mate extract using clarification	microfiltration (0.1 Åμm), ultrafiltration 1 (30–80 kDa), and ultrafiltration 2 (4 kDa)	No significant difference observed between cured and clarified extract. Ultrafiltration process 1 (vinylidene polychloride) produced extracts with 18% loss in phenolic compounds. Reverse osmosis on yerba mate extracts decreased turbidity upto 30 days	[34]

a wide range of microorganisms in the food industry due to its significant oxidative and antibacterial properties [37, 41]. Even at low concentration, O_3 shows strong antiseptic capacity.

10.2.2.1 Properties of O_3

O_3 is a colorless to pale blue gas having a pungent aroma, and it is an allotropic form of O_2 [35]. It has a strong oxidizing property, and the disinfecting property is formed due to ultraviolet (UV) irradiation [42]. O_3 molecules are formed during high-energy inputs that separate the oxygen molecules into singlet oxygen, reacting with the available oxygen molecules to produce O_3 [35, 43]. O_3 gas remains unstable and decomposes rapidly at room temperature, and it has a longer half-life in the gaseous state than in the aqueous state. The half-life of O_3 in distilled water is 20–30 min at 20°C, and an increase in the temperature decreases the half-life of O_3 [44]. The positive and negative destruction rate of O_3 is well correlated with the temperature and purity of water. O_3 quickly gets decompose in hot water, and it is easily soluble in low-temperature water. The solubility rate of O_3 depends on product pH, temperature, pressure, purity of water, O_3 contact time, size of O_3 bubbles, and flow rate [35, 38, 43]. O_3 has three oxygen atoms arranged chemically in the chain with a low molecular mass of 48 g/mol, boiling point −111.9°C and melting point 192.57°C at 1 atm pressure [38]. O_3 density (2.14 kg m^{-3}) is 1.5 times higher than the air density (1.43 kg m^{-3}). The capacity of O_3 oxidation (−2.07 V) is higher compared to chlorine (−1.36 V). O_3 is generated at the time when high-energy sources contact with atmospheric air. The production of O_3 is mainly accomplished from UV radiation, electrolysis, and corona discharge method. The corona discharge process consumes high electricity.

10.2.2.2 Principle of O_3 Generation

The oxygen molecules traveled through two electrodes of oxygen divided from a dielectric substance having a negative charge will bond with another oxygen molecule to generate O_3 (Figure 10.2). UV meter is utilized to measure the concentration of O_3 gas. The UV radiation process is based on the generation of O_3 during exposure of O_2 to UV wavelength ranges from 140 to 190 nm. The recent studies related to O_3 with different compositions are illustrated in Table 10.2.

292 Food Chemistry

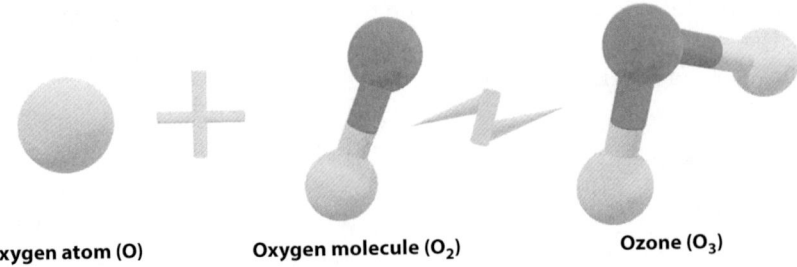

Figure 10.2 Generation of O_3.

10.2.2.3 Challenges and Future Scope of O_3 Processing

The main drawback of this process is the toxicity of various food products. To overcome this difficulty, there is a need for research to monitor the doses of O_3 with different food products.

10.2.3 High Hydrostatic Pressure Treatment

HPP is a non-thermal cold pasteurization technique adopted by food industries due to its numerous advantages over conventional preservation methods [45]. HPP is very well known as ultra-high pressure processing or pascalization in which high pressure (100 to 1,000 MPa) is subjected to the food product for a specific time period to inactivate the growth of harmful microorganism enzymes. HPP process may operate at a temperature below 0°C to more than 100°C in which food product treated at a lower temperature is more sensitive to heat injury, whereas higher temperature eliminates the growth of microbial spores. HPP process does not produce any adverse effect on the nutritional profile and organoleptic properties of food products, which is the key advantage over thermal processing [45].

The high-pressure system consists of a pressure vessel, pressure-transmitting fluid, hydraulic compressors and heating/cooling unit, and material handling system. There are two types of high-pressure system: a batch type system and the semi-continuous process system. In a batch type system, solid foods or bulk foods are processed; they are packed and sealed before processing. In the semi-continuous type system, liquid food products like fruit juices are treated [46, 47]. In HPP, uniform pressure will be applied to all directions, which will not squeeze out the packaged food product (Figure 10.3); this is the key advantage of HPP over thermal processing [48].

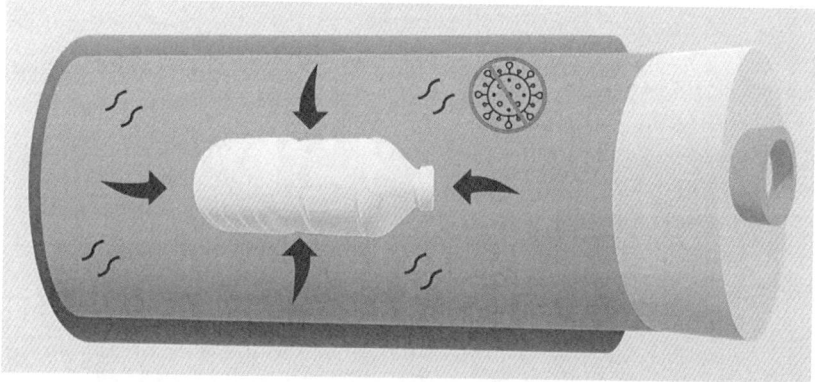

Figure 10.3 HPP processed food product.

High-pressure processing is usually accompanied by a moderate temperature increase called adiabatic heating, which is dependent on the composition of the food product being processed. The water temperature in the food increases by 3°C per 100 MPa, whereas the fats and oils' temperature increases about 8–9°C per 100 MPa [46, 48, 49].

The result of HPP on microorganisms has been widely investigated and found that the microorganisms vary in their response to the HPP, and indeed, there can be vast high-pressure sensitivity among strains and bacterial species [50, 51]. As compared to vegetative cells, endospores tend to be extremely HPP, requiring a combination of high-pressure treatment at a pressure exceeding 1,000 MPa and heat treatment with a temperature of more than 80°C [52, 53]. Yeasts and molds are inactivated at 300–400 MPa at room temperature for a few minutes because yeasts and molds are sensitive to HPP. A wide range of sensitivity of HPP is found in viruses. The recent studies related to HPP treatment O_3 with different level of pressure are illustrated in Table 10.2.

The most common packaging material utilized during HPP is polypropylene (PP), polyester tubes, polyethylene (PE) pouches, and nylon cast PP pouches due to their unique properties like reversible, flexibility, and resiliency of packaging material [45, 48, 49]. While in the HPP process, the utilized packaging co-polymers are biaxially oriented PP film (BOPP), ethylene–vinyl alcohol (EVOH), high-density PE (HDPE), low-density PE (LDPE), linear low-density PE (LLDPE), PE, PE terephthalate (PET), PP, and polytetrafluoroethylene (PTFE).

10.2.3.1 Principles of HPP Treatment

HPP treatment operated on four main principles under high pressure given as follow:

a. Le Chatelier's principle: It states that a chemical reaction or physical process under an equilibrium system associated with a decrease in volume is enhanced by pressure and vice versa [54].
b. Isostatic (Pascal's) principle: It states that the applied pressure is transmitted uniformly throughout every direction and later returns to its initial shape when pressure is realized [54].
c. Principle of microscopic ordering: Microscopic ordering states that at a perpetual temperature, the degree of ordering of the molecules of a given substance increases with increasing pressure. Higher pressure and temperature utilize the antagonistic forces on molecules and chemical reactions [54, 55].
d. Transition state theory: According to transition state theory, the volume of the molar in the intermediate state changes from its reacting element, the changes in pressure increase or decrease the reacting velocity, and according to transitional state, it is more or less voluminous [45].

10.2.3.2 HPP Time

a. Come-up time is the time required for the pressure of the treated sample to increase the atmospheric pressure (P1) to a processing pressure (P2). The pressure comes uptime is the time interval between P1 and P2, and it depends on product rate of compression, fluid rate of compression, the power supply of pressure pump, and target pressure. The commercial HPP equipment may require 1 to 3 min of pressure come up time [45, 48, 49].
b. Pressure holding time is the time period between the come up and decompression time and its time period of 3 to 10 min.
c. Decompression time is the time period between process pressure and atmospheric pressure (0.1 MPa) to carry the

product. High-pressure equipment requires a few seconds for depressurizing.

10.2.3.3 Challenges and Future Scope of HPP Treatment

HPP's hurdle effect with other non-thermal processing, including irradiation and ultrasound, has shown a synergistic effect on microbial lethality. There is a need to change the pressure transmitting fluid, improving the textural property of food products.

10.2.4 Ultrasound Treatment

Ultrasound is an effective non-thermal preservation technique utilized in food process industries, like crystallization, extraction, extrusion, fermentation, filtration, heat transfer, homogenization, and inactivation of microbes and enzymes [56]. It is employed with high-pressure sound waves ranging from 20 kHz to 10 MHz (Figure 10.4). The frequencies are divided into high frequency, low intensity, and diagnostic ultrasound. High frequency is also known as "power ultrasound" utilized in food industries to inactivate microbial growth and operate at the lower frequency range from 16 to 100 kHz. Low-intensity ultrasound operated at high frequency between 100 kHz and 1 MHz for analytical, non-invasive imaging, and sensing. Diagnostic ultrasound utilizes frequency from 1 to 10 MHz for medical imaging [56, 57].

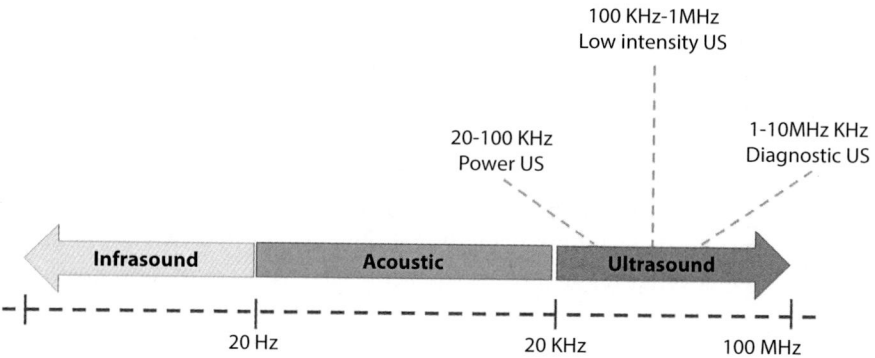

Figure 10.4 Different frequencies of ultrasound.

10.2.4.1 Principle of Ultrasound Treatment

Ultrasound technique generates vibration of high frequency of electric field due to molecules' motion in the medium through which the waves are induced and create regions of rarefaction (negative pressure) and compression (positive pressure). At higher frequencies, the negative pressure causes tiny bubbles with subsequent increasing temperature and produces cavitation due to the tensioning effect. At the compression cycle, bubble sizes reach a critical size, and consequent collapse of bubbles occurs at a higher pressure and temperature [58, 59]. The recent studies related to the ultrasound process with different frequencies are illustrated in Table 10.2.

10.2.4.2 Challenges and Future Scope of Ultrasound Treatment

Nowadays, in food process industries, significant attention toward high-powered ultrasound technique is considered for the preservation of food products owing to its considerable properties, like low operating and maintenance charge, high product output, shorter processing duration, and inactive microbial growth at lower temperature and improves flavor, taste, texture, and color.

10.2.5 Pulsed Electric Field Treatment

Pulsed electric field (PEF) is one of the emerging non-thermal food preservation techniques which retain color, flavor, and texture of treated food product by eliminating microbial population present in liquid and semi-liquid food products [56]. In PFE treatment, most vegetative cells are eliminated, but bacterial spores are more resistant and can be eliminated under certain circumstances. The effectiveness of PEF can be affected by the chemical and physical properties of food, products, processing factors and different types of microorganisms [60].

The significant components of PEF equipment are as follows: 1) power supply of high voltage; 2) capacitor for energy storage; 3) switches for transferring electrical energy; 4) treatment chamber containing two conductive electrodes; 5) cooling refrigerators; 6) oscilloscope to control and monitor current, electric field, and voltage; and 7) packaging system.

PEF utilizes high voltage pulses for a brief span of time; the generated energy is in the form of electric energy stored in the capacitor. While treating food products in PFE, numerous phenomenons, like ohmic heating (OH), electrolysis, electroporation, arc discharge, and disruption of the cell membrane, eliminate microorganisms' growth [60, 61]. According to Amit *et al.* [62], the effectiveness of PEF will be more effective in eliminating the growth of gram-negative bacteria rather than gram-positive bacteria. However, at low-voltage pulses, the plant cell can survive, and an increase in voltage pulses up to 15 kV/cm killed the cells. The high voltage pulses up to 30 kV/cm may be applied for inactivating the growth of bacteria and fungi. According to Di Benedetto *et al.* [56], the high voltage pulses are applied at equal to ambient temperature in oscillatory, bipolar, square wave, or exponentially decaying.

10.2.5.1 Principle of PEF Treatment

In PFE, the treating food product is placed between two conductive electrodes placed on a nonconductive substance and high voltage pulses of 10–80 kV/cm are applied for few microseconds (µs) to serval seconds to induce microbial inactivation (Figure 10.5). Further, treated food products are packed aseptically and stored in refrigerated conditions [56, 60, 62]. The PEF technique has been effectively utilized for milk, juice, soup, liquid, egg, and yogurt. This technology's limitation is that it is utilized for food products without air bubbles while treating the gas bubble burns and produces some unwanted carcinogenic materials. The recent studies related to PFE technology with different voltage are illustrated in Table 10.2.

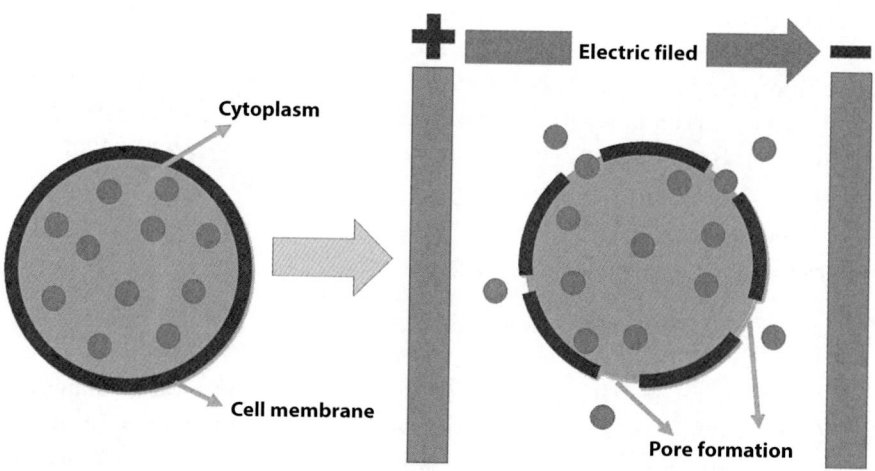

Figure 10.5 Operating mechanism of PEF.

10.2.5.2 Challenges and Future Scope of PEF Treatment

This technique's main drawback is that it is not suitable for solid food products due to the high electric field applied to solid food products, which may change its structure. The effect of PFE with different non-thermal processing will be beneficial to eliminate the microbial growth from food products and help extend the storage period of treated food products.

10.2.6 Cold Plasma Treatment

Cold plasma (CP) processing is another emerging non-thermal technology that has expanded its popularity in the agro-food process industries for extending the shelf life and safety of agricultural food products by inactivating microbial growth. The application of CP in the food industry is presented in Figure 10.6. It is also utilized in medicine to sanitize delicate, polymer and electronic substances for adhesion and printing, improving glass and paper properties, better finishing quality of products, and extending the surface energy of substances in the textile industry [2, 63].

Plasma technology discharges bright fluorescent light, which is considered the fourth state of matter, next to solids, liquids, and gases. The method of generating plasma can be categorized into two classes: thermal plasma and low-temperature plasma. Thermal plasma is an equilibrium plasma (20,000 K) comprised of electrons, gas molecules, and ions. Low-temperature plasma sub-divided into quasi-equilibrium (373–423 K) and non-equilibrium (<333 K) plasma. Non-equilibrium plasma is also referred to as CP, cold atmospheric plasma, and non-thermal plasma [64]. CP is a mixture of ionized gas containing reactive species, such as positive and negative ions, electrons, free radicals, photons, and gas molecules at ambient temperature [2, 36]. CP technology is eco-friendly, utilizing green air as a carrier gas, diminished water for treatment, and no chemical residue [64]. In CP, the utilized carrier gases are air, oxygen, nitrogen, or helium.

10.2.6.1 Generation of CP Treatment

CP utilizes electrical energy applied through electrodes on carrier gases (Figure 10.7). The applied electric energy between two electrodes is cathode and anode separated (at a distance of 150 mm) in the air at a voltage of 30 kV/cm, which causes the breakdown of the electric current and ignition of carrier gas to generate the plasma. As a result of molecules absorbing the energy, matter of phase changes from solid through the liquid, followed by

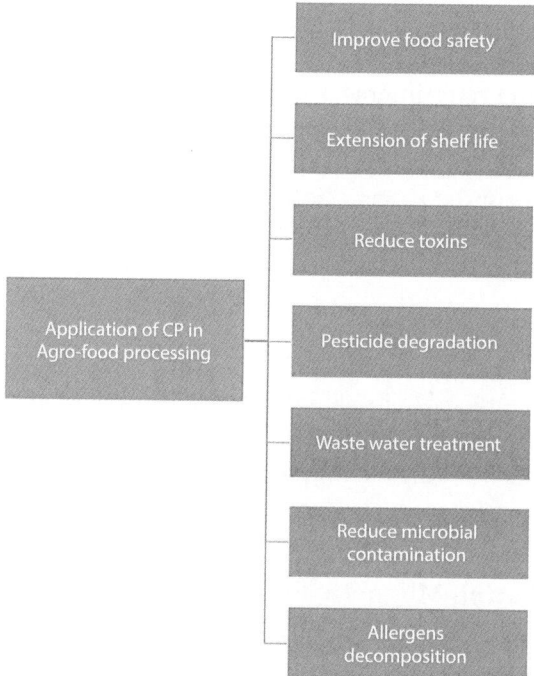

Figure 10.6 Application of cold plasma in agro-food processing industries.

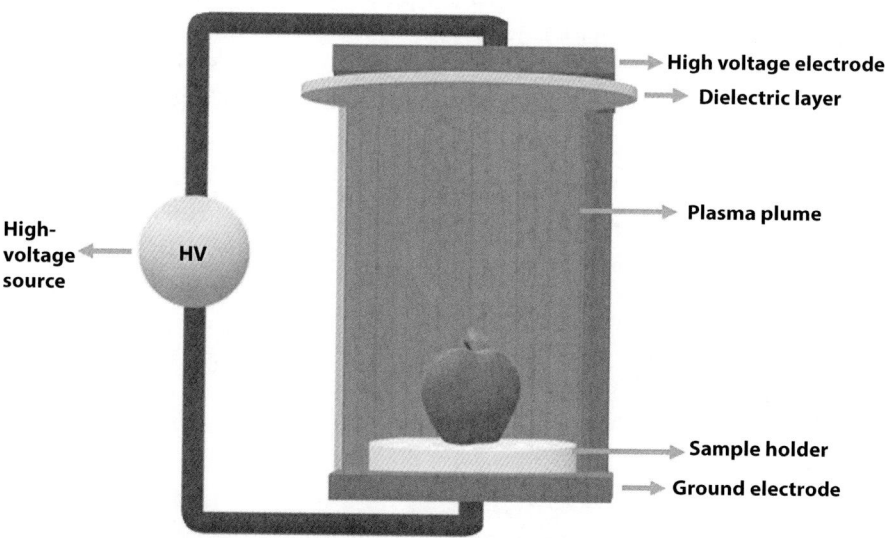

Figure 10.7 Working mechanism of CP.

gas. In the gaseous state, the energy increases beyond its threshold limits that cause gas molecules ionization, which yields a separate plasma state [2]. The cell membrane interacts with radicals, reactive molecules, and charged particles. The reactive species formed due to the breakdown of carrier gas, such as O_2, O_3, hydroxyl, peroxides, and superoxide radicals, play a crucial role in the inactivation growth of microorganisms [65]. The recent studies related to CP with different level of plasma are illustrated in Table 10.2

10.2.6.2 Challenges and Future Scope of CP

The plasma generated using noble gases will be preferable for preventing the oxidation of food products, but this process may increase its final cost. The main drawback of this technique is that this technique is still not utilized on a commercial scale.

10.2.7 Oscillating Magnetic Field

The electromagnetic field utilizes stored energy in the capacitor (up to 8 kJ), charged through a high voltage current (460 V a.c.). Later, the circuit is shut down using an ignitron switch, and the oscillating current produces in between two plates of the capacitor and generates oscillating magnetic field (OMF) for inactivation of microorganisms (Figure 10.8) [66]. The changes in the direction of the current and the polarity have been changed by a magnetic field. The magnetic field frequency ranges from 10 to 15 kHz, and the duration for each pulse varied from 20 to 40 μs. The oscillating current decays rapidly, and the frequency of OMF is measured using capacitance.

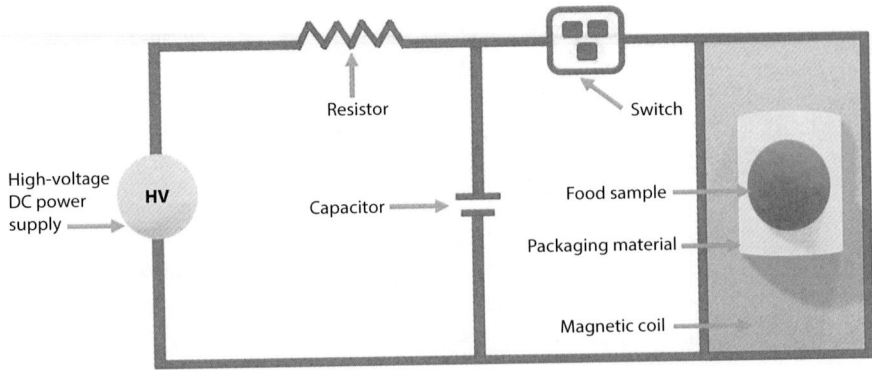

Figure 10.8 Electrical circuit for generation of OMF.

A field shaper is made up of aluminium cylinders with fiberglass insulation for increasing intensity (14–18 Tesla) of the OMF [66].

10.2.7.1 Challenges and Future Scope of OMF

The OMF is an emerging area for research and less utilized to preserve food products, which is the main drawback of this technique. However, less research has been conducted on the high intensity of the magnetic field, and the majority of research carried out with a low magnetic field intensity on biological systems. The recent studies related to OMF with different frequencies are illustrated in Table 10.2.

10.2.8 Membrane Filtration Process

Membrane processing is an important technology utilized in the food process industry to concentrate dairy products, fruit juices and clarification of juices [67, 68]. The membrane filtration process is operated physically and requires less heating, which retains the flavor, color, and quality of food products. The membrane utilized for the filtration process is made from organic polymers, such as cellulose acetate, polyamides, polyolefins, and polysulfones. The inorganic membrane is developed through the oxide of aluminium, silicium, titanium, and zirconium [69].

10.2.8.1 Principle of the Membrane Filtration Process

The basic principle of membrane processing is to act as a barrier to segregate solute and a liquid stream. A membrane is selective, which permits the transfer mixture of one bulk phase to another bulk phase during the driving force utilized across the barrier. The separation of two or more components is categorized based on size, shape, and particle [67].

The membrane filtration processes such as microfiltration (MF), ultrafiltration (UF), nanofiltration (NF), and reverse osmosis (RO) are categorized based on their molecular size of solutes (Table 10.3). The driving force for transferring materials through those four membranes is pressure difference [70]. The recent studies related to membrane process with filtration method and size differences are illustrated in Table 10.2.

10.2.8.2 Microfiltration

MF separates the particle, size ranges from 0.005 to 0.05 μm and permeates some macromolecules to pass through permeation and retained larger

Table 10.3 Typical ranges of membrane filtration.

Filtration process	Membrane pore size (μm)	Molecular weight (kDa)	Operating pressure (psi)	Filtrate
Microfiltration (MF)	0.05–5.0	>1,000	5–50	Amylopectin, bacteria, blood, indigo dye, latex, paint pigment, plant gums, yeast
Ultrafiltration (UF)	0.001–0.1	1–1,000	10–200	Amylose, colloidal silica, enzymes, gelatin, protein, virus
Nanofiltration (NF)	0.0005–0.01	300–500	100–500	Amino acids, antibiotics, colorant, sugars, synthetic dye
Reverse osmosis (RO)	0.0001–0.001	<1	200–1200	Atoms, flavors, fragrance, metal ions, salts

Source: [68, 70, 71].

macromolecules, fat globules, and colloidal structure. Most of the food industry prefers MF to clarify fruit juices, drinking water, beverages water, beer, wine, sugar syrup, and vinegar.

10.2.8.3 Ultrafiltration

UF is mainly utilized in the dairy industry for the separation process. The lower molecular weight species between 0.001 and 0.1 μm can pass through the permeation, while fats globules, proteins, and polysaccharides can be retained.

10.2.8.4 Nanofiltration

NF separates UF and RO components and permits separation for very low molecular weight components [71]. The sizes of separation of particle components are illustrated in Table 10.3. NF is utilized for the purification, fractionation and concentration.

10.2.8.5 Reverse Osmosis

RO permits very low molecular weight components size ranges from 0.0001 to 0.001 μm. RO is mainly utilized for purifying drinking water, which retains all the substances and permits only water.

10.2.8.6 Challenges and Future Scope of the Membrane Filtration Process

The membrane filtration process is a cost-effective process over other filtration and centrifugation processes in the food industry to separate solid-liquid components and preserve food products. The main drawback of this process is the decline of flux during processing, and to improve the flux rate of membrane processing may be considered a future perspective.

10.3 Novel-Thermal Preservation Technique

Novel-thermal processing is meant to generate heat with the food product and direct implications for enhancing the heating efficiency [72]. The use of this process entirely depends on the inherent and complexity of food product properties. These properties are electrical resistance, thermal conductivity, pH, moisture content, porosity, and rheological characteristics. The popularly utilized novel-thermal processing technologies are summarized below:

10.3.1 Ohmic Heating Treatment

OH is a novel-thermal preservation process that utilizes electric current to generate heat. OH is also referred to as Joule heating, electro conductive heating, and electro heating [73, 74]. The food product which contains enough water is treated with OH to generate heat in the form of internal energy within the food product. In the food industry, OH offers several potential applications, including blanching, dehydration, evaporation,

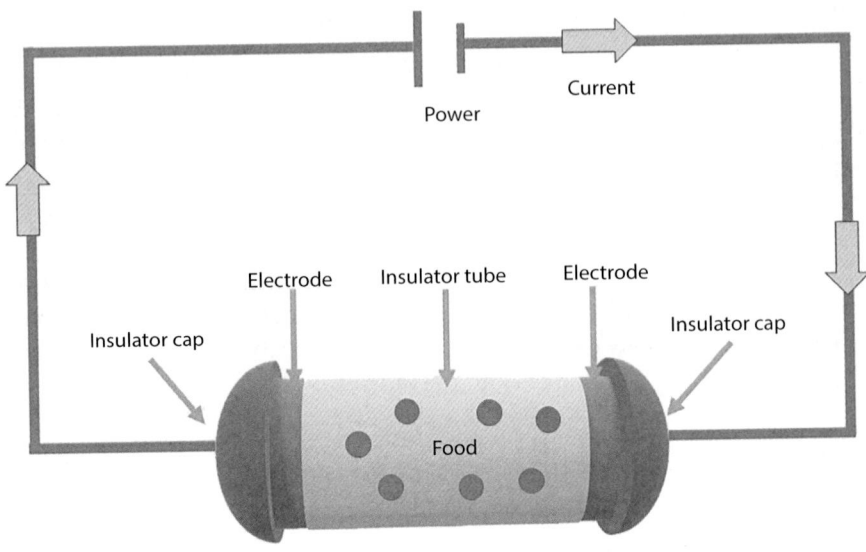

Figure 10.9 Operating mechanism of OH.

extraction, heating, fermentation, pasteurization, and food product sterilization. OH also increases the product gelatinization temperature, enthalpy and extraction rate, and when an electric field is applied in OH, it causes electroporation of the cell membrane.

The heating rate of OH can be influenced by concentration, electrical conductivity, electrodes, filed strength, ionic concentration, and particle size [74]. The higher the electric field intensity, the higher the hating rate and electrical conductivity, which is the key parameter for designing a useful OH (Figure 10.9) [73]. To ensure the proper OH treatment to food products, the cold spot needs to be identified. Once the cold spot is identified, then the temperature can be measured [72]. OH resulted in a faster and uniform heating process and improved quality with minimal structural loss, nutritional value, color, and organoleptic properties.

10.3.1.1 Application of OH Treatment

In OH, the process can be applied on liquid, solid, and mixture of solid-fluid food products. However, OH's successful use was carried out with various food products, including fruit juices, fruits and vegetables, liquid egg products, meat and meat products, pasta, sauces, seafood, soups, and stews. The recent studies related to OH with applied voltage are illustrated in Table 10.4.

Table 10.4 Application of novel-thermal technologies for different food products.

Application	Purpose	Treatment	Results	References
OH treatment				
Infant milk	Evaluate the effect of OH treatment on the physicochemical characteristics of infant milk	8, 12, 16, 20, and 24 V cm^{-1}, 72–75°C/15 s	OH treatment on infant milk significantly lowered the lower hydroxymethylfurfural and viscosity, as well as improved whiteness index and aroma compounds, including butanoic acid, isoamylol, isovaleric acid, heptan-2-one, octan-3-one, and nonan-2-one.	[75]
Grape	Determine the effect of OH on aqueous extract of phytochemical in grape skin	–	OH treatment boosts the extraction process and significantly improved color, soluble solids and total phenol compounds in grape skin. Also, increased in anthocyanin content from 756 to 1349 µg/g	[76]
Orange juice	Evaluate the effect of OH treatment on pectinesterase activity of orange juice	60 Hz, 32–36 V/cm, 60–90°C/0–200 s	OH treatment significantly affect the pectinesterase inactivation in orange juice	[77]

(*Continued*)

Table 10.4 Application of novel-thermal technologies for different food products. (*Continued*)

Application	Purpose	Treatment	Results	References
MW/IRH/RF treatment				
Orange juice-milk beverage (OJMB)	Evaluate the effect of MW heating on bioactive compounds, fatty acid profile, and volatile compounds of OJMB during 28 days of storage at 4°C	MH, 65 and 75°C for 60 s	MW heating on OJMB significantly improved antioxidant activity, ascorbic acid, α-amylase, α-glucosidase, carotenoids, and ACE inhibitory activity, as well as lowered the browning index compared to pasteurized product. MW heating also showed higher retention of bioactive compounds	[78]
Rice bran	Determine the effect of IR heating on biological and chemical profile and deterioration of rice bran during short-term storage	85°C	IR heating significantly reduced hydroperoxides and oxidation, and maintain the linoleic acid (C18:2), palmitic acid (C16:0), and oleic acid (C18:1) content in rice bran during short-term storage	[79]

(*Continued*)

Table 10.4 Application of novel-thermal technologies for different food products. (Continued)

Application	Purpose	Treatment	Results	References
Cumin, paprika, white pepper powder	Evaluate the efficiency of RF heating on thermal resistance of *Enterococcus faecium* and *Salmonella* in Cumin, paprika, white pepper powder samples	27.12–MHz, 6 kW	RF heating reduced thermal resistance capacity of both the microorganisms in paprika than cumin and white pepper powder. RF heating helped to control the growth of foodborne pathogens in spices	[80]
Soybean	Determine the electrode powder voltages associated with the free running on RF heating of soybeans	27.12 MHz, 6 kW	The relative error of 5% of was observed between improved and predicted model. The developed model for estimating voltage is more accurate for RF heating of soybean	[81]
Freezing treatment				
Spinach	Determine the changes in polyphenols of spinach sample and evaluate the bio accessibility of by-product at frozen storage	−80°C	The frozen storage of by-product significantly lowered the flavonoids (98%) and phenolic acid (90%) content, as well as increased in bio accessible flavonoids (15%) and phenolic acids (16%) compared to fresh spinach	[82]

(Continued)

Table 10.4 Application of novel-thermal technologies for different food products. (*Continued*)

Application	Purpose	Treatment	Results	References
Snakehead fish muscle	Evaluate the effect of immersion freezing on formation of ice crystal in fish muscle at different temperatures	20, −30, and −40°C	Immersion freezing at lower temperature showed formation of larger number of smaller ice crystal and improved microstructures of frozen snakehead fish muscle	[83]
Dehydration				
Papad (Indian snack food)	Evaluate the effect of microwave-assisted hot air drying (MAHD) on fried papad quality	653 W, 56 s, 43°C	MAHD reduced oil content (7.90 ± 0.02%), oil uptake ratio (1.50 ± 0.03), porosity (16.33 ± 0.29%), and expansion (7.97 ± 0.02%) of fried papad compared to conventional sun dried papad	[84]
Avocado wastewater	Determine the yield of Avocado wastewater into spray dried powder for the use of food preservative	160°C, 5.8g/min flow rate	Spry drying process significantly improved 49% yield of avocado wastewater powder	

(*Continued*)

Table 10.4 Application of novel-thermal technologies for different food products. (*Continued*)

Application	Purpose	Treatment	Results	References
Extrusion process				
Buckwheat flour	Evaluate the effect of extrusion process parameters on physicochemical properties of buckwheat flour (MBF)	70–100°C	Extrusion process significantly increased water absorption index and retention rate of MBF	[85]
Rice Dough	Investigate the suitable formulation rice types including indica rice, japonica rice and waxy rice, and water content for 3-D printing of rice-based food	Travel speed 20 mm/s, initial layer thickness 0.3 mm, shell thickness 1.52 mm, layer height and nozzle size 0.76 mm, nozzle height 2.0 mm, and infill density 100%	The formulation of waxy rice flour to water 100:90 (3DWRD-3) showed acceptable shape, microstructure and precision using 3-D printing	[86]

(*Continued*)

Table 10.4 Application of novel-thermal technologies for different food products. (*Continued*)

Application	Purpose	Treatment	Results	References
Cereal snack	Determine the ability of 3-D printed snack food texture	Travel speed 30 mm/s, initial layer thickness 0.6 mm, shell thickness 0.84 mm, layer height 0.8 mm, nozzle size 0.84 mm, and infill density 100%	3-D printing showed significant increase in porosity from 5% to 25% and reduced length of pores owing to crushing of dough filament. Significantly reduced hardness and relative density from 289 N to 84 N and 0.569 to 0.401	[87]

10.3.1.2 Challenges and Future Scope of OH Treatment

The OH process' initial cost is relatively higher, but the products processed with OH technology have superior quality compared to available conventional technologies. Enormous research work has been carried out for OH with various food products, but this technology requires some advancement to identify the cold-spot and measurement of complex food processing.

10.3.2 Microwave Heating

MW heating is electromagnetic radiation with frequency ranges from 300 MHz to 300 GHz. According to the Federal Communications Commission (FCC), MW utilized in food processing operates at 915 and 2,450 MHz (Figure 10.10). The oven utilized in the home operates at 2,450 MHz [72, 88]. MW travels as light waves and functions mainly on two mechanisms: dipole rotation and ionic polarization. MW heating can achieve the desired temperature in one-quarter of the time or less than conventional heating required. The distribution of temperature in MW heating is more uniform compared to conventional heating methods [72]. The penetration depth of MW into the food product ranges from 2 to 4 cm and travel through all directions to cook the food product.

10.3.2.1 Principle of MW Heating

MW is reflected by metals, absorbed by dielectric substances (like fats, water, and sugars) and transmitted from other dielectric substances. When they are absorbed, they convert into atomic motion, converted into heat [56, 88]. According to Vicente and Castro [72], the MW heating can be influenced by large numbers of parameters including, frequency, moisture content, temperature, specific heat, and product parameters (mass, density, size, and shape). The recent studies related to MW with different power levels applied to different food products are illustrated in Table 10.4.

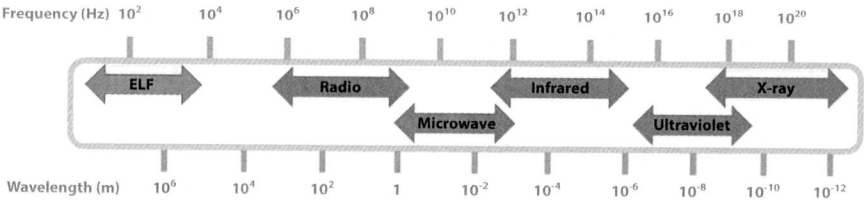

Figure 10.10 Spectrum of electromagnetic wavelength and frequencies.

Table 10.5 MW frequencies and power level utilized for different food products.

Application	Frequency (MHz)	Power (kW)	Food products
Baking	915	2–10	Bread, donuts, pastry
Blanching	2,450	10–30	Corn, fruits, potatoes
Cooking	915	50–240	Bacon, meat patties, potatoes, sausages
Dying	915	30–50	Grains, onion, pasta, snacks
Freeze-drying	2,450	30–50	Fruits and vegetables, meat products
Pasteurization	2,450	10–30	Dairy products, prepared foods
Sterilization	2,450	10–30	Fruit juices, milk, pastry
Tempering	915	30–70	Butter, fish, meat, poultry

Source: [88, 89].

10.3.2.2 Applications of MW Heating

In the food processing industry, MW heating is utilized for various applications, including baking, blanching, cooking, drying, freeze-drying, pasteurization, sterilization, and tempering. The application of MW heating is illustrated in Table 10.5. However, the utilization of industrial MW is cost-effective over conventional methods.

10.3.2.3 Challenges and Future Scope of MW Heating

MW heating utilizes high-energy costs and the difficulties of setting up industrial MW heating and controlling processes. The use of MW heating and food product changes due to dielectric properties are still not understandable to the consumers.

10.3.3 Infrared Heating (IRH)

IRH is electromagnetic radiation of thermal energy generated through a quartz lamp or quartz tube. IRH spectra categorized based on wavelength

Table 10.6 IRH spectra categorized based on wavelength and appearing temperatures.

IR Spectra	Wavelength (μm)	Temperature (°C)
Short wave	0.7–2.0	>1,000
Medium wave	2.0–4.0	400–1,000
Long wave	4.0–1,000	<400

Source: [72].

are illustrated in Table 10.6. Heat is produced owing to the absorption of radiation energy [72]. IRH system produces heat instantly, and there is no need to build-up heat as required by other conventional processes. The surrounding air does not absorb the main advantage of IRH as it does not heat up.

10.3.3.1 Application of IRH

The applications of IRH are boiling, cooking, drying, food warming, heating, melting, and toasting. The main advantage of IRH is reducing processing time, up to 70%.

The factor influencing IRH is the absorption of IRH, penetration properties, radiator efficiency, and radiator temperature [72]. The heat transfer rate may depend on the shape of the absorbing material, the surface temperature of the material, and the material's surface properties. The recent studies related to IRH with different wavelengths applied to a food product are illustrated in Table 10.4.

10.3.3.2 Challenges and Future Scope of IRH

The combination of IRH with other novel techniques can reduce the processing time, which will increase productivity. The main drawback of this process is that it is still developing the food industry's condition, and there is a need to understand the interaction between product properties and the heating process.

10.3.4 Radio Frequency Heating

Radio frequency (RF) heating is also termed dielectric heating, which involves transferring electromagnetic energy ranging from 10 and 300

MHz [90]. In RF heating, the wavelength frequency is more extensive compared to sample dimensions (Figure 10.10). The penetration depth of RF heating is more profound, and it provides uniform heating throughout the food product.

10.3.4.1 Principle of RF Heating

The main principle of IR heating is to generate electromagnetic energy when the food product is placed on the applicator between two electrodes; the electric energy is absorbed and reflected by the food product to cause dielectric heating [72, 88]. RF heating consists of a generator and applicator; the generator responsible for generating RF waves, and the applicator is the main part of electrodes responsible for RF power. The three main configurations of RF electrodes are through the field, staggered through the field and fringe field. The through field contains two electrodes in between food products that are treated at high-frequency voltage. This configuration is mainly applied to thick food products. The staggered through the field is the modified version of through field, which is utilized for the intermediate thickness of food products. The fringe field reduces the electric field required for generating power density within the food products. It contains a set of electrodes that are connected to the RF generator. This configuration reduces the risk of variation in the electric field when applied to thin food products [72]. The recent studies related to RF heating with different frequencies applied to a different food product are illustrated in Table 10.4.

10.3.4.2 Factor Influencing of RF Heating

The factor influencing RF heating is electric conductivity, relative dielectric constant and relative dielectric loss factor. Electric conductivity has a close relationship with the electric loss factor, which influences RF hating [72]. RF heating has several applications and is illustrated in Table 10.5.

10.3.4.3 Challenges and Future Scope of RF Heating

Various researches have been carried out with IR heating on various applications that help preserve the food product. This technology's main obstacles affect the texture of the treated food product, and this area of research needs to be studied. IR heating can be utilized with other non-thermal or novel-thermal technologies to improve the properties of food products.

10.4 Other Alternate Preservation Techniques

10.4.1 Freezing

Freezing is a preservation technique that exposes the food product to a low temperature, which removes latent heat and sensible heat from the product and results in lower product temperature [69, 70]. The temperature of freezing at which the vapor pressure of water is identical to that of ice crystal.

The freezing process reduces food products' water activity by converting most of the water molecules into a solid state, such as ice, which slows down the biochemical and physicochemical reaction and inhibits pathogenic microorganisms [70]. Due to ice crystal formation, less water is available within the food product to sustain the deteriorative reactions [69]. Typically, fish and ice-cream are frozen at −25°C, and fruits and vegetables are frozen at −18°C. However, frozen food may have 10% of the water in the liquid state [69].

The freezing process is categorized as indirect and direct contact freezing and is illustrated in Table 10.7. In indirect contact freezing, the food

Table 10.7 Types of freezing process operated at different temperatures.

Freezing types	Freezing process	Freezing temperature (°C)	Food products
Indirect contact freezing	Plate freezers	−40	Pastries, fish fillets, beef patties, asparagus, cauliflower, spinach, and broccoli
	Air-bast freezers	−35 to −45	Meat and poultry products, fruits and vegetables, fish, cheese
Direct contact freezing	Individual quick freezing	−30 to −50	Green peas, cut beans, cauliflower pieces, shrimps, meat chunks, and fish
	Immersion freezing	Below −18°C	Meat and poultry products, fruits, and vegetables

Source: [69, 70].

product is distinct from the refrigerant with some barrier. In direct contact freezing, the food product is surrounded with refrigerant [70].

Freezing time is the time required to change the initial product temperature to a given temperature. The freezing rate has a significant effect on the quality of the frozen product [70]. Typically, slow freezing of food tissue produces larger ice crystals than rapid freezing, which affects the freezing rate by damaging the texture and quality of end products [62]. Factors influencing the freezing time are enthalpy, heat transfer coefficient, refrigerating medium, the shape of the product, temperature, and thermal conductivity. Freezing can also affect the color, flavor, texture, and nutritional value of the end product. The recent studies related to different freezing processes at different freezing temperatures applied to various food products are illustrated in Table 10.4.

10.4.1.1 Challenges and Future Scope of Freezing

The freezing process helps preserve food products for a longer duration, but it increases the end product's cost. The main obstacle using this process is that it damages the texture of the food product. To overcome this problem, the freezing process can be utilized with other novel techniques to accelerate freezing, which will retain the end product's color, flavor, and texture.

10.4.2 Dehydration

The process of eradicating solid and liquid food due to evaporation is known as drying or dehydration. This oldest preservation method is focused on attaining a solid product with less water content [70]. During drying, the moisture content is lowered to a level where the microorganisms' actions are ceased. This is the cheapest preservation method and has various advantages like reducing weight and volume, storage, packaging, and transportation of food products. However, this method also has some limitations such as loss of smell, flavor, vitamins, proteins, and lipids drying, which is of three types: convective, conductive and radiative among which convective drying can dehydrate 90%, making it the most popular, drying method [62]. Drying is classified as thermal drying, osmotic dehydration, and mechanical dewatering based on the water removal method. Further, thermal drying is classified into air drying, low air environment drying, and modified atmosphere drying. Various factors are type of product, expected properties of the desired product, temperature, heat susceptibility of the product, pretreatment, environment, processing, and capital cost. For drying food products electric heating or flue gas heating is followed conventionally. Electromagnetic wavelength spectrum is used as energy to increase the rate of heating by MW, infrared, RF, refractance window, and dielectric heating. A frequency of 1 to 100 MHz is used by radiofrequency, but in MW drying, frequencies range from 0 MHz to 300 GHz. Moreover, MW energy applied in a vacuum condition is advantageous for improving food products' energy efficiency and quality. Thus, in food industries, electrotechnology is attaining popularity for drying [91].

10.4.3 Frying

Frying is a cooking method where fat or oil is used directly as a medium of heat transfer. Fat is used for frying as it has a boiling point higher than water, and heat can be applied until it reaches its boiling point without smoking. Frying is broadly classified into pan-frying, stir-frying and deep-frying. Pan-frying is usually done for flat, wide, and thin food products like omelettes, fillets, and patties [92–94]. Food products from small or medium-size cooked in a small amount of very hot oil is followed for stir-frying. Sauteing is an example of stir-frying. When the food particles are completely immersed in oil or fat, the uniform frying is carried out through all directions, referred to as deep frying. In deep frying, usually a temperature range from 150°C to above 200°C [95]. Fried potato products, fried snacks, ready to eat meals like fish sticks, vegetarian patties, and poultry or veal schnitzels are some deep-fried foods commercialized as frozen heat and serve foods. Several processes like cyclic fatty acids and polymerization take place during frying. The type of oil or fat for frying plays a vital role in avoiding food safety issues. The main aim of frying is to cook food to encourage several thermal phenomena like gelatinization of starch, hydrolysis, Millard reaction, protein denaturation, flavor development, and caramelization. Dehydration occurs as a result of frying along with moisture loss. Fat content may change during frying as deep-frying oil uptake occurs, but fat loss occurs in pan-frying. As a result of frying, heat transfer occurs, and a temperature gradient is created due to which texture and structure may be changed [96]. Batch or continuous fryers are used for frying, and industries usually use continuous fryers for deep frying. To avoid overheating the oil, a large area must be provided for more heat transfer [97]. Both vegetable and animal origin fats are used in frying [98]. Acrylamide issue of frying has become a recent trend of research [99].

10.4.4 Chilling

Chilling is a process where the foods' temperatures are maintained at −1 and 8°C [62]. The products' initial and final temperatures are maintained to extend food products' shelf life during chilling. The shelf lives of fresh and processed foods are extended by lowering the biochemical and microbiological aspects. Generally, the process of freezing is referred to as chilling when it is carried out below 15°C. However, the application of partial freezing in modern industries prolongs the shelf life of fresh food products [62]. Super chilling is another method of cooling food products to a cooling temperature below the initial freezing temperature. Various equipment are air cooler, jacket heat exchanger, cryogenic chamber, ice bank cooler, ice implementation system, plate heat exchanger, and vacuum attribution system. The chilling rate depends on different factors such as thermal conductivity, density, presence or absence of a lid on the storage vessel, the initial temperature of foods, moisture content, and size and weight of foods. Chilling is mostly used for short term preservation. Due to chilling, microbial proliferation is inhibited and prevents post-harvest loss. Some disadvantages like reducing the crispness, dehydration of unwrapped foods, loss of nutrients, and oxidative browning are accompanied during chilling. This process is advantageous over complete freezing as most water is in a liquid state [70].

10.4.5 Extrusion

Extrusion is a Latin word, which means pushing out, and it is used since the 1950s [70]. It is an operation of forcing a food product through a narrow space. Extrusion's principal function is to pass on a shape or structure to food products without disturbing the food products' properties; pasta presses and feedstuff pelletizers are some extruded food products. When the thermochemical processes where heat transfer, pressure change, mass transfer, and shear are united to produce drying, cooling, cooking, melting, sterilization, conveying, texturizing, mixing, foaming, puffing, freezing, etc. Extruder-cooker is a pump, heat exchange, and a continuous high-pressure, high-temperature reactor, and all these are collectively put up for single equipment. Since ancient times before its use in food industries, this method has been used in metallurgy and plastic polymers [70]. This method has been used for the texturization of protein-rich food products that are of plant origin. This method also has been utilized to produce biodegradable and nanocomposite films with good barrier properties. This method is also used for disinfection, sterilization of foods, and inhibiting heat resistant toxins like aflatoxins [100]. In extrusion, cooking energy is less spent as energy delivery direct and not through any intermediate source. A single screw is the first cooking extruder used, but nowadays, twin-screw extruder is used. Single screw consists of a hopper, barrel, screw or worm, flow channel, compression section, feed section, metering section, and dies [101]. The twin-screw consists of a parallel screw, barrel, cross-section, and die [102]. Pumping efficiency, mixing, heat exchange, and residence time are more efficient in a twin-screw extruder than a single screw extruder. The modular configuration of the twin-screw is more flexible. The high cost and complexity of the twin-screw extruder cause some disadvantages [103]. The recent studies related to the extrusion process with different food products are illustrated in Table 10.4.

10.4.6 Three-Dimensional (3-D) Printing

Nowadays, extrusion-based printing such as 3-D printing is popularly utilized for designing new food products. This is an emerging novel-thermal processing technology utilized in various fields like aerospace, construction, food, material science and medicine. 3-D printing is operated using computer-based controls that add materials (gel, liquid or powdered) layer-by-layer to obtain a specific 3-D object [104, 105]. In the food industry, extrusion-based methods are most studied for 3-D printing. 3-D printing can create food with unique shapes, textures, flavors, and color that are more popular among kids and the elderly [105].

10.4.6.1 Principle of 3-D Printing

While 3-D printing, the pressure is forced on the material to squeeze out through the nozzle where deformation of material occurs, and a subsequent layer of material is printed over the previous layer. This process creates the desired shape [105]. The recent studies related to 3-D printing with different food products are illustrated in Table 10.4.

10.4.6.2 Factor Influencing 3-D Printing

The factor influencing 3-D printing is the material's viscosity, particle structure, pump size for the flow of material, the syringe geometry for printing, and total printing duration [104]. To determine the efficiency of 3-D printability, it is essential to study the food matrix's textural and rheological characteristics, including adhesiveness, hardness, shear modulus, shear recovery, and springiness viscosity and yield strength, which play a crucial role in 3-D printing [104, 105]. This technology has many food industry applications, including bakery, confectionery, chocolate, cheese, and meat and poultry products.

10.4.7 Blanching

Blanching is a process for solid foods similar to pasteurization that inactivates an enzyme system within the temperature range of 70–100°C for a short duration of 20 s to 15 min [106]. Such a high temperature leads to cause structural damage and losses of the firmness of fruits and vegetable tissues; therefore, blanching at a lower temperature of 55–72° may increase the firmness and reduce physical damage of fruit and vegetables [106]. Conveying systems are used to carry food products since process is built for solid foods. Steam and hot water blanching are mainly preferred to heat the products in the system. Design and speed are responsible for setting the thermal treatment for the inactivating enzyme system. In the first stage of the steam or hot water treatment process, the foods' slowest heating portion will increase the temperature. In the second stage, cold air or water is exposed to food products. The process considers temperature-time profile at slowest heating or cooling portions of the food products, and this process is established in both the stages through the speed of the conveyor [69].

10.5 Hurdle Technology for Preservation of Food

Nowadays, the use of a single preservation technique is not so feasible to obtain high quality, safety, and stability of food product; therefore, the combination of two or more preservation technologies is more feasible. Combining a non-thermal preservation process with other non-thermal or conventional preservation processes may enhance the lethal effect, reduce the severity, and prevent the proliferation of non-thermal processing. The use of non-thermal, novel thermal, or conventional processing is developed to maximize the synergistic effect on the inactivation of microorganisms.

Hurdle technology was mainly developed due to increasing consumer demands for developing fresh and safe, minimally processed food products. According to Raso *et al.* [107], hurdle processed food has natural and fresh attributes and requires minimum effort and processing time. Before applying the hurdle effect, proper knowledge about individual factors and combination of mechanisms is required. The understanding of the hurdle effect provides cost-effective, safe, and stable food. However, combinations of preservation processes are more advantageous over a single preservation process that improves quality and stability and improves the nutritional value and sensorial effect of food products [107].

During hurdle processing, the appropriate hurdle may require particular products, and all hurdles may not be used for food preservation. The hurdle process may depend on its microbial growth, target shelf life and type of hurdle. While preservation of food, the most commonly utilized hurdles are acidity (pH), competitive microorganisms (lactic acid bacteria), preservatives (nitrite, sorbate, sulfite, etc.), redox potential (Eh), temperature (high or low), and water activity (aw) [108]. Hurdle processing is categorized into four groups: physical hurdle, physical non-thermal hurdle, physicochemical hurdle, and microbiological hurdles [109]. Physical hurdle involves baking, blanching, chilling, freezing, MW, packaging, pasteurization, radiation, etc. Physical non-thermal hurdle utilizes HPP, electric pulse field, pulse light, etc. Physicochemical hurdle involves carbon dioxide, ethanol, lactic acid, low pH and low water activity, O_2, O_3, smoking, spices sulfite, etc. Simultaneously, microbiological hurdles use competitive flora, microbial produce, and protective cultures [109]. The appropriate hurdle may require a particular product, and all hurdle process may not be used for food preservation. Factor influencing the hurdle process is that high concentration and strength of single hurdles may affect food products' quality (color, texture, and nutrient loss). The recent studies related to the hurdle process with different technology and food products are illustrated in Table 10.8.

10.6 Irradiation Process for Preservation of Food

Irradiation is one of the existing non-thermal processes in which food substances are directly or indirectly exposed to ionizing radiation (IR), and it comprises of X-ray, electron beams (β) and gamma (γ) rays [62, 115]. The dose level for IR food products is measured in kilo gray (kGy). The limits for IR doses for food products are set by the regulatory bodies and represented in Table 10.9, and the limits are expressed as low, medium, and high doses. About 40 countries have approved the low doses of irradiation for quarantine treatment on more than 60 food products. The low IR doses up to 1 kGy may be applicable for disinfecting plant food products. The medium IR doses from 1 to 10 kGy are utilized for decontaminating animal origin food products. The high IR doses above 10 kGy can decontaminate spices and herbs and purify low acid food products. The IR doses at a high level may cause micronutrients from food, but it will not lose the

Table 10.8 Application of hurdle technology for different food products.

Application	Purpose	Treatment	Results	References
Foodborne pathogens	Investigate the effect of γ-irradiation on the microbial inactivation of foodborne pathogens (Bacillus cereus, Escherichia coli O157:H7, Listeria monocytogenes, Salmonella typhimurium, and Staphylococcus aureus)	2.5% sodium citrate, 0.5% sodium carbonate, 0.75% citric acid and γ-irradiation of 0.25 to 10 kGy	γ-irradiation process effective against pathogens in frozen IF. A hurdle effect with sodium carbonate generated high radiosensitization and lower radiosensitization observed for all pathogens, compared to other additives. The use of combine effect of food additives—mainly sodium carbonate with γ-irradiation reduced time of irradiation process	[110]
Apple juice	Evaluate the combine effect of efficiency of ultrasound (US) and fumaric acid (FA) treatment against Escherichia coli O157:H7, Listeria monocytogenes and Salmonella typhimurium in apple juice	40 kHz (US) and 0.05, 0.1, and 0.15% (FA)	The combined effect of US and FA (0.15%) for 5 min showed 5.67, 3.47, and 6.35 log reductions in E. coli O157:H7, L. monocytogenes and S. typhimurium, respectively	[111]

(Continued)

Table 10.8 Application of hurdle technology for different food products. (*Continued*)

Application	Purpose	Treatment	Results	References
Ready-to-eat ham	Evaluate the effect of atmospheric cold plasma (ACP) treatment on MAP stored ready-to-eat ham at 4°C	20% O_2 + 40% CO_2 + 40% N_2	ACP treated on MAP stored (after 24 h) ready-to-eat ham showed >2 CFU/cm^2 and 4-6 CFU/cm^2 log reduction of *Listeria monocytogenes*. The inside gas composition of ready-to-eat ham influences the efficiency of ACP treatment	[112]
Cherry	Evaluate the effect of alternating magnetic field (AMF) and permanent magnetic field (PMF) during freezing of cheery	1.26 mT AMF and 10 mT PMF, −30°C and 4°C/min.	Freezing temperature significantly improved AMF compared to control and use of magnetic field reduced drip loss. The applied magnetic field showed homogeneous ice crystal and reduced area of ice crystal of 78% and 67% in AMF and PMF	[113]
Mushroom chips	Determine the optimal condition of microwave (MW) and ultrasound (US) treatment on frying condition to improve quality of fried mushroom chips	3,000 W (MW) and 600W (US), frying at 90°C and vacuum pressure 0.01 MPa	The combine effect of MW and US significantly increased moisture evaporation rate and flavor, as well as reduced oil content from fried mushroom chips	[114]

Table 10.9 Regulatory bodies limitation for IR doses applied on different food products.

Type of doses	Level of doses (kGy)	Target food product	Purpose
Low doses up to 1 kGy	0.05–0.15	Onions, potatoes, ginger garlic, etc.	Inhibition of sprouting
	0.15–0.5	Grains, pulses, fruits, dry fruits, meat, and meat products	Insect disinfectant
	0.25–1.0	Fruits and vegetables	Delay ripening
Medium doses 1 –10 kGy	1.0–3.0	Strawberries, mushroom, fish, etc.	Shelf life extension
	1.0–7.0	Frozen meat, poultry and seafood, etc.	Reduce the growth of pathogenic microorganism
	2.0–7.0	Dehydrated vegetable, grapes, etc.	Improve food properties
High doses above 10 kGy	30–50	Meat, poultry, and seafoods	Sterilization
	10–50	Spices, herbs, etc.	Decontamination

Source: [56, 62].

treated food product's nutritional quality. As per the FDA, the food treated with the IR technique has a similar nutritional quality compared to other conventional food preservation techniques [62]. The maximum doses of irradiation up to 10 kGy for any food product can be considered safe [56].

During IR processing of any food product, the energy sources from IR damage the bacterial cell (chromosomal DNA) to prevent microorganisms' growth. IR doses may depend on the food product characteristics and external factors, including moisture, temperature, density and respiratory gas composition. The recent studies related to irradiation with different processes applied to different food products are illustrated in Table 10.10.

Table 10.10 Application of IR for different food products.

Application	Purpose	Treatment	Results	References
Horseshoe crab chitosan	Investigate the effect of γ-irradiation on structural changes and antioxidant capacity of horseshoe crab chitosan	CO-60 at 10–20 kGy/hour	The antioxidant properties of horseshoe crab-derived chitosan were evaluated *in vitro*. The 20 kGy γ-irradiation showed changes in chitosan structure with increase in antioxidant activity and decrease in molecular weight	[116]
Red raspberries (*Rubus idaeus* L.)	Evaluate the effect of E-beam on microbial inactivation and bioactive properties of red raspberries (*Rubus idaeus* L.)	3 kGy	Treatment of E-beam with 3 kGy successfully reduced 2 log CFU/g of mesophilic bacteria and 3 log CFU/g on filamentous fungi and nil foodborne pathogens (3 log CFU/g) was achieved at 7 days of refrigerated storage. E-beam preserved antioxidant activity and phenolic content, as well as decreased ascorbic acid concentration (80%)	[117]

(Continued)

Table 10.10 Application of IR for different food products. (*Continued*)

Application	Purpose	Treatment	Results	References
Calf liver	Detection of γ-irradiated calf liver through measurements FT-IR spectra of their extracted DNA	3 kGy	FT-IR spectra showed significant differences between irradiated and control calf liver. The significant characteristic of guanine nucleoside observed at band range of 1,491 cm^{-1}	[118]
Ricotta cheese	Identify the sanitizing effects of X-rays on ricotta cheese	0.5, 2, and 3 kGy	All the applied intensities of X-ray treatment on ricotta cheese can significantly prolong the shelf life upto 84 days	[119]

10.6.1 Electron Beam

Electron beam (e-beam) is also known as cathode rays, which produce high energy in electron guns equipped with the cathode (e−) and anode (e+) electrodes [56]. In tubes, the cathodes travel through the filament to generate the current and make the filament red hot, knock the surface electron, and increase the filament's random heat motion. This process is termed as thermionic emission, which reduces cathode and anode voltage required to obtain current.

During irradiating of food products, the depth of e-beam is limited up to 3–5 cm, and the energy limit is set up to 10 MeV, and for liquid products, it can be applicable up to 0.50 cm/MeV [115, 120]. While processing cartons foods, the doses of energy levels are reduced to provide a two-sided irradiation effect. The accelerators' power output may vary from 10 to 100 kW, and penetration depth may increase to 8 cm [120]. The drawback of the e-beam process is the depth of penetration applicable to food products. Due to the limited penetration capacity, it is used for food surface decontamination. It is also utilized for sterilization of equipment, medical, and pharmaceutical goods [56].

10.6.2 X-Radiation (X-Ray)

X-ray is also termed as Röntgen radiation, and this modern IR technique utilizes high-energy photons for food application, and it has higher penetration power than e-beam and similar penetration to gamma (γ) rays from Cobalt-60. About 4–6% are utilized from total electrons to produce the X-ray technique, and it operated at a maximum energy level of 5 MeV [56]. X-ray technique is mainly utilized for sanitizing packaged products and decontaminating meat and poultry products.

10.6.3 Gamma Rays

Gamma (γ) radiation is the easiest form of irradiation, which utilizes radionuclides of Cobalt (60 Co) and Caesium (137 Cs) for food irradiation [115, 120]. When the neutron of Co59 is bombarded, the generation of Co60 takes place inside a reactor and produces 1.17 and 1.33 MeV of gamma rays. Cs137 is atomic fission by-products produced by separation processing, and it produces 0.66 MeV of gamma rays. The half-life of Co60 is about 5.27 years and 30 years for Cs137. The penetration depth of gamma rays is up to several feet that eliminate microorganism's growth without affecting food characteristics. The main drawback of this process is that the emitted radiation cannot be control.

10.7 Food Additives for the Preservation of Food

In ancient times, for the preservation of food, salt and spices were utilized. Nowadays, consumers demanding for fresh and safe food added with fewer additives. To satisfy consumer demands, there is a need to modify the traditionally utilized food preservation processes.

According to the FDA, food additives are utilized to deliver a technical effect in the food product. More than 2,500 additives were added to the food product to improve food products' shelf life [121] in European Union (labeled the food additives are labeled with "E" latter and mentioned with specific number to identify the additive. Preservation is one of the functions of food additives that control microorganisms' growth, improve flavor, appeal, and nutritional value. FDA is strictly monitoring and regulating the use of food additives for safety purpose. The intended use of additives in the food products and their consumption can cause several chronic diseases [121, 122].

Additives are categorized into six different groups: preservatives, coloring agents, flavoring agents, texturing agents, nutritional additives and miscellaneous additives. The additives used in food processing may be either natural additives or synthetic additives [121]. The recent studies related to food additives with different food products are illustrated in Table 10.11.

10.7.1 Natural Additives

Natural additives are compounds of chemicals extracted from animal and plant sources. Natural additives consider healthier to provide various functions, including antimicrobial, antioxidant, and coloring agents, and gain more attention from the public [122]. The derived sources of natural additives compounds are illustrated in Table 10.12.

10.7.2 Synthetic Additives

Synthetic additives are not extracted and produced from the chemical and enzymatic reaction. The overconsumption of synthetic additives causes serious health issues, but it has antimicrobial property and antioxidant property and acts as coloring and flavoring agents [122]. Synthetic agents and their properties are illustrated in Table 10.13.

Table 10.11 Application of food additives for different food products.

Application	Purpose	Treatment	Results	References
Natural additives				
Egg yolk	Evaluate the rheological and thermal properties of frozen and unfrozen thawed egg yolk with and without additives	5% hydrolyzed egg yolk protein treated with 10% salt or sugar	The egg yolk with freeze-thaw showed better stability compared to conventional antigelation method	[123]
Ready-to-cook (RTC) idli batter	Investigate the use of natural colorant (annatto and pomegranate extract powder) for identification of RTC idli batter spoilage	Dye concentration 1 mg/ml	The use of natural colorants as a dye solution did not show any significant effect on properties of RTC idli batter. Natural day could be used as indicator for monitoring freshness of RTC stored idli batter	[12]

(*Continued*)

Table 10.11 Application of food additives for different food products. (*Continued*)

Application	Purpose	Treatment	Results	References
Synthetic additives				
Saffron	Determine the adulteration in saffron sample	-	Thin-Layer Chromatography (TLC) showed adulteration of saffron with magenta-III and pink-colored dye. Although, Mass spectrometry (MS) identified magenta III and pink-colored dye seen to be carcinogenic	[124]

Table 10.12 Derived sources and compounds of natural additives.

Natural additives	Compounds	Derived sources
Natural antioxidant	Ascorbic acid, carotenoids, polyphenols, tocopherol	Plant
Natural antimicrobial	Bacteriocins, natamycin, reuterin	Animal or microorganisms
	Essential oil, polyphenols	Plant
	Lactoferrin, lactoperoxidase, lysozyme, poly-L-Lysine	Animal
Natural colorant	Annatto, anthocyanins, betalains, β-carotene, caminic acid, carotenoids, chlorophylls, curcumin, lutein, paprika	Plant

Source: [125].

Table 10.13 Synthetic agent and properties of synthetic additives.

Synthetic agent	Properties
Ascorbic acid, butylated hydroxyanisole (BHA), butylated hydroxytoluene (BHT), citric acid, sulfites, tertiary butylhydroquinone (TBHQ), tocopherols	Antioxidant
Acetic acid, benzoic acid, natamycin, nisin, nitrates, nitrites, propionic acid, sorbic acid, sulfites and sulfur dioxide	Antimicrobial
Allura red, brilliant blue, erythrosine, fast green, indigo carmine, sunset yellow, tartrazine	Coloring agent

10.7.3 Challenges and Future Scope of Additives

Food additives play a significant role in the food industry to preserve food and improve the taste and flavor of food products. The overconsumption of additives will create a severe health issue that needs to be controlled and reduce its health impact. Nowadays, nanotechnology has focused on utilizing additives for various purposes such as packaging, encapsulation, indicators and sensors, etc., which will be the future research area.

10.8 Conclusion

Alternative food preservation technology can inactivate pathogenic microorganism's growth and deliver safe and stable food products with improved quality. However, these technologies are advantageous over existing thermal preservation technology in terms of high-energy efficiency, a shorter duration for treatment, and higher safety of the product with improved shelf life. These technologies have few limitations to widespread implementation at an industrial scale due to high investment capital, regulatory affairs, lack of knowledge about process variables, and utilization. As these issues get sorted, non-thermal and novel-thermal technologies will get more attention in the food industries.

References

1. Ortega-Rivas, E., *Non-thermal processing: Steam Vacuuming*, Encyclopedia of Food Microbiology (Second Edition), Academic Press, Cambridge, MA, 2014.
2. Mandal, R., Singh, A., Singh, A.P., Recent developments in cold plasma decontamination technology in the food industry. *Trends Food Sci. Technol.*, 80, 93–103, 2018.
3. Ezeh, O., Yusoff, M.M., Niranjan, K., Nonthermal processing technologies for fabrication of microstructures to enhance food quality and stability, in: *Food Microstructure and Its Relationship with Quality and Stability*, pp. 239–274, Woodhead Publishing, Cambridge, UK, 2018.
4. Han, J.H., A review of food packaging technologies and innovations, in: *Innovations in Food Packaging*, pp. 3–12, Academic Press, Cambridge, MA, 2014.
5. Restuccia, D., Spizzirri, U.G., Parisi, O.I., Cirillo, G., Curcio, M. et al., New EU regulation aspects and global market of active and intelligent packaging for food industry applications. *Food Control*, 21, 11, 1425–1435, 2010.
6. Vaclavik, V.A. and Christian, E.W., Packaging of Food Products, in: *Essentials of Food Science*, pp. 472–500, Springer, New York, 2008.
7. Brody, A.L., Zhuang, H., Han, J.H., *Modified atmosphere packaging for fresh-cut fruits and vegetables*, John Wiley & Sons, United Kingdom, 2010.
8. Ooraikul, B., Modified atmosphere packaging (MAP), in: *Food preservation Techniques*, pp. 338–359, Woodhead Publishing, Cambridge, MA, 2003.
9. Densensor, https://www.ametekmocon.com/, 2018.
10. Robertson, G.L., Food Packaging Principles and Practice, in: *Food Science and Technology*, Second Edition, p. 152, 2006.

11. Djenane, D. and Roncalés, P., Carbon monoxide in meat and fish packaging: advantages and limits. *Foods*, 7, 2, 12, 2018.
12. Gaikwad, P.S., Yadav, B.K., Sugumar, A., Fabrication of natural colorimetric indicators for monitoring freshness of ready-to-cook idli batter. *Packag. Technol. Sci.*, 34, 4, 211–218, 1–8, 2020.
13. Fellows, P.J., *Food Processing Technology: Principles and Practice*, Elsevier, Cambridge, UK, 2009.
14. Ghosh, T. and Dash, K.K., Modeling on respiration kinetics and modified atmospheric packaging of fig fruit. *J. Food Meas. Charact.*, 14, 2, 1092–1104, 1–13, 2020.
15. Minh, N.P., Influence of modified atmospheric packaging and storage temperature on the physico-chemical, microbial and organoleptic properties of cantaloupe (Cucumis melo) fruit. *Res. Crops*, 21, 3, 506–511, 2020.
16. Baswal, A.K., Dhaliwal, H.S., Singh, Z., Mahajan, B.V.C., Influence of Types of Modified Atmospheric Packaging (MAP) Films on Cold-Storage Life and Fruit Quality of 'Kinnow'Mandarin (Citrus nobilis Lour X C. deliciosa Tenora). *Int. J. Fruit Sci.*, 20, sup3, S1552–S1569, 1–18, 2020.
17. Junior, M.M., Castanha, N., Dos Anjos, C.B.P., Augusto, P.E.D., Sarmento, S.B.S., Ozone technology as an alternative to fermentative processes to improve the oven-expansion properties of cassava starch. *Food Res. Int.*, 123, 56–63, 2019.
18. Alexandre, A.P.S., Castanha, N., Costa, N.S., Santos, A.S., Badiale-Furlong, E., *et al.*, Ozone technology to reduce zearalenone contamination in whole maize flour: degradation kinetics and impact on quality. *J. Sci. Food Agric.*, 99, 15, 6814–6821, 2019.
19. Sroy, S., Fundo, J.F., Miller, F.A., Brandão, T.R., Silva, C.L., Impact of ozone processing on microbiological, physicochemical, and bioactive characteristics of refrigerated stored Cantaloupe melon juice. *J. Food Process. Preserv.*, 43, 12, e14276, 2019.i.
20. Chou, C.H., Wang, C.Y., Shyu, Y.T., Wu, S.J., The effect of high-pressure processing on reducing the glycaemic index of atemoya puree. *J. Sci. Food Agric.*, 101, 4, 1546–1553, 2020.
21. Chuang, S., Sheen, S., Sommers, C.H., Sheen, L.Y., Modeling the Reduction of Salmonella and Listeria monocytogenes in Ground Chicken Meat by High Pressure Processing and trans-Cinnamaldehyde. *LWT - Food Sci. Technol.*, 139, 110601, 2020.
22. Raj, A.S., Chakraborty, S., Rao, P.S., Thermal assisted high-pressure processing of Indian gooseberry (Embilica officinalis L.) juice–Impact on colour and nutritional attributes. *LWT - Food Sci. Technol.*, 99, 119–127, 2019.
23. Yuan, S., Li, C., Zhang, Y., Yu, H., Xie, Y. *et al.*, Degradation of parathion methyl in bovine milk by high-intensity ultrasound: degradation kinetics, products and their corresponding toxicity. *Food Chem.*, 327, 127103, 2020.

24. Meurer, M.C., de Souza, D., Marczak, L.D.F., Effects of ultrasound on technological properties of chickpea cooking water (aquafaba). *J. Food Eng.*, 265, 109688, 2020.
25. Lu, C., Ding, J., Park, H.K., Feng, H., High intensity ultrasound as a physical elicitor affects secondary metabolites and antioxidant capacity of tomato fruits. *Food Control*, 113, 107176, 2020.
26. Jeong, S.H., Kim, E.C., Lee, D.U., The Impact of a Consecutive Process of Pulsed Electric Field, Sous-Vide Cooking, and Reheating on the Properties of Beef Semitendinosus Muscle. *Foods*, 9, 11, 1674, 2020.
27. Fauster, T., Giancaterino, M., Pittia, P., Jaeger, H., Effect of pulsed electric field pretreatment on shrinkage, rehydration capacity and texture of freeze-dried plant materials. *LWT - Food Sci. Technol.*, 121, 108937, 2020.
28. Rybak, K., Samborska, K., Jedlinska, A., Parniakov, O., Nowacka, M., et. al., The impact of pulsed electric field pretreatment of bell pepper on the selected properties of spray dried juice. *Innov. Food Sci. Emerg. Technol.*, 65, 102446, 2020.
29. Akasapu, K., Ojah, N., Gupta, A.K., Choudhury, A.J., Mishra, P., An innovative approach for iron fortification of rice using cold plasma. *Food Res. Int.*, 136, 109599, 2020.
30. Rashid, F., Bao, Y., Ahmed, Z., Huang, J.Y., Effect of high voltage atmospheric cold plasma on extraction of fenugreek galactomannan and its physicochemical properties. *Food Res. Int.*, 138, 109776, 2020.
31. Kang, T., Hoptowit, R., Jun, S., Effects of an oscillating magnetic field on ice nucleation in aqueous iron-oxide nanoparticle dispersions during supercooling and preservation of beef as a food application. *J. Food Process Eng.*, 43, 11, e13525, 2020.
32. Jarulertwattana, P., Siriwattanayotin, S., Asavasanti, S., Effect of Oscillating Magnetic Field on Freezing Rate, Phase Transition Time and Supercooling of Deionized Water. *Appl. Sci. Eng. Prog.*, 13, 1, 32–37, 2020.
33. Al-Hatim, R.R., Al-Rikabi, A.K., Ghadban, A.K., The Physico-chemical properties of bovine and buffalo whey proteins milk by using ultrafiltration membrane Technology. *Basrah J. Agric. Sci.*, 33, 1, 122–134, 2020.
34. dos Santos, L.F., Vargas, B.K., Bertol, C.D., Biduski, B., Bertolin, T.E. et al., Clarification and concentration of yerba mate extract by membrane technology to increase shelf life. *Food Bioprod. Process. Trans. Inst. Chem. Eng. Part C*, 122, 290, 2020.
35. Pandiselvam, R., Subhashini, S., Banuu Priya, E.P., Kothakota, A., Ramesh, S.V., Shahir, S., Ozone based food preservation: a promising green technology for enhanced food safety. *Ozone Sci. Eng.*, 41, 1, 17–34, 2019.
36. Porto, E., Alves Filho, E.G., Silva, L.M.A., Fonteles, T.V., do Nascimento, R.B.R. et al., Ozone and plasma processing effect on green coconut water. *Food Res. Int.*, 131, 109000, 2020.

37. Brodowska, A.J., Nowak, A., Śmigielski, K., Ozone in the food industry: Principles of ozone treatment, mechanisms of action, and applications: An overview. *Crit. Rev. Food Sci. Nutr.*, 58, 13, 2176–2201, 2018.
38. Pandiselvam, R., Sunoj, S., Manikantan, M.R., Kothakota, A., Hebbar, K.B., Application and kinetics of ozone in food preservation. *Ozone Sci. Eng.*, 39, 2, 115–126, 2017.
39. Khadre, M.A., Yousef, A.E., Kim, J.G., Microbiological aspects of ozone applications in food: a review. *J. Food Sci.*, 66, 9, 1242–1252, 2001.
40. Güngör, F.Ö., Ocak, Ö.Ö., Ünal, M.K., Effect of ozone treatment on the physical, microbiological and sensorial properties of Spanish-style table olives. *Grasas Aceites*, 71, 1, 348, 2020.
41. Varol, K., Koc, A.N., Atalay, M.A., Keles, I., Antifungal Activity of Olive Oil and Ozonated Olive Oil Against Candida Spp. and Saprochaete Spp. *Ozone Sci. Eng.*, 39, 6, 462–470, 2017.
42. Mohammadi, H., Mazloomi, S.M., Eskandari, M.H., Aminlari, M., Niakousari, M., The Effect of Ozone on Aflatoxin M1, Oxidative Stability, Carotenoid Content and the Microbial Count of Milk. *Ozone Sci. Eng.*, 39, 6, 447–453, 2017.
43. Miller, F.A., Silva, C.L., Brandão, T.R., A review on ozone-based treatments for fruit and vegetables preservation. *Food Eng. Rev.*, 5, 2, 77–106, 2013.
44. Cullen, P.J., Tiwari, B.K., O'Donnell, C.P., Muthukumarappan, K., Modelling approaches to ozone processing of liquid foods. *Trends Food Sci. Technol.*, 20, 3–4, 125–136, 2009.
45. Balasubramaniam, V.M., Barbosa-Cánovas, G.V., Lelieveld, H. (Eds.), *High Pressure Processing of Food: Principles, Technology and Applications*, Springer, New York, 2016.
46. Hogan, E., Kelly, A.L., Sun, D.W., High pressure processing of foods: an overview, in: *Emerging Technologies for Food Processing*, pp. 3–32, Academic Press, Cambridge, MA, Cambridge, MA, 2005.
47. Mertens, B., Hydrostatic pressure treatment of food: equipment and processing, in: *New Methods of Food Preservation*, pp. 135–158, Springer, Boston, MA, 1995.
48. Balasubramaniam, V.M. and Farkas, D., High-pressure food processing. *Food Sci. Technol. Int.*, 14, 5, 413–418, 2008.
49. Balasubramaniam, V.M., Ting, E.Y., Stewart, C.M., Robbins, J.A., Recommended laboratory practices for conducting high-pressure microbial inactivation experiments. *Innovative Food Sci. Emerg. Technol.*, 5, 3, 299–306, 2004, 2004.
50. Alpas, H.A.M.İ., Kalchayanand, N., Bozoglu, F., Sikes, A., Dunne, C.P., Ray, B., Variation in resistance to hydrostatic pressure among strains of foodborne pathogens. *Appl. Environ. Microbiol.*, 65, 9, 4248–4251, 1999.
51. Benito, A., Ventoura, G., Casadei, M., Robinson, T., Mackey, B., Variation in resistance of natural isolates ofEscherichia coli O157 to high hydrostatic

pressure, mild heat, and other stresses. *Appl. Environ. Microbiol.*, 65, 4, 1564–1569, 1999.
52. Rastogi, N.K., Raghavarao, K.S.M.S., Balasubramaniam, V.M., Niranjan, K., Knorr, D., Opportunities and challenges in high pressure processing of foods. *Crit. Rev. Food Sci. Nutr.*, 47, 1, 69–112, 2007.
53. Smelt, J.P.P.M., Recent advances in the microbiology of high pressure processing. *Trends Food Sci. Technol.*, 9, 4, 152–158, 1998.
54. Balasubramaniam, V.B., Martinez-Monteagudo, S.I., Gupta, R., Principles and application of high pressure–based technologies in the food industry. *Annu. Rev. Food Sci. Technol.*, 6, 435–446, 2015.
55. Chauhan, O.P. (Ed.), *Non-Thermal Processing of Foods*, CRC Press, Boca Raton, FL, 2019.
56. Di Benedetto, N., Perricone, M., Corbo, M.R., Alternative non thermal approaches: microwave, ultrasound, pulsed electric fields, irradiation, in: *Application of Alternative Food-Preservation Technologies to Enhance Food Safety and Stability*, pp. 143–160, Bentham Science Publishers, Sharjah, UAE, 2010.
57. Torley, P.J. and Bhandari, B.R., Ultrasound in food processing and preservation, in: *Handbook of Food Preservation*, pp. 731–758, CRC Press, Boca Raton, FL, 2007.
58. Gallego-Juárez, J.A., Basic principles of ultrasound, in: *Ultrasound Food Process*, pp. 1–26, John Wiley & Sons, Chichester, UK, 2017.
59. Mason, T.J., Paniwnyk, L., Chemat, F., Ultrasound as a preservation technology, in: *Food Preservation Techniques*, pp. 303–337, Elsevier, Cambridge, UK, 2003.
60. Deeth, H.C., Datta, N., Ross, A.I.V., Dam, X.T., Pulsed Electric Field Technology: Effect on Milk and Fruit Juices', in: *Advances in Thermal and Non-Thermal Food Preservation*, Blackwell Publishing, Hoboken, NJ, USA, 2007.
61. Picart, L. and Cheftel, J.C., Pulsed electric fields, in: *Food Preservation Techniques*, pp. 360–427, Woodhead Publishing, Boca Raton, FL, 2003.
62. Amit, S.K., Uddin, M.M., Rahman, R., Islam, S.R., Khan, M.S., A review on mechanisms and commercial aspects of food preservation and processing. *Agric. Food Secur.*, 6, 1, 51, 2017.
63. Gavahian, M. and Khaneghah, A.M., Cold plasma as a tool for the elimination of food contaminants: Recent advances and future trends. *Crit. Rev. Food Sci. Nutr.*, 60, 9, 1581–1592, 2020.
64. Misra, N.N., Schlüter, O., Cullen, P.J. (Eds.), *Cold Plasma in Food and Agriculture: Fundamentals and Applications*, Academic Press, Cambridge, MA, 2016.
65. Feizollahi, E., Misra, N.N., Roopesh, M.S., Factors influencing the antimicrobial efficacy of Dielectric Barrier Discharge (DBD) Atmospheric Cold Plasma (ACP) in food processing applications. *Crit. Rev. Food Sci. Nutr.*, 1–24, 2020.

66. Barbosa-Cánovas, G.V., Swanson, B.G., Harte, F., Use of magnetic fields as a nonthermal technology, in: *Novel Food Processing Technologies*, pp. 443–451, CRC Press, Boca Raton, FL, 2005.
67. Grandison, A.S., *Membrane filtration techniques in food preservation*, Woodhead Publishing, Cambridge, UK, 2003.
68. Sablani, S.S., Food preservation and processing using membranes, in: *Food Preservation*, p. 365, CRC Press, Boca Raton, FL, 2007.
69. Singh, R.P. and Heldman, D.R., *Introduction to Food Engineering*, Gulf Professional Publishing, Houston, Texas, 2001.
70. Berk, Z., *Food Process Engineering and Technology*, Academic Press, London, UK, MA 2018.
71. Ramaswamy, H.S. and Marcotte, M., *Food Processing: Principles and Applications*, CRC Press, Boca Raton, FL, 2005.
72. Vicente, A.A. and Castro, I., *Novel Thermal Processing Technologies*, Blackwell Publishing, Ames, Iowa, 2007.
73. Knirsch, M.C., Dos Santos, C.A., de Oliveira Soares, A.A.M., Penna, T.C.V., Ohmic heating–a review. *Trends Food Sci. Technol.*, 21, 9, 436–441, 2010.
74. Kaur, N. and Singh, A.K., Ohmic heating: concept and applications—a review. *Crit. Rev. Food Sci. Nutr.*, 56, 14, 2338–2351, 2016.
75. Pires, R.P., Cappato, L.P., Guimarães, J.T., Rocha, R.S., Silva, R. *et al.*, Ohmic heating for infant formula processing: Evaluating the effect of different voltage gradient. *J. Food Eng.*, 280, 109989, 2020.
76. Pereira, R.N., Coelho, M.I., Genisheva, Z., Fernandes, J.M., Vicente, A.A., Pintado, M.E., Using Ohmic Heating effect on grape skins as a pretreatment for anthocyanins extraction. *Food Bioprod. Process.*, 124, 320–328, 2020.
77. Funcia, E.S., Gut, J.A., Sastry, S.K., Effect of electric field on pectinesterase inactivation during orange juice pasteurization by ohmic heating. *Food Bioprocess Technol.*, 13, 7, 1206–1214, 2020.
78. Martins, C.P., Cavalcanti, R.N., Cardozo, T.S., Couto, S.M., Esmerino, E.A. *et al.*, Effects of microwave heating on the chemical composition and bioactivity of orange juice-milk beverages. *Food Chem.*, 345, 128746, 2020.
79. Yan, W., Liu, Q., Wang, Y., Tao, T., Liu, B. *et al.*, Inhibition of Lipid and Aroma Deterioration in Rice Bran by Infrared Heating. *Food Bioprocess Technol.*, 13, 10, 1677–1687, 2020.
80. Ozturk, S., Kong, F., Singh, R.K., Evaluation of Enterococcus faecium NRRL B-2354 as a potential surrogate of Salmonella in packaged paprika, white pepper and cumin powder during radio frequency heating. *Food Control*, 108, 106833, 2020.
81. Qu, Y., Ramaswamy, H., Li, R., Guana, X., Cheng, T., Wang, S., Determining the top electrode voltage in free-running oscillator radio frequency heating of soybeans under different electrode configurations, in: *2020 ASABE Annual International Virtual Meeting*, American Society of Agricultural and Biological Engineers, p. 1, 2020.

82. Kamiloglu, S., Industrial freezing effects on the content and bioaccessibility of spinach (Spinacia oleracea L.) polyphenols. *J. Sci. Food Agric.*, 100, 11, 4190–4198, 2020.
83. Liu, S., Zeng, X., Zhang, Z., Long, G., Lyu, F. et al., Effects of Immersion Freezing on Ice Crystal Formation and the Protein Properties of Snakehead (Channa argus). *Foods*, 9, 4, 411, 2020.
84. Gaikwad, P.S., Pare, A., Sunil, C. K. Effect of process parameters of microwave-assisted hot air drying on characteristics of fried black gram papad. *J. Food Sci. Technol.*, 1–11, 2021.
85. Cheng, W., Gao, L., Wu, D., Gao, C., Meng, L. et al., Effect of improved extrusion cooking technology on structure, physiochemical and nutritional characteristics of physically modified buckwheat flour: Its potential use as food ingredients. *LWT - Food Sci. Technol.*, 133, 109872, 2020.
86. Liu, Y., Tang, T., Duan, S., Qin, Z., Zhao, H. et al., Applicability of Rice Doughs as Promising Food Materials in Extrusion-Based 3D Printing. *Food Bioprocess Technol.*, 13, 3, 548–563, 2020.
87. Derossi, A., Caporizzi, R., Paolillo, M., Severini, C., Programmable texture properties of cereal-based snack mediated by 3D printing technology. *J. Food Eng.*, 289, 110160, 2020.
88. Tewari, G., Microwave and radio-frequency heating, in: *Advances in Thermal and Non-Thermal Food Preservation*, pp. 91–98, John Wiley & Sons, Ames, Iowa, 2008, 2007.
89. Fito, P., Chiralt, A., Martin, M.E., Current state of microwave applications to food processing, in: *Novel Food Processing Technologies*, CRC Press, Boca Raton, FL, 2005.
90. Tang, J., Wang, Y., Chan, T.V.C.T., Radio frequency heating in food processing, in: *Novel Food Processing Technologies*, pp. 501–524, CRC Press, Boca Raton, FL, 2005.
91. Pare, A. and Mandhyan, B.L., *Food Process Engineering and Technology*, New India Pub. Agency, New Delhi, IN, 2011.
92. Ufheil, G. and Escher, F., Dynamics of oil uptake during deep-fat frying of potato slices. *LWT - Food Sci. Technol.*, 29, 7, 640–644, 1996.
93. Sioen, I., Haak, L., Raes, K., Hermans, C., De Henauw, S. et al., Effects of pan-frying in margarine and olive oil on the fatty acid composition of cod and salmon. *Food Chem.*, 98, 4, 609–617, 2006.
94. Haak, L., Sioen, I., Raes, K., Van Camp, J., De Smet, S., Effect of pan-frying in different culinary fats on the fatty acid profile of pork. *Food Chem.*, 102, 3, 857–864, 2007.
95. Ramesh, M.N., Cooking and frying of foods. *Food Sci. Technol.*, 167, 625, 2004.
96. van Koerten, K.N., Somsen, D., Boom, R.M., Schutyser, M.A.I., Modelling water evaporation during frying with an evaporation dependent heat transfer coefficient. *J. Food Eng.*, 197, 60–67, 2017.

97. Morton, I.D., Geography and history of the frying process. *Grasas Aceites*, 49, 3–4, 247–249, 1998.
98. Kalogeropoulos, N., Salta, F.N., Chiou, A., Andrikopoulos, N.K., Formation and distribution of oxidized fatty acids during deep-and pan-frying of potatoes. *Eur. J. Lipid Sci.*, 109, 11, 1111–1123, 2007.
99. Arvanitoyannis, I.S. and Dionisopoulou, N., Acrylamide: formation, occurrence in food products, detection methods, and legislation. *Crit. Rev. Food Sci. Nutr.*, 54, 6, 708–733, 2014.
100. Saalia, F.K. and Phillips, R.D., Degradation of aflatoxins by extrusion cooking: Effects on nutritional quality of extrudates. *LWT - Food Sci. Technol.*, 44, 6, 1496–1501, 2011.
101. Meuser, F. and Wiedmann, W., Extrusion plant de-Sign, in: *extrusion cooking*, C. Mercier, P. Linko, J.M. Harper, (Eds.), pp. 128–133, AACC, St. Paul, MIN, 1989.
102. Harper, J.M., Food extruders and their applications, in: *Extrusion cooking*, 1989.
103. Harper, J.M., *The technology of extrusion cooking*, N. Frame (Ed.), Blackie Academic and Professional, London, 1994.
104. Álvarez-Castillo, E., Oliveira, S., Bengoechea, C., Sousa, I., Raymundo, A., Guerrero, A.A., Rheological approach to 3D printing of plasma protein based doughs. *J. Food Eng.*, 288, 110255, 2021.
105. Pulatsu, E., Su, J.W., Kenderes, S.M., Lin, J., Vardhanabhuti, B., Lin, M., Effects of Ingredients and Pre-heating on the Printing Quality and Dimensional Stability in 3D Printing of Cookie Dough. *J. Food Eng.*, 294, 110412, 2020.
106. Abu-Ghannam, N. and Crowley, H., The effect of low temperature blanching on the texture of whole processed new potatoes. *J. Food Eng.*, 74, 3, 335–344, 2006.
107. Raso, J., Pagán, R., Condón, S., Nonthermal technologies in combination with other preservation factors, in: *Novel Food Processing Technologies*, pp. 453–475, CRC Press, Boca Raton, FL, 2005.
108. Leistner, L. and Gould, G.W., *Hurdle Technologies: combination treatments for food stability, safety and quality*, Springer Science and Business Media, New York, USA, 2012.
109. Erkmen, O. and Bozoglu, T.F., *Food Microbiology, 2 Volume Set: Principles into Practice*, John Wiley & Sons, New York, USA, 2016.
110. Robichaud, V., Bagheri, L., Salmieri, S., Aguilar-Uscanga, B.R., Millette, M., Lacroix, M., Effect of γ-irradiation and food additives on the microbial inactivation of foodborne pathogens in infant formula. *LWT*, 139, 110547, 2020.
111. Park, J.S. and Ha, J.W., Ultrasound treatment combined with fumaric acid for inactivating food-borne pathogens in apple juice and its mechanisms. *Food Microbiol.*, 84, 103277, 2019.
112. Yadav, B., Spinelli, A.C., Misra, N.N., Tsui, Y.Y., McMullen, L.M., Roopesh, M.S., Effect of in-package atmospheric cold plasma discharge on microbial

safety and quality of ready-to-eat ham in modified atmospheric packaging during storage. *J. Food Sci.*, 85, 4, 1203–1212, 2020.
113. Tang, J., Zhang, H., Tian, C., Shao, S., Effects of different magnetic fields on the freezing parameters of cherry. *J. Food Eng.*, 278, 109949, 2020.
114. Devi, S., Zhang, M., Ju, R., Bhandari, B., Water loss and partitioning of the oil fraction of mushroom chips using ultrasound-assisted vacuum frying. *Food Biosci.*, 38, 100753, 2020.
115. Odueke, O.B., Farag, K.W., Baines, R.N., Chadd, S.A., Irradiation applications in dairy products: a review. *Food Bioprocess Technol.*, 9, 5, 751–767, 2016.
116. Pati, S., Chatterji, A., Dash, B.P., Raveen Nelson, B., Sarkar, T. *et al.*, Structural Characterization and Antioxidant Potential of Chitosan by γ-Irradiation from the Carapace of Horseshoe Crab. *Polymers*, 12, 10, 2361, 2020.
117. Elias, M.I., Madureira, J., Santos, P.M.P., Carolino, M.M., Margaça, F.M.A., Verde, S.C., Preservation treatment of fresh raspberries by e-beam irradiation. *Innovative Food Sci. Emerg. Technol.*, 66, 102487, 2020.
118. Hamad, A.M., Fahmy, H.M., Elshemey, W.M., FT-IR spectral features of DNA as markers for the detection of liver preservation using irradiation. *Radiat. Phys. Chem.*, 166, 108522, 2020.
119. Ricciardi, E.F., Lacivita, V., Conte, A., Chiaravalle, E., Zambrini, A.V., Del Nobile, M.A., X-ray irradiation as a valid technique to prolong food shelf life: The case of ricotta cheese. *Int. Dairy J.*, 99, 104547, 2019.
120. Moy, J.H., Food irradiation—an emerging technology, in: *Novel Food Processing Technologies*, CRC Press, Boca Raton, FL, 2005.
121. Branen, A.L., Davidson, P.M., Salminen, S., Thorngate, J. (Eds.), *Food Additives*, CRC Press, New York, USA, 2001.
122. Carocho, M., Barreiro, M.F., Morales, P., Ferreira, I.C., Adding molecules to food, pros and cons: A review on synthetic and natural food additives. *Compr. Rev. Food Sci. Food Saf.*, 13, 4, 377–399, 2014.
123. Primacella, M., Acevedo, N.C., Wang, T., Effect of freezing and food additives on the rheological properties of egg yolk. *Food Hydrocolloids*, 98, 105241, 2020.
124. Bhooma, V., Nagasathiya, K., Vairamani, M., Parani, M., Identification of synthetic dyes magenta III (new fuchsin) and rhodamine B as common adulterants in commercial saffron. *Food Chem.*, 309, 125793, 2020.
125. Carocho, M., Morales, P., Ferreira, I.C., Natural food additives: Quo vadis? *Trends Food Sci. Technol.*, 45, 2, 284–295, 2015.

11

Green Solvents for Food Processing Applications

A. Surendra Babu[1], A. Sangeetha[2] and R. Jaganmohan[3]*

[1]Department of Food Technology, School of Liberal Arts and Applied Science, Hindustan Institute of Technology and Science (Deemed to be University), Chennai, Tamil Nadu, India
[2]Department of Nutrition and Dietetics, Jamal Mohamed College, Trichy, Tamil Nadu, India
[3]Department of Food Product Development, National Institute of Food Technology, Entrepreneurship and Management (NIFTEM) - Thanjavur (formerly Indian Institute of Food Processing Technology - IIFPT), Ministry of Food Processing Industries (MoFPI), Government of India, Tamil Nadu, Thanjavur

Abstract

Food bioactive compounds are known to be heat sensitive and vulnerable to chemical, physical, and microbiological changes. Losses of some nutrients, low production efficiency, and time- and energy-consuming are limitations of various conventional food-processing methods. These shortcomings have led to the use of new sustainable "green" techniques which typically involve less time, water, and energy. The advantages of using green solvents created a revolution for the demand of ionic liquids (ILs), to replace the traditional organic solvents that are highly volatile and toxic in nature. Nonetheless, their green aspects have been questioned in the recent times due to their poor biocompatibility and biodegradability. Hence, to overcome this lacuna, deep eutectic solvents (DESs) came into existence since 2004, as a green alternative to ILs. Natural deep eutectic solvents (NADESs), a new class of DESs, may be regarded as designer solvents due to their numerous structural variations and possibility to design their physicochemical properties. These ideal alternative solvents are extremely recommended for extraction of food bioactive compounds as they have high solvency, low toxicity and low environmental impacts, easily biodegradable, obtained from renewable resources at a reasonable price, and should be easy to recycle without any deleterious effect to the environment.

---

*Corresponding author: jagan@iifpt.edu.in

Mousumi Sen (ed.) Food Chemistry: The Role of Additives, Preservatives and Adulteration, (341–374)
© 2022 Scrivener Publishing LLC

Keywords: Green solvents, natural deep eutectic solvents, phenolic compounds, flavonoids, antioxidants, super critical carbon dioxide

11.1 Introduction

Bioactive compounds are non-nutritive components that are present in minor quantities in fruits, vegetables, milk, whole grains, and their other by-products [1]. They are known to have demonstrated physiological health benefits on human health like antioxidant, anti-carcinogenic, anti-inflammatory, anti-diabetic, and anti-microbial properties. Extraction, characterization, and clinical trials of such bioactive components from plant and animal foods are of great interest in the areas of research [2]. Extraction of potential bioactive compounds such as polyphenols, proteins, pigments, and essential oils and incorporation in various food products in order to improve their nutritional and health benefits are now widely been studied [3].

Consequently, various conventional techniques like hydrodistillation, steamdistillation, maceration or advanced methods using microwave, ultrasound-assisted, pressurized hot water extraction, and supercritical fluids (SCFs) have been adopted for the extraction of bioactive compound and widely been reported [4, 5]. In connection to extraction process, there are a number of factors that might play a vital role in deciding the process efficiency which includes type of solvent, extraction time, temperature, pressure, solid-to-solvent ratio, etc. [6]. Among them, extraction solvents have grabbed more attention for the bad reason since a decade. These solvents are principally volatile organic liquids mostly acquired from non-renewable resources particularly of petroleum-based and are supposed to be harmful to mankind and the environment [7]. Furthermore, the recovery of bioactive compounds from the solvents and retaining their native chemical structure are also found to be problematic in the extraction process [8]. Extraction solvents are only an intermediate of the extraction process and are not directly responsible for the composition of the extracted product or an active component of the formulation. Hence, the use of toxic, flammable, or environment concern solvents seems to be redundant since these qualities have no impact on the function or progress of the system in which the solvent is applied [9].

In the recent times, best alternative solvents that exhibit negligible acute impacts on the human health and environment are being studied to replace the harmful organic solvents. In this regard, upsurge attention is gained toward green extraction alternatives which is based on renewable plant resources, alternative solvent usage (non-petroleum based), and emergent technologies that reduces energy, cost, and time [10]. There is a lot of demand for the research in area of developing green alternatives and extractions which obeys the principles of "green chemistry" that focuses on using environmental friendly and non-toxic solvents [11]. Therefore, replacement of conventional solvents with eco-friendly green alternatives would revolutionize pharmaceutical and food industry.

According to Anastas and Warner [12], the following new 12 principles of green chemistry (Figure 11.1) have to be followed to achieve the sustainable eco-development of the chemical and food industry in the future. The outlines of 12 principles of green chemistry provide a frame work for learning, designing, or improving materials, methods that make a greener chemical, process, or product and are listed as follows:

1. Prevent Waste: It is better to prevent waste than to treat or clean up waste after it has been created.
2. Atom Economy: Synthetic methods should be designed to maximize the incorporation of all materials used in the process into the final product.

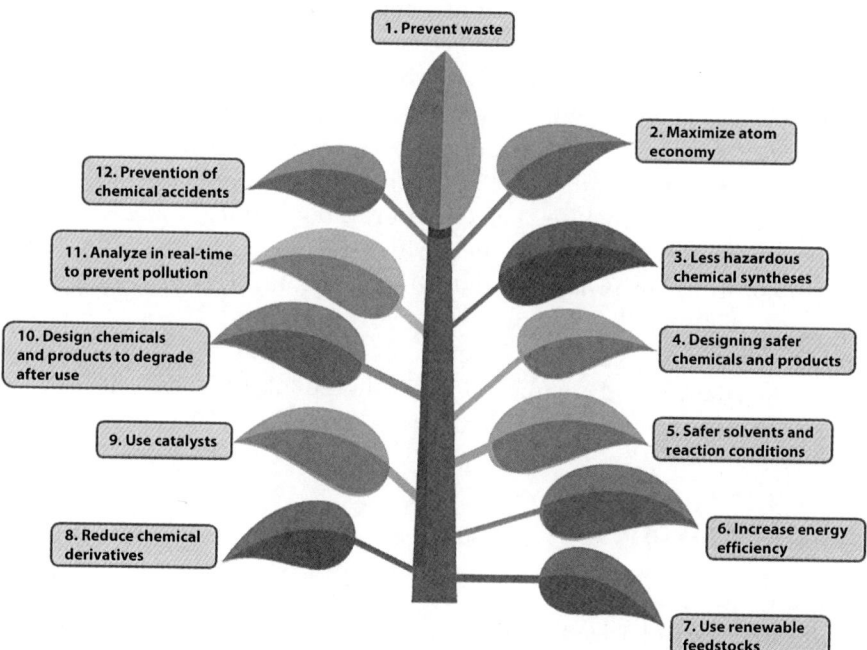

Figure 11.1 New 12 principles of green chemistry postulated by Anastas and Warner [12].

3. Less Hazardous Chemicals: Syntheses, wherever practicable, should be designed to use and generate substances that possess little or no toxicity to human health and the environment.
4. Designing Safer Chemicals: Various chemicals that are employed during chemical processes must be highly efficient in performing desired function while minimizing their toxicity.
5. Safer Solvents and Auxiliaries Substances: Solvents, separation agents, etc., must be safe for workers and the environment.
6. Design for Energy Efficiency: Energy requirements of chemical processes should be recognized for their environmental and economic points of view and should be minimized.
7. Use of Renewable Feedstocks: Raw materials or feedstock should be renewable rather than depleting.
8. Reduce Derivatives: Unnecessary derivatization (blocking groups, protection/deprotection, temporary modification, etc.) should be minimized or avoided if possible.
9. Catalysis and new catalytic reagents (enzymes, as selective as possible) are superior to stoichiometric reagents.
10. Design Products for Degradation: Chemical products should be designed so that at the end of chemical function they perform, they break down into biodegradable degradation products and do not persist in the environment.
11. Real-Time Analysis for Pollution Prevention: Analytical methodologies need to be further developed to allow for real-time, in-process monitoring, and control prior to the formation of hazardous substances.
12. Inherently Safer Chemistry for Accident Prevention: Substances and chemical process should be chosen to minimize the potential for chemical accidents.

11.2 Green Solvents

According to Choi and Verpoorte [13], "What you see is what you extract" which emphasize the prominence of solvent extraction during the extraction of bioactive compounds; nevertheless, the statement holds true to extraction of other industrial products like aromas, colors, fats and oils, and flavors for food and non-food industrial applications. Consequently, solvents have gained a lot of attention as they are applied in a surplus amount during extraction or purification stages [14]. Most of the commonly used solvents are documented to have environmental concerns due to three aspects: the source and synthesis of the solvent, its properties in use, and final disposal [15]. However, considering the health and environment concerns associated with the application of regular organic solvents in the extraction process the term "green solvents" has drawn more popular in the recent times. The following characteristics need to be addressed for a solvent to claim as a green solvent, health and safety issues, indirect impact during their production, application, and disposal which includes energy consumption in their synthesis, recycling of solvent, and waste treatment [16].

According to a great deal of the literature the solvents and solvent classes that are considered to be "green solvents" include water [17, 18], SCFs [19, 20], gas expanded liquids [21], ionic liquids (ILs) [22, 23], solvents derived from biomass [24, 25], and deep eutectic solvents (DESs) [26, 27].

11.2.1 Water as Green Solvent

Besides water being considered as "universal solvent" and is used widely as a medium over centuries for extraction of natural components from food and other natural sources, it is also reported to have a potential application as a green solvent due to the fact that water is inexpensive, eco-friendly, nonflammable, and nontoxic in nature providing opportunities for pollution free clean and green processing conditions. The small size of water molecule with a diameter of 2.75 Å offers a supreme ability to hydrate most of the solutes. The net dipole moment of water is found to 1.85D which is ascribed due to the presence of positive charges on hydrogen (H^+) atoms and negative charge on oxygen (O^-) atom. Whereas the dielectric constant (ε) value (which is a measure of the polarity) of water is 78.3 that is attributed to dipole orientation of the hydrogen bonding. It is considered to be a bad solvent for non-polar or few semi-polar compounds. However, the polarity of water can be reduced at extreme temperature and pressure as the hydrogen bond gets disintegrated [28]. Pure water at ambient temperature and pressure has a ε value about 79, while increasing temperature to 200°C at a pressure of 1.5 MPa (need to maintain the liquid state) its ε value drops to about 35 [29] which is similar to that of methanol at ambient conditions [30]. Thus, water at elevated temperature is a solvent of lower polarity. It also favors water to better wet and penetration of the sample matrix due to reduced surface tension and lesser viscosity. Furthermore, it improves analyte diffusion and mass-transfer kinetics as well [31, 32]. This process of extraction is generally known as subcritical water extraction (SWE), superheated water extraction, pressurized hot water extraction.

11.2.2 Subcritical Water Extraction

SWE runs under different temperature between the boiling point and critical point of water (100°C at 1 bar and 374°C at 221 bar) and high pressure conditions to maintain water in a liquid state. It is reported that temperature plays a vital role on attaining the polarity of subcritical water than pressure. Hence, organic compounds are more soluble in subcritical water state due to its lower polarity. In a recent study, SWE extraction of bioactive compounds from carrot leaves at varying temperatures (110°C–230°C) has been investigated by Song et al. [33]. It was found that there is a significant increase in the extraction of phenolic content. Similarly, Lachos-Perez et al. [34] postulated that SWE is a superior method for recovery of flavanones particularly hesperidin and narirutin form defatted orange peels with enhanced antioxidant activity. The yield accounts to about approximately 21% of total flavonones in the extract leading to burst extracts obtained in SWE. These reports state that SWE is considered to be an efficient method for the extraction of bioactive compounds with high antioxidant activity.

11.2.3 Supercritical Fluids as Green Solvent

If a fluids pressure and temperature crosses beyond its respective critical value (Tc means critical temperature and Pc means critical pressure), it is considered as super critical.

As per Figure 11.2, the critical point situated at the upper end of right and the phase area over this point is regarded as SCF region [35]. A SCF can behave both as a liquid and a gas. SCF is a state obtained by simultaneous exertion of temperature and pressure higher than the critical point. SCF extraction is new separation techniques established as an alternative to regular organic solvents. They exhibit unique properties of both liquids and gases, i.e., being able to penetrate anything and being capable of dissolving compounds into them, respectively. Furthermore, SCF offers an advantage of being able to vary its density to a larger extent in a continuous matter and their gas-like viscosity results in high mass transfer. The physiochemical properties of SFE can be altered by increasing the temperature and pressure levels beyond their critical limits [36]. As a result, carbon dioxide or water in their SCF state can be a suitable alternative for organic

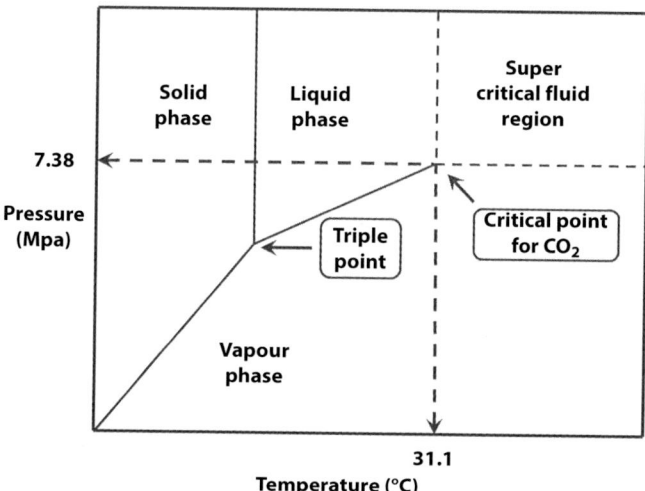

Figure 11.2 Schematic phase diagram of super critical carbon dioxide.

solvents [37]. Among the SCF carbon dioxide (CO_2) is the predominantly used SCF due to inert nature, non-toxic, non-flammable, less expensive, abundance, and easy separation from the product and offers moderate critical properties (Tc = 31.1°C, Pc = 7.38 Mpa). Manipulation of these critical properties can change its density which in turn improve solvating power for selective extractions due to its high volatility and solvent-free depressurization and is widely studied in the literature [36]. Various studies have been reported on extraction of bioactive compounds using supercritical CO_2; however, it has limited industrial application because of high pressure conditions (Pc = 7.38 MPa) [38].

11.2.4 Gas Expanded Liquids as Green Solvent

Although supercritical carbon dioxide system offers profuse advantages in extraction of various compounds, it is operated at high temperature and pressure and in general permits limited solubility of many compounds. To overcome these limitations, gas-expanded liquid (GXL) systems have emerged as promising media for separation and extraction of natural compounds [39]. GXL are mixed solvent system composed primarily an organic solvent and a compressed gas. In general the liquid solvent like hexane is compressed with a gas mostly CO_2 where gas dissolves into the liquid portion to generate a solvent at a pressure lower than vapor pressure of the pure gas and higher than atmospheric pressure (Jossop and

Subramaniam, 2007). The organic liquid enhances the solubility of solutes, while CO_2 increases gas solubility and mass transfer.

Peng and Robinson [40] designed an equation to predict the vapor liquid equilibrium of gas liquid mixtures as follows:

$$\frac{V}{Vo} = \frac{V - Vo}{Vo}$$

where V is the total liquid volume at an applied pressure of gas and Vo is the liquid volume of the organic solvent at atmospheric conditions.

The volume of the liquid phase increases considerably when the pressure of the applied gas increases because of enhanced gas solubility. The physiochemical properties of GXL can be tuned by manipulating properties of the solvent and gas just by varying partial pressure of the gas [41]. GXL is known to be an effective green solvent for extraction of natural compounds. Application of expansion gas (N_2) in high-pressure water in combination with ultrasound had improved the yield of phenolic compound from pomegranate peel to about 26% [42]. Reyes et al. [43] reported a significant improvement of astaxanthin yield from *Haematococcus pluvialis* using CO_2-expanded ethanol with enhanced antioxidant activity.

11.2.5 Ionic Liquids as Green Solvent

ILs are organic salts which contain an organic cation and a polyatomic inorganic anion, which remains as liquids under 100°C. Most of the ionic have negligible vapor pressure, non-flammable, thermal stability, and recognized as green "designer" solvents [44]. They exhibit wide temperature range and negligible vapor pressure. For example, 1-butyl-3-methylimidazolium hexafluorophosphate remains stable even upto 300°C. In addition, they present tremendous solvent properties and can influence chemical reactions without being altered [45].

11.2.5.1 Classification of Ionic Liquids

Generally, simple ILs contain a cation and an anion. The cations are bulky attached with alkyl chains while anions are small. Most common anions for a simple IL solvent are as follows:

Cl^- (Chloride ion)
Br^- (Bromide ion)

BF4⁻ (Tetrafluoroborate ion)
CF3SO3⁻ (Triflate ion)
$N(SO2CF3)_2^-$ [Bis(trifluoromethane)sulfonamide ion]

On the other hand, most common cations for IL systems are 1-alkyl-3-methyl-imidazolium cations, whereas other simple cations include phosphorous or nitrogen containing organic ions bonded with alkyl chains with varied chain length [46]. ILs as per Seddon [47] are described as neoteric solvents with seven desirable properties

- Higher liquid range compared to water
- Thermal stability (upto 200°C)
- In expensive and easy to prepare
- They display Bronsted, Lewis, and other types of acidity
- They possess limited or negligible vapor pressure
- Good solvent for organic, inorganic and polymeric compounds
- Water sensitivity does not affect industrial application

De Oliveira *et al.* [48] reported the extraction of alkaloids, caffeine and theobromine from cocoa beans using 2-hydroxy ethylammonium acetate as IL using ultrasound. The results showed an outstanding recovery of alkaloids from the cocoa sample. Wang *et al.* [49] applied 1-Butyl3-methylimidazolium bromide [(Bmim)Br] as an ionic solvent for the extraction of rutin from *Tamarix Chinensis* and found 15% higher yield than that of ultrasound extraction using methanol (Table 11.1). While Zhao *et al.* [50] demonstrated a twofold higher (5.49 mg/g) yield for the ultrasound and microwave-assisted rutin extraction from *Abutilon theophrasti* using IL 1-butyl-3-methylimidazolium bromide (Bmim)Br.

11.2.6 Solvents Derived From Biomass as Green Solvent

Agro- or bio-based solvents play a crucial role for the replacement of conventional organic solvents. Bio-based solvents are usually manufactured from natural biomasses such as wood, starch, fruit, and vegetables. Some common bio-based solvent include limonene, ethanol, glycerol, and 2-methyltetrahydrofuran. Although they have certain limitations like high viscous, high boiling point, and off flavor, they are far more advantageous owing to their biodegradability, nontoxic, ecofriendly, and nonflammable nature [58]. Bio-based solvents are classified into three types: cereal/sugar based, oleo-proteagineous based, and wood based. Solvents obtained from cereal/sugar are usually produced by natural fermentation

Table 11.1 Ionic liquids as green solvent for phytochemical extraction from various plant sources.

Ionic liquids	Method	Phytochemical	Source	Reference
1-butyl-3-methylimidazolium bromide, 1-butyl-3-methylimidazolium chloride and 1ethyl-3-methylimidazolium tetrafluoroborate,	MAE	Flavonoids (quercetin), phenolic acids (gallic acid)	*Labisia pumila*	[51]
Ethyl acetate 2-hydroxy ethylammonium acetate	UAE	Caffeine and theobromine	*Theobroma cacao* L.	[48]
Imidazolium, pyridinium, pyrrolidynium, and piperidinium-based anions with different anions	-	Quercetin	Red onion (*Allium cepa* L.)	[52]
Imidazolium-based anions with varying alkyl chain length (ethyl – octyl) and different cations	-	Flavonoids (kaempferol, quercetin and myricetin), phenolic acids (gallic acid and caffeic acid)	*Labisia pumila*	[53]
Choline-leucine	-	Flavonoids and pectin	*Citrus reticulata*, Blanco	[54]

(*Continued*)

Table 11.1 Ionic liquids as green solvent for phytochemical extraction from various plant sources. (*Continued*)

Ionic liquids	Method	Phytochemical	Source	Reference
1-alkyl-3-methylimidazolium bromide (with varying alkyl chain length), 1-Ethyl-3-methylimidazolium Tetrafluoroborate	MAE, Heat flux extraction	Flavonoid glycosides	*Chrysanthemum morifolium* Ramat	[55]
1-Butyl-3-methylimidazolium bromide	UAE, MAE	Rutin	*Abutilon theophrasti*	[56]
1-Butyl-3-methylimidazolium tetrafluoroborate, and 1-Butyl-3-methylimidazolium chloride	UAE	Phenolic compounds (3,4-DHBD, 4-HBA, SA, 2,3-DHBA and 4-HBD)	*Laminaria japonica*, Aresch	[57]
1-Butyl-3-methylimidazolium bromide	UAE	Rutin	*Tamarix Chinensis*	[49]

of sugars present in plants such as sugar beet, sugarcane, wheat, and corn. Limonene production from microbial fermentation by *E. coli* or *S. cerevisiae* has an enormous potential in green chemistry applications. This also minimizes bulk usage of citrus fruits and instead utilizes glucose from bio-waste for fermentation process [59]. On the other hand, lignocellulose residues obtained from straw and wood waste is utilized for the purpose of furfural production which is known to be great economic importance. Further, hydrogenation of furfural leads to formation of 2-methyl tetrahydrofuran which can be used for extraction of vegetable oils and aromatic compounds and be a best alternative for hexane [60]. The solvents derived from oleo-proteagineous sector include esters of fatty acids and derivatives of glycerol. This sector covers colza, sunflower, soya, etc., that are a rich source of vegetable oils [61].

A study was carried out to extract fatty acids from soybean, pean nuts, and sunflower using α-pinene as a bio-based solvent presented approx. In addition, 90% solvent recovery with a high efficiency in fatty acid extraction compared to the 50% recovery of n-hexane [62]. Likewise, oil extraction from sandbox seed was demonstrated using ethyl acetate as bio-based solvent. Results have confirmed 1.28% higher oil yield with ethyl acetate than n-hexane [63].

11.2.7 Deep Eutectic Solvents as Green Solvents

ILs which are considered as green solvents possess certain drawbacks due to their poor biocompatibility and biodegradability [64]. To overcome these limitations, DESs have been evolving as a green alternative for ILs since 2004 [65]. These solvents resemble ionic solvents in terms of its designability, stability, and negligible vapor pressure; in addition, they offer numerous advantages with respect to diversity, cost effectiveness, easy source raw materials, green, and simple synthesis [66]. DES is a mixture of a hydrogen bond donor (HBD) and a hydrogen bond acceptor (HBA). DESs are generally composed of carboxylic acids, amides, alcohols as HBD, and a quaternary phosphonium salts as a HBA. Since they are analogous to ILs in most of its characteristics and properties, they are admitted as a new class of IL analogues [67]. If a DES is produced by combining a HBD and an acceptor of a natural origin, then it is termed as natural DES (NADES). Most commonly used HBA are quaternary ammonium salts, quaternary phosphonium salts, imidazolium-based salts, dication-based salts, and molecular imidazole, while HBD includes water, urea, thiourea, amides, indoles, azoles, alcohols, acids, and phenol. Most of these compounds are solid at room temperature.

NADES are a new class of DES which has been emerged in 2011. It was first postulated by Choi *et al.* [68] to represent the solubility of intracellular components which are either hydrophilic or lipophilic in nature. The term NADES was coined for the eutectic mixtures containing two or more natural compounds such as choline chloride, citric acid, malic acid, glucose, and fructose water. They are known to be eco-friendly, non-volatile, safe, and readily biodegradable as they are of natural origin and their reaction products can be safer for food and non-food industrial applications [65].

11.3 Synthesis of NADES

Preparation of NADES does not require solvent and purification of final product is also not necessary for most of the mixtures as no by product formation occurs during the synthesis. They are synthesized by proper selection and mixture of HBA and HBD which then associates by themselves to form a eutectic phase having lower melting points than their respective ones. This decrease in the melting point is owed to formation of hydrogen bonding between the functional groups such as hydroxyl, carbonyl, carboxyl, and amino groups in the mixture [69, 70]. The higher the hydrogen bond between the counterparts, the lower the melting point of the mixture [71]. In addition, other bonding that includes electrostatic, dipole-dipole, and Van der Waals interaction may be possible [72]. Most of these eutectic mixtures are prepared simple agitation at about 80°C with continuous stirring. However, stirring becomes difficult for the sugar-based eutectic mixtures with high viscosity. In that case, to over the problem additional water can be added into the mixture [73]. Other preparative methods of NADES include freeze-drying [71], grinding in a mortar [74], or mixing in an extruder [75] (Table 11.2). Gutierrez *et al.* [71] prepared by mixing pure choline chloride:urea and choline chloride:thiourea deep eutectic mixtures with appropriate concentration (1:2 molar ratio) of choline and urea/thiourea individually in water. Later, the mixtures were freeze dried to obtain pure NADES. Santana *et al.* [76] had synthesized three different NADES based on malic acid-citric acid-water (MA-CA), xylitol-citric acid-water (Xyl-CA), and xylitol-malic acid-water (Xyl-MA) with a molar concentration of 1:1:10 employed by ultrasound and microwave. The authors proved that NADES synthesis using ultrasound and microwave was fast and efficient and recommended as a new advanced synthesis techniques.

Table 11.2 Various methods and components for the synthesis of NADESs.

Method of synthesis	NADESs components	Molar concentration	Reference
Controlled heating in water bath with agitation	ChCl-urea ChCl-glycerol ChCl-LA ChCl-EG ChCl-CA	1:2 1:1 1:1 1:1 1:1	[77]
Controlled heating and stirring, UAS, and MAS	MA-CA-water Xyl-CA-water Xyl-MA-water	1:1:10	[76]
Freeze drying, Vacuum evaporation, and Heating and stirring	Betaine-TA B-Alanine-MA Glucose-sucrose CA-glucose	2:1 3:2 1:1 2:1	[78]
Vortexing, Heating and mixing (thermal treatment, MAS, UAS, shaking incubation), Evaporation technique (freeze drying and rotary evaporation)	Sixteen different NADESs with varying components (ChCl, Acetyl ChCl, urea, glycerol)	Different ratios	[79]
Extrusion	ChCl-urea	1:2	[75]
Controlled heating and stirring	Betaine-urea-water Betaine-methylurea-water Betaine-glucose-water Betaine-sorbitol-water Betaine-glycerol-water Betaine-EG-water	1:2:1 1:3:1 1:1:2 1:2:1 1:1 1:2	[80]

(Continued)

Table 11.2 Various methods and components for the synthesis of NADESs. (*Continued*)

Method of synthesis	NADESs components	Molar concentration	Reference
Heating method, Grinding method	ChCl - carboxylic acids (LeA, GA, MoA, OA, and GyA)	Different ratios	[74]
Controlled heating and vacuum drying	Imidazole, halide, zinc halide and amide-based DESs	-	[81]
Freeze drying	ChCl -urea and ChCl - thiourea	1:2	[71]

UAS, ultrasound-assisted synthesis; MAS, microwave-assisted synthesis; ChCl, choline chloride; MA, malic acid; CA, citric acid; Xyl, Xylitol; TA, tartaric acid; EG, ethyle glycol; LeA, levulinic acid; GA; glutaric acid; MoA, malonic acid; OA, oxalic acid; GyA, glycolic acid; DES, deep eutectic solvent.

11.3.1 NADES for Extraction of Phenolic Compounds

Table 11.3 represents the list of NADESs used for extraction of phenolic compounds from the natural resources in the course of last 5 years (2015–2020). The studies dealing with extraction of bioactive compounds from plant sources using NADESs as an extraction solvent has been increased in a rapid proportion which is evident from the Table 11.3. The principle reason which makes them an ideal choice as an extraction solvent is because of its natural composition, high stability, and high solubility that can be applied for the extraction of desired phytochemicals of our choice such as proteins, peptides, phenolic compounds, flavonoids, and alkaloids. For an example, Grillo *et al.* [82] indicated that choline chloride–lactic acid used in 1:1 molar ratio had a greater yield of anthocyanins from blueberry peels with higher efficiency and lesser cost. Other study demonstrated about 112% higher extraction efficiency of choline chloride–tartaric acid (molar ratio of 1:2) along with 44% water as NADES for polyphenols from winery by-product (red grape pomace) [83]. Shikov *et al.* [84] optimized the extraction of phenyletanes and phenylpropanoids from *Rhodiola rosea* rizhomes using L-lactic acid, glucose fructose, and water as NADES. The authors reported a greater recovery of phenolic compounds from L-lactic acid-fructose-water with a molar ratio of 5:1:11.

Table 11.3 NADES as green solvent for phenolic compound extraction from various food samples.

NADES system	Molar ratio	Targeted analyte	Source	Reference
ChCl-LA (TST: 1)	1:1	Anthocyanins	Blueberry peels (*Vaccinium myrtillus* L.)	[82]
3 HBA (ChCl, betaine, L-carnitine), 4 HBD groups (TEG, EG, propanediol, glycerol, butanediol) (TST: 15)	Different ratios	Gingerol	Ginger (*Zingiber officinale* Roscoe)	[86]
ChCl-urea ChCl-sorbital ChCl-sucrose ChCl-glycerol SA-glycerol ChCl-LA SA-LA ChCl-MA (TST: 8)	Different ratios	Polyphenols	Ripe mango peel (*Mangifera indica* L.)	[87]

(Continued)

Table 11.3 NADES as green solvent for phenolic compound extraction from various food samples. (Continued)

NADES system	Molar ratio	Targeted analyte	Source	Reference
ChCl-urea ChCl-LA ChCl-LA1 ChCl-fructose-water ChCl-glucose-water (TST:5)	1:1 1:1 1:2 2:1:1 2:1:1	Gallic acid, protocatehuic acid, chlorogenic acid, vanillic acid, caffeic acid syringic acid, epicatechin p-coumaric acid, ferulic acid, and transcinnamic acid	Chokeberry (*Aronia Melanocarpa*)	[88]
ChCl-TA (TST: 1)	1:2	Polyphenols	Red grape pomace	[83]
L-LA-fructose-water (TST: 1)	5:1:11	Phenyletanes and phenylpropanoids	*Rhodiola rosea* rizhomes	[84]
2 HBA (ChCl, and betaine), 4 HBD groups (Glycerol, sucrose, butanediol, propanediol, levulinic acid, GA, OA) (TST: 7)	Different ratios	Cyanidin-3-glucoside, pelargonidin-3-glucoside, pelargonidin-3-rutinoside, Quercetin-glucoside, ellagic acid and its derivative, kaemferol-3-glucoside, kaemferol, anthocyanins	Strawberry extrudate, Raspberry extrudate	[89]

(Continued)

Table 11.3 NADES as green solvent for phenolic compound extraction from various food samples. (Continued)

NADES system	Molar ratio	Targeted analyte	Source	Reference
Betaine- propanediol (TST:1)	3:1	α-mangostin	Rind of mangosteen (Garcinia mangostana L.)	[90]
ChCl-sucrose ChCl-urea ChCl-OA ChCl-sorbitol (TST: 4)	4:1 1:2 1:1 3:1	Various phenolic compounds	Onion peels (Allium cepa L.)	[91]
ChCl-CA ChCl-LA ChCl-maltose ChCl-glycerol (TST: 4)	1:2	Various phenolic compounds	Virgin olive pomace (Olea europaea L.)	[92]
ChCl-glycerol ChCl-glycerol ChCl-glycerol ChCl-EG ChCl-EG ChCl-EG (TST: 6)	1:2 1:3 1:4 1:2 1:3 1:4	Gallic acid, ferulic acid, and p-coumaric acid	Orange peel	[93]

(Continued)

Table 11.3 NADES as green solvent for phenolic compound extraction from various food samples. (*Continued*)

NADES system	Molar ratio	Targeted analyte	Source	Reference
ChCl-propanediol ChCl-butanediol ChCl-LA ChCl-CA **(TST:4)**	2:1,1:1,1:2, 1:3,1:4,1:5	Various anthocyanins	*Lycium ruthenicum* Murr. Fruit	[94]
ChCl-EG ChCl-glycerol **(TST: 2)**	1:2	Ferulic, caffeic and cinnamic acids	Olive, almond, sesame, cinnamon oil	[95]
ChCl-CA ChCl-maltose ChCl-fructose CA-maltose **(TST: 4)**	1:1 4:1 5:2 4:1	Various anthocyanins	Grape skin	[96]

ChCl, choline chloride; LA, lactic acid; TEG, triethylene glycol; HBA, hydrogen bond acceptor; HBD, hydrogen bond donor; TA, tartaric acid; EG, ethylene glycol; SA, sodium acetate; MA, malic acid; GyA, glycolic acid; OA, oxalic acid; CA, citric acid; EG, ethylene glycol; TST, total samples tested.

Similar study was conducted by Tsvetov et al. [85], where cinnamic alcohol extracted from *Rhodiola rosea* using NADESs prepared using choline chloride–glycerol-water with a molar ratio of 1:2:1 showed twofold higher extraction efficiency than conventional organic solvents. The results of the studies (listed in Table 11.3) prove that NADES has remarkable efficiency in the extraction of phenolic compounds from plant/food sources and can be regarded as the best green alternatives to the conventional solvents.

11.3.2 NADES for Extraction of Flavonoids

Flavonoids are a group of bioactive compounds naturally present in fruits, vegetables, etc. Increased interest in flavonoid research has been evident in the literature which is ascribed to its reported health benefits. Various studies have confirmed the potential efficacy and promising results of NADES in the extraction of flavonoid from different plant materials. The overview of NADES application in the extraction of flavonoid extractions from plant food sources are presented in Table 11.4. In a recently study, choline chloride and betaine-based NADES were applied as a green solvent for the extraction of quercetin from onion and broccoli. The recovery rate was found to be 99% which proves that the developed natural solvents are more efficient than organic solvent [97]. A novel attempt was made by Jokic et al. [98] to isolate hesperidin from four varieties of mandarin peels. The authors have used 15 different choline chloride–based NADES mixtures with varying molar concentrations. The results of the screening test confirmed the highest and lowest extraction efficiency for hesperidin to be choline chloride:acetamide and choline chloride:citric acid solvent mixtures, respectively. In another study, different flavonoids were extracted from a variety of fruits, vegetables, and spices using a wide range of NADES solvent mixtures (17 types of DES based on choline chloride, acetylcholine chloride, choline tatrate, betaine, and carnitine mixtures) [99]. Bajkacza and Adamek [100] isolated isoflavones from soy products using choline chloride:citric acid with a molar ratio of 1:1. The results of the study indicated that the developed NADES system could be an alternative sustainable media for the extraction of isoflavones compared to organic solvents. The findings of all the studies reported in the Table 11.4 would provide an insight of obtaining potential bioactive flavonoids using natural green solvents and further to minimize the substantial environmental issues of waste disposal.

Table 11.4 NADES as green solvent for flavonoid extraction from various food samples.

NADES system	Molar ratio	Targeted analyte	Source	Reference
TMACl-LA Borneol-decanoic acid (**TST:3**)	1:2, 1:3 1:4	Morin, kaempferol, and quercetin	Dark tea, chocolate, tomato juice, pineapple juice, and orange juice	[101]
ChCl as HBA, and 11 HBD groups (propanediol, glycerol, EG, MA, MoA, TSA, LeA, OA, resorcinol, Xyl, urea (**TST: 11**)	Different ratios	Myricetin, Morin, rutin, luteolin, hyperoside, Quercitrin, apigenin	*Lycium barbarum L. fruits*	[102]
2 HBA groups (ChCl, and betaine) and 9 HBD groups (carbohydrate based) (**TST:18**)	Different ratios	Quercetin, isorhamnetin kaempferol	Onion and broccoli	[97]
ChCl as HBA, and 10 HBD groups (Acetyl amide, TEG, propanediol, butanediol, urea, EG, glycerol, OA, MoA (**TST:18**)	Different ratios	Orientin, isoorientin, vitexin, isovitexin, quercetin-3-O-robinobioside, and rutin	Buckwheat sprouts	[103]

(*Continued*)

Table 11.4 NADES as green solvent for flavonoid extraction from various food samples. (Continued)

NADES system	Molar ratio	Targeted analyte	Source	Reference
ChCl-EG TMACl-EG TEACl-EG **(TST:3)**	1:4 1:3 1:3	Quercetin, and morin	Onion, apple, orange, and pineapple juices	[104]
3 HBA groups (ChCl, betaine, L-proline) 4 HBD groups (carboxylic acid, alcohol, sugar and amine based) **(TST:20)**	Different ratios	Ononin, sissotrin, formononetin, biochanin A	Chickpea (*Cicer arietinum* L.) sprouts	[105]
ChCl as HBA, 15 HBD groups (Amides, carboxylic acid, alcohols, and sugars) **(TST:15)**	Different ratios	Hesperidin	Four varieties of Croatian mandarian peels (*Zorica rana, Chahara, Okitsu, Kuno*)	[106]

(*Continued*)

Table 11.4 NADES as green solvent for flavonoid extraction from various food samples. (*Continued*)

NADES system	Molar ratio	Targeted analyte	Source	Reference
5 HBA groups (ChCl, acetlyChCl, ChT, betaine, and carnitine based) 5 HBD groups (CA, MA, TA, LA, betaine groups) (TST: 17)	Different ratios	Rutin, hesperidin, naringin, quercetin and chrysin	Fruits (Cranberry, fruits of *Lycium barbarum* L., grape, plum, and orange peel) Vegetables (onion, and broccoli), and spices (mustard, rosemary, and black pepper)	[107]
17 different NADES systems (TST:17)	Different ratios	Isoflavones (daidzin, genistin, genistein, daidzein)	Soy-based products (Soy flour, pasta, breakfast cereals, cutlets, tripe, soy drink, soy nuts, soy cubes, and three different dietary supplements)	[108]
2 HBA groups (ChCl, glycerol) 13 HBD groups (alcohol, sugar and amino acids based) (TST:13)	Different ratios	Rutin	Tartary buckwheat hull	[109]

LA, lactic acid; TMACl, tetramethylammonium chloride, ChCl, choline chloride; HBA, hydrogen bond acceptor; HBD, hydrogen bond donor; EG, ethylene glycol, MA, malic acid; MoA, malonic acid; TSA, toluenesulfonic acid; LeA, levulinic acid, OA, oxalic acid, Xyl, Xylitol; TEG, triethylene glycol; TEACl, tetraethylammonium chloride, ChT, choline tatrate; TA, tartaric acid; TST, total samples tested.

11.3.3 NADES for Extraction of Other Polar Compounds

11.3.3.1 Ferulic Acid Extraction From Ligusticum Chuanxiong Hort and NADES

Xie et al. [110] had extracted ferulic acid from L. Chuanxiong using NADES-MAE. The usage of NADES with microwave assistance (1:2 choline chloride-1,2-propanediol and 30% water as solvent) had presented a highest extraction efficiency, higher yield, shorter extraction time, and energy. The results have demonstrated that these optimum extraction conditions would have potential industrial applications.

11.3.4 NADES for Extraction of Food Samples

11.3.4.1 Extraction of Vanillin With NADES

Traditional method of vanillin extraction is a tedious process which required huge amount of organic solvents. Applications of NADES such as malic acid-fructose-glucose and malic acid-glucose have shown to produce a greater extraction efficiency of 16.3 and 16.7 mg/g, respectively, than traditional organic solvent like ethanol. These results confirmed that NADES can be used as an alternative solvent for extraction of food flavors and has been recommended for various food applications [111].

11.3.4.2 Extraction of Anthocyanins With NADES

An attempt was made to extract anthocyanin from grape pomace using eight different NADES. A combination of ultrasound/microwave-assisted extraction using choline chloride–citric acid for 10 min with water as solvent (30%) was found to be the best combination for efficient recovery of anthocyanins (1.77 mg/g) [112].

11.3.4.3 Extraction of Phenolic Compounds With NADES

A study was conducted to isolate phenolic metabolites from safflower using NADEs solvents with varying polarity, viscosity, and composition. The study had demonstrated that the extraction efficiency of the phenolic metabolites was significantly higher in sucrose-choline chloride, proline-malic acid, and lactic acid-glucose solvents compared to their counter ones. The authors have noted that the viscosity of the NADES solvents negatively affected the extraction efficiency. To counter the affect, extraction conditions were manipulated to reduce the viscosity of NADES. The overall yield of the phenolic compounds was ranged between 75% and 97% [26].

11.3.5 General Considerations Using NADES as Extraction Solvents

The research on NADES was started a decade back and the early publication on NADES was appeared in 2011. The novel DES and NADES solvents investigated are proved to have potential green approach in extracting bioactive compounds from various natural sources. The potential components applied for NADES synthesis are carefully selected according to their ability to form a mixture with melting point lower than their individual constituents. The list of components for the synthesis of NDES is presented in Table 11.5. Further, certain parameters needs to be considered while choosing the components for NADES synthesis which includes time and

Table 11.5 Components for the synthesis of natural deep eutectic solvent (NADES).

Name of the components	Natural deep eutectic solvent
Amides	Dimethylurea, urea
Amines	Choline
Amino acids	Alanine, arginine, glycine, histidine, lysine, proline, serine, threonine
Diols	1, 4-butanediol, ethylene glycol, glycol, 1,2-propanediol
Organic acids	Acetic acid, aconitic acid, ascorbic acid, carboxylic acid, citric acid, fumaric acid, glycolic acid, lactic acid, levulinic acid, maleic acid, malic acid, malonic acid, oxalic acid, propionic acid, succinic acid, succinic acid, tartaric acid
Polyols	Adonitol, erythritol, glycerol, inositol, mannitol, ribitol, sorbitol, xylitol
Sugars	Arabinose, fructose, galactose, glucose, maltose, mannose, raffinose, rhamnose, sacarose, sorbose, sucrose, trehalose, xylose
Quaternary ammonium/ sulfonium/ phosphonium salts	betaine, choline acetate, choline bitartrate, choline chloride, cholinium chloride, ethylammonium chloride, methyltriphenylphosphonium bromide, N, N-diethyl ethanol ammonium chloride

Source: Kua and Gan [114].

energy required for synthesis to decrease the cost, environment aspects, and tendency to solubilize the target compounds [113, 26]. The solubility of some NADES solvents such as lactic acid-propanediol (620 mg/ml) and lactic acid-fructose (320 mg/ml) was on par with organic solvents (methanol, 632 mg/ml; ethanol, 375 mg/ml). Polarity is reported to be the principal property of NADES. In general, NADES are hydrophilic in nature, due to the presence of water as solvent, which further lowers the melting point and changes the properties of solvation [115].

11.4 Conclusion and Future Trends

NADES are emerging green solvents for extraction of natural compounds. While the extraction efficiency of these green solvents are associated to the nature of the individual constituent and their concentration and further on the extraction method. Basing on the literature, it is clear that NADES are found to be an ideal green solvent for the extraction of a variety of bioactive compounds such as alkaloids, phenolic compounds, essential oils, and proteins. On the other hand, the industrial application of DES and NADES as extraction solvents are limited and are still under viability evaluation. The tractability of NADES in connection to the environment impact, stability, solute recovery, toxicity, and operational costs is under consideration. Bottom-line is that several studies confirmed NADES as eco-friendly, low cost, and ideal solvents for food and non-food applications for extraction and recovery of natural products.

References

1. Gökmen, V., *Acrylamide in food: analysis, content and potential health effects*, p. 532, Academic Press, London, UK, 2015.
2. Radojković, M., Zeković, Z., Jokić, S., Vidović, S., Lepojević, Ž., Milošević, S., Optimization of solid-liquid extraction of antioxidants from black mulberry leaves by response surface methodology. *Food Technol. Biotechnol.*, 50, 167–176, 2012.
3. Jablonský, M., Škulcová, A., Malvis, A., Šima, J., Extraction of value-added components from food industry based and agro-forest biowastes by deep eutectic solvents. *J. Biotechnol.*, 282, 46–66, 2018.
4. Corrales, M., Toepfl, S., Butz, P., Knorr, D., Tauscher, B., Extraction of anthocyanins from grape by-products assisted by ultrasonics, high hydrostatic pressure or pulsed electric fields: A comparison. *Innovative Food Sci. Emerg. Technol.*, 9, 1, 85–91, 2008.

5. Ghafoor, K., Hui, T., Choi, Y.H., Optimization of ultrasound-assisted extraction of total anthocyanins from grape peel. *J. Food Biochem.*, 35, 735–746, 2011.
6. Hernández-Carranza, P., Ávila-Sosa, R., Guerrero-Beltrán, J.A., Navarro-Cruz, A.R., Corona-Jiménez, E., Ochoa-Velasco, C.E., Optimization of antioxidant compounds extraction from fruit by-products: Apple pomace, orange and banana peel. *J. Food Process. Preserv.*, 40, 1, 103–115, 2016.
7. Chemat, F., Abert Vian, M., Ravi, H.K., Khadhraoui, B., Hilali, S., Perino, S., Fabiano Tixier, A.S., Review of alternative solvents for green extraction of food and natural products: Panorama, principles, applications and prospects. *Molecules*, 24, 16, 3007, 2019.
8. Zainal-Abidin, M.H., Hayyan, M., Hayyan, A., Jayakumar, N.S., New horizons in the extraction of bioactive compounds using deep eutectic solvents: A review. *Anal. Chim. Acta*, 979, 1–23, 2017.
9. Ashcroft, C.P., Dunn, P.J., Hayler, J.D., Wells, A.S., Survey of solvent usage in papers published in Organic Process Research Development 1997–2012. *Org. Process Res. Dev.*, 19, 7, 740–747, 2015.
10. Li, Y., Fabiano-Tixier, A.S., Tomao, V., Cravotto, G., Chemat, F., Green ultrasound-assisted extraction of carotenoids based on the bio-refinery concept using sunflower oil as an alternative solvent. *Ultrason. Sonochem.*, 20, 1, 12–18, 2013.
11. Bubalo, M.C., Ćurko, N., Tomašević, M., Ganić, K.K., Redovniković, I.R., Green extraction of grape skin phenolics by using deep eutectic solvents. *Food Chem.*, 200, 159–166, 2016.
12. Anastas, P.T. and Warner, J., *Green Chemistry: Theory and Practice*, Oxford University Press, London, 1998.
13. Choi, Y.H. and Verpoorte, R., Metabolomics: What you see is what you extract. *Phytochem. Anal.*, 25, 4, 289–290, 2014.
14. Pena-Pereira, F., Kloskowski, A., Namieśnik, J., Perspectives on the replacement of harmful organic solvents in analytical methodologies: a framework toward the implementation of a generation of eco-friendly alternatives. *Green Chem.*, 17, 7, 3687–3705, 2015.
15. Welton, T., Solvents and sustainable chemistry. *Proc. R. Soc. A: Proc. Math. Phys. Eng. Sci.*, 471, 2183, 20150502, 2015.
16. Gałuszka, A., Migaszewski, Z., Namieśnik, J., The 12 principles of green analytical chemistry and the Significance mnemonic of green analytical practices. *TrAC Trends Anal. Chem.*, 50, 78–84, 2013.
17. Jiménez-González, C., Constable, D.J., Ponder, C.S., Evaluating the "Greenness" of chemical processes and products in the pharmaceutical industry—a green metrics primer. *Chem. Soc. Rev.*, 41, 4, 1485–1498, 2012.
18. McElroy, C.R., Constantinou, A., Jones, L.C., Summerton, L., Clark, J.H., Towards a holistic approach to metrics for the 21st century pharmaceutical industry. *Green Chem.*, 17, 5, 3111–3121, 2015.
19. Howdle, S., Supercritical fluids: A clean route to polymer synthesis and polymer processing. *Green Chem.*, 4, 3, G29–G31, 2002.

20. Beckman, E.J., Supercritical and near-critical CO2 in green chemical synthesis and processing. *J. Supercrit. Fluids*, 28, 2–3, 121–191, 2004.
21. Scurto, A.M., Hutchenson, K., Subramaniam, B., Gas-expanded liquids: fundamentals and applications. *ACS Symp. Ser.*, 1006, 3–37, 2009.
22. Wasserscheid, P. and Stark, A. (Eds.), *Handbook of Green Chemistry, Green Solvents, Vol. 6, Ionic Liquids*, Wiley-VCH, Weinheim, Germany, 2010.
23. Olivier-Bourbigou, H., Magna, L., Morvan, D., Ionic liquids and catalysis: Recent progress from knowledge to applications. *Appl. Catal. A-Gen.*, 373, 1–2, 1–56, 2010.
24. Lomba, L., Giner, B., Bandrés, I., Lafuente, C., Pino, M.R., Physicochemical properties of green solvents derived from biomass. *Green Chem.*, 13, 8, 2062–2070, 2011.
25. Azadi, P., Inderwildi, O.R., Farnood, R., King, D.A., Liquid fuels, hydrogen and chemicals from lignin: A critical review. *Renewable Sustainable Energy Rev.*, 21, 506–523, 2013.
26. Dai, Y., Witkamp, G.J., Verpoorte, R., Choi, Y.H., Natural deep eutectic solvents as a new extraction media for phenolic metabolites in *Carthamus tinctorius* L. *Anal. Chem.*, 85, 13, 6272–6278, 2013.
27. Xin, R., Qi, S., Zeng, C., Khan, F.I., Yang, B., Wang, Y., A functional natural deep eutectic solvent based on trehalose: Structural and physicochemical properties. *Food Chem.*, 217, 560–567, 2017.
28. Mustafa, A. and Turner, C., Pressurized liquid extraction as a green approach in food and herbal plants extraction: A review. *Anal. Chim. Acta*, 703, 1, 8–18, 2011.
29. Uematsu, M. and Frank, E.U., Static dielectric constant of water and steam. *J. Phys. Chem. Ref. Data*, 9, 4, 1291–1306, 1980.
30. Carr, A.G., Mammucari, R., Foster, N.R., A review of subcritical water as a solvent and its utilisation for the processing of hydrophobic organic compounds. *Chem. Eng. J.*, 172, 1, 1–17, 2011.
31. Richter, B.E., Jones, B.A., Ezzell, J.L., Porter, N.L., Avdalovic, N., Pohl, C., Accelerated solvent extraction: a technology for sample preparation. *Anal. Chem.*, 68, 1033–1039, 1996.
32. Jessop, P.G., Mercer, S.M., Heldebrant, D.J., CO 2-triggered switchable solvents, surfactants, and other materials. *Energy Environ. Sci.*, 5, 6, 7240–7253, 2012.
33. Song, R., Ismail, M., Baroutian, S., Farid, M., Effect of subcritical water on the extraction of bioactive compounds from carrot leaves. *Food Bioprocess Technol.*, 11, 10, 1895–1903, 2018.
34. Lachos-Perez, D., Baseggio, A.M., Mayanga-Torres, P.C., Junior, M.R.M., Rostagno, M.A., Martínez, J., Forster-Carneiro, T., Subcritical water extraction of flavanones from defatted orange peel. *J. Supercrit. Fluids*, 138, 7–16, 2018.
35. Brennecke, J.F. and Eckert, C.A., Phase equilibria for supercritical fluid process design. *AIChE J.*, 35, 9, 1409–1427, 1989.

36. Lumia, G., Extraction par fluides supercritiques, in: *Eco-Extraction du Végétal*, F. Chemat, (Ed.), pp. 231–258, Dunod, Paris, France, 2011.
37. Rehman, I.U., Darr, J., Moshaverinia, A., Supercritical fluid processing, in: *Encycl. Biomater. Biomed. Eng. Second Ed.-Four Vol. Set*, pp. 2522–2530, CRC Press, California, USA, 2008.
38. Chemat, F., Vian, M.A., Fabiano-Tixier, A.S., Nutrizio, M., Jambrak, A.R., Munekata, P.E., Lorenzo, J.M., Barba, F.J., Binello, A., Cravotto, G., A review of sustainable and intensified techniques for extraction of food and natural products. *Green Chem.*, 22, 8, 2325–2353, 2020.
39. Jessop, P.G. and Subramaniam, B., Gas-expanded liquids. *Chem. Rev.*, 107, 6, 2666–2694, 2007.
40. Peng, D.Y. and Robinson, D.B., A new two-constant equation of state. *Ind. Eng. Chem. Fundam.*, 15, 1, 59–64, 1976.
41. Siougkrou, E., Galindo, A., Adjiman, C.S., On the optimal design of gas-expanded liquids based on process performance. *Chem. Eng. Sci.*, 115, 19–30, 2014.
42. Santos, M.P., Souza, M.C., Sumere, B.R., da Silva, L.C., Cunha, D.T., Bezerra, R.M.N., Rostagno, M.A., Extraction of bioactive compounds from pomegranate peel (Punica granatum L.) with pressurized liquids assisted by ultrasound combined with an expansion gas. *Ultrason. Sonochem.*, 54, 11–17, 2019.
43. Reyes, F.A., Mendiola, J.A., Ibanez, E., del Valle, J.M., Astaxanthin extraction from Haematococcus pluvialis using CO2-expanded ethanol. *J. Supercrit. Fluids*, 92, 75–83, 2014.
44. Pacheco-Fernández, I. and Pino, V., Green solvents in analytical chemistry. *Curr. Opin. Green Sustainable Chem.*, 18, 42–50, 2019.
45. Welton, T., Room-temperature ionic liquids. Solvents for synthesis and catalysis. *Chem. Rev.*, 99, 8, 2071–2084, 1999.
46. Bernhem, K., *How Ionic are Ionic Liquids?* (M.Sc. thesis), KTH Royal Institute of Technology, Department of Chemistry, Stockholm, 2013.
47. Seddon, K.R., Room-temperature ionic liquids: neoteric solvents for clean catalysis. *Kinet. Catal.*, 37, 5, 693–697, 1995.
48. De Oliveira, N.S., Carlos, A.S., Mattedi, S., Soares, C.M.F., De Souza, R.L., Fricks, A., Lima, A.S., Ionic liquid-based ultrasonic-assisted extraction of alkaloids from cacao (Theobroma cacao). *Chem. Eng. Trans.*, 64, 49–54, 2018.
49. Wang, Y.G., OuYang, X.K., Yang, L.Y., Li, Q.L., Microwave-assisted extraction of Rutin from Tamarix Chinensis with ionic liquid. *Chem. Eng. Chin.*, 3, 411–415, 2011.
50. Zhao, C., Lu, Z., Li, C., He, X., Li, Z., Shi, K., Zu, Y., Optimization of ionic liquid based simultaneous ultrasonic-and microwave-assisted extraction of rutin and quercetin from leaves of velvetleaf (Abutilon theophrasti) by response surface methodology. *Chem. Eng. Trans.*, 2014, 283024, 2014.
51. Rahman, N.R.H.A., Idris, A., Yunus, N.A., Mustaffa, A.A., Optimization of ionic liquid-based microwave extraction of flavonoid and phenolic acid from

Labisia Pumila, in: *AIP Conference Proceedings*, vol. 2124, AIP Publishing LLC, p. 020027, 2019.
52. Domanska, U., Wisniewska, A., Dabrowski, Z., Extraction of Quercetin from Red Onion (Allium cepa L.) with Ionic Liquids. *Chromatogr. Sep. Tech. J.*, 1, 2, 116, 2018.
53. Rahman, N.R.A., Yunus, N.A., Mustaffa, A.A., Selection of optimum ionic liquid solvents for flavonoid and phenolic acids extraction, in: *IOP Conf. Ser., Mater. Sci. Eng.*, vol. 206, p. 012061, 2017.
54. Wang, R., Chang, Y., Tan, Z., Li, F., Applications of choline amino acid ionic liquid in extraction and separation of flavonoids and pectin from ponkan peels. *Sep. Sci. Technol.*, 51, 7, 1093–1102, 2016.
55. Zhou, Y., Wu, D., Cai, P., Cheng, G., Huang, C., Pan, Y., Special effect of ionic liquids on the extraction of flavonoid glycosides from *Chrysanthemum morifolium* Ramat by microwave assistance. *Molecules*, 20, 5, 7683–7699, 2015.
56. Zhao, C., Lu, Z., Li, C., He, X., Li, Z., Shi, K., Zu, Y., Optimization of ionic liquid based simultaneous ultrasonic-and microwave-assisted extraction of rutin and quercetin from leaves of velvetleaf (Abutilon theophrasti) by response surface methodology. *Sci. World J.*, 2014, 283024, 2014.
57. Han, D., Zhu, T., Row, K.H., Ultrasonic extraction of phenolic compounds from Laminaria japonica Aresch using ionic liquid as extraction solvent. *Bull. Korean Chem. Soc.*, 32, 7, 2213, 2011.
58. Chemat, F., Vian, M.A., Cravotto, G., Green extraction of natural products: concept and principles. *Int. J. Mol. Sci.*, 13, 7, 8615–8627, 2012.
59. Wu, W. and Maravelias, C.T., Synthesis and techno-economic assessment of microbial-based processes for terpenes production. *Biotechnol. Biofuels*, 11, 1, 294, 2018.
60. Filly, A., Fabiano-Tixier, A.S., Fernandez, X., Chemat, F., Alternative solvents for extraction of food aromas. Experimental and COSMO-RS study. *LWT*, 61, 1, 33–40, 2015.
61. Vian, M., Breil, C., Vernes, L., Chaabani, E., Chemat, F., Green solvents for sample preparation in analytical chemistry. *Curr. Opin. Green Sustainable Chem.*, 5, 44–48, 2017.
62. Bertouche, S., Tomao, V., Hellal, A., Boutekedjiret, C., Chemat, F., First approach on edible oil determination in oilseeds products using alpha-pinene. *Curr. Opin. Green Sustainable Chem.*, 25, 6, 439–443, 2013.
63. Ibrahim, A.P., Omilakin, R.O., Betiku, E., Optimization of microwave-assisted solvent extraction of non-edible sandbox (*Hura crepitans*) seed oil: A potential biodiesel feedstock. *Renewable Energy*, 141, 349–358, 2019.
64. Paiva, A., Craveiro, R., Aroso, I., Martins, M., Reis, R.L., Duarte, A.R.C., Natural deep eutectic solvents–solvents for the 21st century. *ACS Sustainable Chem. Eng.*, 2, 5, 1063–1071, 2014.
65. Florindo, C., Lima, F., Ribeiro, B.D., Marrucho, I.M., Deep eutectic solvents: overcoming 21st century challenges. *Curr. Opin. Green Sustainable Chem.*, 18, 31–36, 2019.

66. Aroso, I.M., Paiva, A., Reis, R.L., Duarte, A.R.C., Natural deep eutectic solvents from choline chloride and betaine–Physicochemical properties. *J. Mol. Liq.*, 241, 654–661, 2017.
67. Abbott, A.P., Capper, G., Davies, D.L., Rasheed, R.K., Tambyrajah, V., Novel solvent properties of choline chloride/urea mixtures. *Chem. Commun.*, 1, 70–71, 2003.
68. Choi, Y.H., van Spronsen, J., Dai, Y., Verberne, M., Hollmann, F., Arends, I.W., Verpoorte, R., Are natural deep eutectic solvents the missing link in understanding cellular metabolism and physiology? *Plant Physiol.*, 156, 4, 1701–1705, 2011.
69. Wei, Z., Qi, X., Li, T., Luo, M., Wang, W., Zu, Y., Fu, Y., Application of natural deep eutectic solvents for extraction and determination of phenolics in *Cajanus cajan* leaves by ultra performance liquid chromatography. *Sep. Purif. Technol.*, 149, 237–244, 2015.
70. Dai, Y., Rozema, E., Verpoorte, R., Choi, Y.H., Application of natural deep eutectic solvents to the extraction of anthocyanins from *Catharanthus roseus* with high extractability and stability replacing conventional organic solvents. *J. Chromatogr. A*, 1434, 50–56, 2016.
71. Gutierrez, M.C., Ferrer, M.L., Mateo, C.R., del Monte, F., Freeze-drying of aqueous solutions of deep eutectic solvents: a suitable approach to deep eutectic suspensions of self-assembled structures. *Langmuir*, 25, 5509–5515, 2009.
72. Zhang, Q., Vigier, K.D.O., Royer, S., Jerome, F., Deep eutectic solvents: syntheses, properties and applications. *Chem. Soc. Rev.*, 41, 21, 7108–7146, 2012.
73. Yiin, C.L., Yusup, S., Quitain, A.T., Uemura, Y., Physicochemical properties of low transition temperature mixtures in water. *Chem. Eng.*, 45, 1525–1530, 2015.
74. Florindo, C., Oliveira, F.S., Rebelo, L.P.N., Fernandes, A.M., Marrucho, I.M., Insights into the synthesis and properties of deep eutectic solvents based on cholinium chloride and carboxylic acids. *ACS Sustainable Chem. Eng.*, 2, 10, 2416–2425, 2014.
75. Crawford, D.E., Wright, L.A., James, S.L., Abbott, A.P., Efficient continuous synthesis of high purity deep eutectic solvents by twin screw extrusion. *Chem. Commun.*, 52, 22, 4215–4218, 2016.
76. Santana, A.P., Mora-Vargas, J.A., Guimarães, T.G., Amaral, C.D., Oliveira, A., Gonzalez, M.H., Sustainable synthesis of natural deep eutectic solvents (NADES) by different methods. *J. Mol. Liq.*, 293, 111452, 2019.
77. Bonacci, S., Di Gioia, M.L., Costanzo, P., Maiuolo, L., Tallarico, S., Nardi, M., Natural Deep Eutectic Solvent as Extraction Media for the Main Phenolic Compounds from Olive Oil Processing Wastes. *Antioxidants*, 9, 6, 513, 2020.
78. Meneses, L., Santos, F., Gameiro, A.R., Paiva, A., Duarte, A.R.C., Preparation of Binary and Ternary Deep Eutectic Systems. *J. Vis. Exp.*, 2019, 152, e60326, 2019.

79. Degam, G., *Deep eutectic solvents synthesis, characterization and applications in pretreatment of lignocellulosic biomass*, p. 1156, South Dakota State University, Theses and Dissertations, South Dakota, USA, 2017.
80. Li, N., Wang, Y., Xu, K., Huang, Y., Wen, Q., Ding, X., Development of green betaine-based deep eutectic solvent aqueous two-phase system for the extraction of protein. *Talanta*, 152, 23–32, 2016.
81. Liu, Y.T., Chen, Y.A., Xing, Y.J., Synthesis and characterization of novel ternary deep eutectic solvents. *Chin. Chem. Lett.*, 25, 1, 104–106, 2014.
82. Grillo, G., Gunjević, V., Radošević, K., Redovniković, I.R., Cravotto, G., Deep Eutectic Solvents and Nonconventional Technologies for Blueberry-Peel Extraction: Kinetics, Anthocyanin Stability, and Antiproliferative Activity. *Antioxidants*, 9, 11, 1069, 2020.
83. Sapone, V., Cicci, A., Franceschi, D., Vincenzi, S., Bravi, M., Antioxidant Extraction and Bioactivity Preservation from Winery By-products by Natural Deep Eutectic Solvents (nades). *Chem. Eng. Trans.*, 79, 157–162, 2020.
84. Shikov, A.N., Kosman, V.M., Flissyuk, E.V., Smekhova, I.E., Elameen, A., Pozharitskaya, O.N., Natural Deep Eutectic Solvents for the Extraction of Phenyletanes and Phenylpropanoids of *Rhodiola rosea* L. *Molecules*, 25, 8, 1826, 2020.
85. Tsvetov, N.S., Mryasova, K.P., Korotkova, G.V., Asming, S.V., Nikolaev, V.G., Extraction of cinnamic alcohol from Rhodiola rosea using deep eutectic solvents, in: *IOP Conf. Ser. Earth Environ. Sci.*, vol. 315, p. 042006, 2019.
86. Hsieh, Y.H., Li, Y., Pan, Z., Chen, Z., Lu, J., Yuan, J., Zhang, J., Ultrasonication-assisted synthesis of alcohol-based deep eutectic solvents for extraction of active compounds from ginger. *Ultrason. Sonochem.*, 63, 104915, 2020.
87. Pal, C.B.T. and Jadeja, G.C., Microwave-assisted extraction for recovery of polyphenolic antioxidants from ripe mango (Mangifera indica L.) peel using lactic acid/sodium acetate deep eutectic mixtures. *Food Sci. Technol. Int.*, 26, 1, 78–92, 2020.
88. Islamčević Razboršek, M., Ivanović, M., Krajnc, P., Kolar, M., Choline Chloride Based Natural Deep Eutectic Solvents as Extraction Media for Extracting Phenolic Compounds from Chokeberry (Aronia melanocarpa). *Molecules*, 25, 7, 1619, 2020.
89. Vázquez-González, M., Fernández-Prior, Á., Oria, A.B., Rodríguez-Juan, E.M., Pérez-Rubio, A.G., Fernández-Bolaños, J., Rodríguez-Gutiérrez, G., Utilization of strawberry and raspberry waste for the extraction of bioactive compounds by deep eutectic solvents. *LWT*, 109645, 130, 2020.
90. Mulia, K., Yoksandi, Y., Kurniawan, N., Pane, I.F., Krisanti, E.A., 1, 2-Propanediol-Betaine as Green Solvent for Extracting α-Mangostin from the Rind of Mangosteen Fruit: Solvent Recovery and Physical Characteristics. *J. Phys. Conf. Ser.*, 1198, 6, 062003, 2019.
91. Pal, C.B.T. and Jadeja, G.C., Deep eutectic solvent-based extraction of polyphenolic antioxidants from onion (Allium cepa L.) peel. *J. Sci. Food Agric.*, 99, 4, 1969–1979, 2019.

92. Chanioti, S. and Tzia, C., Extraction of phenolic compounds from olive pomace by using natural deep eutectic solvents and innovative extraction techniques. *Innovative Food Sci. Emerg. Technol.*, 48, 228–239, 2018.
93. Ozturk, B., Parkinson, C., Gonzalez-Miquel, M., Extraction of polyphenolic antioxidants from orange peel waste using deep eutectic solvents. *Sep. Purif. Technol.*, 206, 1–13, 2018.
94. Sang, J., Li, B., Huang, Y.Y., Ma, Q., Liu, K., Li, C.Q., Deep eutectic solvent-based extraction coupled with green two-dimensional HPLC-DAD-ESI-MS/MS for the determination of anthocyanins from *Lycium ruthenicum* Murr. fruit. *Anal. Methods*, 10, 10, 1247–1257, 2018.
95. Khezeli, T., Daneshfar, A., Sahraei, R., A green ultrasonic-assisted liquid–liquid microextraction based on deep eutectic solvent for the HPLC-UV determination of ferulic, caffeic and cinnamic acid from olive, almond, sesame and cinnamon oil. *Talanta*, 150, 577–585, 2016.
96. Jeong, K.M., Zhao, J., Jin, Y., Heo, S.R., Han, S.Y., Lee, J., Highly efficient extraction of anthocyanins from grape skin using deep eutectic solvents as green and tunable media. *Arch. Pharmacal Res.*, 38, 12, 2143–2152, 2015.
97. Dai, Y. and Row, K.H., Application of natural deep eutectic solvents in the extraction of quercetin from vegetables. *Molecules*, 24, 12, 2300, 2019.
98. Jokić, S., Šafranko, S., Jakovljević, M., Cikoš, A.M., Kajić, N., Kolarević, F., Babić, J., and Molnar, M., Sustainable Green Procedure for Extraction of Hesperidin from Selected Croatian Mandarin Peels. *Processes*, 7, 7, 469, 2019.
99. Bajkacz, S. and Adamek, J., Development of a method based on natural deep eutectic solvents for extraction of flavonoids from food samples. *Food Anal. Methods*, 11, 5, 1330–1344, 2018.
100. Bajkacz, S. and Adamek, J., Evaluation of new natural deep eutectic solvents for the extraction of isoflavones from soy products. *Talanta*, 168, 329–335, 2017.
101. Majidi, S.M. and Hadjmohammadi, M.R., Development of magnetic dispersive micro-solid phase extraction based on magnetic agarose nanoparticles and deep eutectic solvents for the isolation and pre-concentration of three flavonoids in edible natural samples. *Talanta*, 222, 121649, 2020.
102. Ali, M.C., Chen, J., Zhang, H., Li, Z., Zhao, L., Qiu, H., Effective extraction of flavonoids from *Lycium barbarum* L. fruits by deep eutectic solvents-based ultrasound-assisted extraction. *Talanta*, 203, 16–22, 2019.
103. Mansur, A.R., Song, N.E., Jang, H.W., Lim, T.G., Yoo, M., Nam, T.G., Optimizing the ultrasound-assisted deep eutectic solvent extraction of flavonoids in common buckwheat sprouts. *Food Chem.*, 293, 438–445, 2019.
104. Nia, N.N. and Hadjmohammadi, M.R., The application of three-phase solvent bar microextraction based on a deep eutectic solvent coupled with high-performance liquid chromatography for the determination of flavonoids from vegetable and fruit juice samples. *Anal. Methods*, 11, 40, 5134–5141, 2019.
105. Shang, X., Dou, Y., Zhang, Y., Tan, J.N., Liu, X., Zhang, Z., Tailor-made natural deep eutectic solvents for green extraction of isoflavones from chickpea (*Cicer arietinum* L.) sprouts. *Ind. Crops Prod.*, 140, 111724, 2019.

106. Jokić, S., Šafranko, S., Jakovljević, M., Cikoš, A.M., Kajić, N., Kolarević, F., Molnar, M., Sustainable Green Procedure for Extraction of Hesperidin from Selected Croatian Mandarin Peels. *Processes*, 7, 7, 469, 2019.
107. Bajkacz, S. and Adamek, J., Development of a method based on natural deep eutectic solvents for extraction of flavonoids from food samples. *Food Anal. Methods*, 11, 5, 1330–1344, 2018.
108. Bajkacz, S. and Adamek, J., Evaluation of new natural deep eutectic solvents for the extraction of isoflavones from soy products. *Talanta*, 168, 329–335, 2017.
109. Huang, Y., Feng, F., Jiang, J., Qiao, Y., Wu, T., Voglmeir, J., Chen, Z.G., Green and efficient extraction of rutin from tartary buckwheat hull by using natural deep eutectic solvents. *Food Chem.*, 221, 1400–1405, 2017.
110. Xie, Y., Liu, H., Lin, L., Zhao, M., Zhang, L., Zhang, Y., Wu, Y., Application of natural deep eutectic solvents to extract ferulic acid from *Ligusticum chuanxiong* Hort with microwave assistance. *RSC Adv.*, 9, 39, 22677–22684, 2019.
111. González, C.G., Mustafa, N.R., Wilson, E.G., Verpoorte, R., Choi, Y.H., Application of natural deep eutectic solvents for the "green" extraction of vanillin from vanilla pods. *Flavour Fragr. J.*, 33, 1, 91–96, 2018.
112. Panić, M., Gunjević, V., Cravotto, G., Redovniković, I.R., Enabling technologies for the extraction of grape-pomace anthocyanins using natural deep eutectic solvents in up-to-half-litre batches extraction of grape-pomace anthocyanins using NADES. *Food Chem.*, 300, 125185, 2019.
113. Crespo, E.A., Silva, L.P., Martins, M.A., Fernandez, L., Ortega, J., Ferreira, O., Sadowski, G., Held, C., Pinho, S.P., Coutinho, J.A., Characterization and modeling of the liquid phase of deep eutectic solvents based on fatty acids/alcohols and choline chloride. *Ind. Eng. Chem. Res.*, 56, 42, 12192–12202, 2017.
114. Kua, Y.L. and Gan, S., Natural deep eutectic solvent (NADES) as a greener alternative for the extraction of hydrophilic (polar) and lipophilic (non-polar) phytonutrients, in: *Key Eng. Mater*, vol. 797, pp. 20–28, Trans Tech Publications Ltd, Bäch SZ, Switzerland, 2019.
115. Mišan, A.N., The perspectives of natural deep eutectic solvents in agri-food sector. *Crit. Rev. Food Sci. Nutr.*, 60, 1–29, 2019.

12
Technological Advancement in Food Additives and Preservatives

Shikha Pandhi, Arvind Kumar* and Akansha Gupta

Department of Dairy Science and Food Technology, Institute of Agricultural Sciences, Banaras Hindu University, Varanasi, Uttar Pradesh, India

Abstract

Food additives and preservatives are purposefully added to food in minute quantity to perform a defined technological or sensory function like enhancing shelf-life, imparting, and restoring color, to maintain palatability and wholesomeness, to enhance or preserve nutritive value, or to enhance flavor. Apart from the advantages offered by these substances during processing and storage, certain safety and health issues hinder their proficient use for food application. This has caused a leaning trend towards the utilization of compounds from a natural source such as phytochemicals and plant essential oils as natural antioxidants and food preservatives. The utilization of metallic nanoparticles as an effective antimicrobial agent has also been a rising trend in the research field. Novel techniques such as nanoencapsulation have further encouraged the development of these novel food ingredients for promising food applications. This chapter provides a succinct overview of various conventional food additives and preservatives with a major emphasis on their types, functionality, mode of action, and significance for food applications. Further, it tends to discuss various novel alternative ingredients together with discussing the opportunities shaped by the advent of technologies like nanoencapsulation. It also provides a concise review of different techniques employed for the determination of food additives.

Keywords: Food additives, functionality, preservation, processing, technological advancement

---

*Corresponding author: arvind1@bhu.ac.in

Abbreviations

%	Percentage
3-D	3-Dimensional
ADIs	Acceptable Daily Intakes
a_w	Water Activity
BHA	Butylated Hydroxyanisole
BHT	Butylated Hydroxytoluene
Eos	Essential Oils
EU	European Union
FAO	Food and Agriculture Organization
FAPs	Food Additive Petitions
FDA	Food and Drug Administration
FSSAI	Food Safety and Standard Authority of India
GC	Gas Chromatography
GMP	Good Manufacturing Practices
GRAS	Generally Recognized as Safe
HPLC	High-pressure Liquid Chromatography
INS	International Numbering System
IR	Infrared
JECFA	Joint Expert Committee on Food Additives
LC	Liquid Chromatography
LD_{50}	Lethal Dose, 50%
mm	Millimetre
MS	Mass Spectroscopy
NIR	Near-infrared
nm	Nanometer
NOAEL	No observed Adverse Effect Level
°C	Degree Celsius
QSAR	Quantitative Structure–Activity Relationship
TBHQ	Tertiary Butylhydroquinone
TBQ	Tertiary Butylquinone
UV	Ultraviolet
UV/Vis	Ultraviolet/Visible
WHO	World Health Organization

12.1 Introduction

Hectic lifestyles, changing food practices, and global expansion are the key aspects for rapid augmentation of processed and convenience food products in the market. Using escalated consumption of processed food, the food additive intake has also shown an extensive rise [1]. Food additives and preservatives have been known to be an invariable part of various food preparations since immemorial times. The utilization of salt for food preservation to hinder the growth of microbes by reducing water activity (a_w) has been known since antiquity. The food additives are the constituents which on addition to the food tends to maintain its flavor and to improve its appearance, taste, and other characteristics. With the advent of many processed food products and preparations, numerous newer food additives both of natural and synthetic origin have been introduced to serve various technological functions during several stages of the food chain starting from manufacturing to packaging and ultimately to transportation and distribution [2]. As per the Codex Alimentarius, it is defined as a non-nutritional constituent not primarily added as a food ingredient but it is deliberately incorporated during the preparation, processing, packaging, storage, and distribution stage to exercise a technological advantage [3]. The use of food additives enhances food quality and shelf life on the retail shelves. Generally, the food industry incorporates only those food additives in their food formulations that propose a fair technological advantage which cannot be attained through other low-cost feasible methods. The key technological function of food additives includes pH regulation, uniformity, stability, viscosity, and shelf life extension, prevents degradation and extends shelf life, and retains sensory attributes like color, flavor, and texture. The concentration of food additive to be added, the stage of addition, means of addition, and the type of food can be accustomed based on the required technological function [4]. The usage of additives in food formulations can be acceptable only when their usage does not deceive consumers, has the technical requirement, and performs a distinct technological function, to enhance the food stability and preserve the nutrient value of the product. The safety assessments conducted by the Joint Expert Committee on Food Additives (JECFA) are generally utilized by the Codex Alimentarius Commission, which is a joint intergovernmental organization of the FAO and WHO that forms food standards and establishes maximum usage level of additives in food and beverages [2]. Food additives offer numerous advantages such as improved consistency, palatability, wholesomeness, nutritional value, shelf-stability, and improved sensory attributes. Apart from these advantages, there are certain health risks associated with the consumption of certain food additives that may cause immediate or delayed health implications in the long run. Various synthetic food additives react with the body's cellular constituents and lead to numerous health complications. Given this, the use of food additives of natural origin and additives that have been provided the status of generally recognized as safe (GRAS) within the limits of acceptable daily intakes (ADIs) has been encouraged to safeguard the health of the consumers [5]. Various natural constituents such as plant essential oils (EOs) and phytochemicals have been used to perform similar functions to other food additives (like preservatives and antioxidants). Plant secondary metabolites extracted from the agri-food waste offer a remarkable and economical source of various bioactive constituents such as phytochemicals to serve the function of an antimicrobial, antioxidant agent, and food colorant to endorse a circular economical idea [6]. Nanoencapsulation has been used as an advanced approach to increase the antimicrobial activity of these phytoconstituents like EOs [7] and bioactive [8]. Among these, nanoliposome, solid lipid nanoparticles, and nanoemulsion are among the advanced nanoencapsulation techniques. Further, the use of novel greener approaches for the synthesis of antimicrobial agents such as silver [9], gold [10], copper [11], and other metallic nanoparticles using biological sources has also received great potential for future food preservation approaches. This chapter tends to provide a bird-eye insight into various aspects of food additives such as their types, functions, and significance in foods. Also, it provides a comprehensive overview of various technological advancements made over time with the main emphasis on novel food additives and advanced techniques like nanoencapsulation. Various analytical methods of determination along with associated safety and regulatory aspect will also be addressed in brief.

12.2 Food Additives and Preservatives

Food additives and preservatives are substances that are added to food to perform an explicit function such as imparting color, flavor, and texture, to improve consistency and shelf life extension by inhibiting the growth of microorganisms and also as a thickening, gelling, emulsifying, stabilizing agent, etc. [12]. The European Union's definition for food additives is "any constituent purposely incorporated in a food during the stage of preparation, processing, packaging, storage or distribution to exert a technological advantage". Food additives usually have the following three features: (a) they are the constituents added to food but cannot be solely used for consumption as food themselves; (b) they can be either of natural or synthetic origin; and (c) the purpose of its incorporation into food is to enhance food quality, color, fragrance, and the flavor of food and to fulfill the need of fresh and wholesome processed food product [13]. Food additives have been classified into different classes discussed in the next section.

12.2.1 Classes of Food Additives

Food additives used in food formulation essentially relies on the following basic functions [14]:

- Food defense: Use of antimicrobials and antioxidants to maintain safety and wholesomeness of food product
- Food colors or flavors: To enhance the aesthetic appeal and flavor of the product
- Use of artificial sweeteners as a substitute for sugar (calorie reduction)
- Structural function: Use of thickeners, gelling agents, stabilizers, etc.

The obligatory declaration of these additives on the food label stands mandatory and is linked to their categorization in agreement with the International Numbering System (INS). All the EU-approved additives have been assigned a particular identification code entailing the letter "E" followed by three or four digits. The chemical name of the additive may also be specified on the product label. The major classes of food additives are discussed below.

Food preservatives are the substances employed for extending the product life and to assure their safety throughout the extended time. They work primarily by suppressing the growth of microorganisms that may cause degradation leading to the release of various toxins that can cause foodborne illnesses. Thus, they bid a strong consumer advantage by rendering food safe for consumption for extended periods on the shelves and to support modern lifestyles and occasional bulk buying [15]. Antioxidants, on the other hand, diminish the chances of oxidative deterioration that may result in oxidative rancidity, loss of flavor, nutritive value, and aesthetic quality. Fats, oils, vitamins, and colors are most susceptible to oxidative degradation on exposure to oxygen. They can either be natural or synthetic in nature. Ascorbic acid and tocopherol are among the natural antioxidants and butylated hydroxyl anisole (BHA), butylated hydroxytoluene (BHT), and propyl gallate are the examples of synthetic antioxidant. Further, the use of emulsifiers and stabilizers as structural additives has aided the process of mixing two immiscible phases. The use of stabilizers for maintaining the stability of two-phase aqueous and lipid in products like ice-creams, mayonnaise, chocolates, and fat spreads has been widely practiced. Emulsifiers, plasticizers, surface active agents, and dispersing agents are other types belonging to this class of additives. Aesthetic agents like food colorants are also widely used to enhance product appeal. Similar to other food additives, their use in food formulations is also strictly monitored and regulated. Flavor enhancers are used to improve the flavor attribute of a product. They can either be isolated from natural sources or synthetic origin [15]. Food sweeteners are also segmented into natural and artificial sweeteners. These sweeteners are mainly employed to provide a sweet taste in food and beverage formulations with lower calories. They are mainly used in food formulations that are prepared exclusively for people who cannot consume sugar (sucrose) [14]. As per the commercial purpose, the food additives are broadly categorized into 25 different categories, which comprises of 230 different compounds. Several compounds showcase multiple functions which are solely dependent on their usage quantity and processing method of food. The major technological function comprises of pH control, stability, microbial inhibition, homogeneity, viscosity, and improved sensorial properties like color, flavor, and smell. The amount of additive to be incorporated, the step of their incorporation, the processing method, and type of food greatly affect the desired technological function of the additive and can be adjusted. Among these food additives, the major ones are listed as acidity regulators (controls and modifies the pH), anti-caking agents (prevents agglutination and lump formation in powder products), raising agents (promote expansion of dough or batter), stabilizers (prevent phase separation), humectants (prevent moisture loss), preservatives (prevent spoilage), and antioxidants (prevent oxidative degradation) [4].

12.2.2 Significance in Food Processing and Preservation

Food additives perform numerous technological functions at different stages. Food processing and preservation are among the two major areas where food additives have to serve as a multi-functional assets. Food preservation is a science that deals with the processes and techniques employed to surpass the effect of inherent and external parameters that can cause spoilage of food. The key objective of food preservation is to enhance the shelf life of food products along with retaining their original nutritional values, color, texture, and flavor. Various traditional food preservation methods are dehydration, freezing, heating, and chemical preservation, etc. [16]. The use of chemical preservatives has repeatedly augmented among these conventional preservation methods and serves as the most economical candidate by a processor to guarantee a fair product shelf life. Antioxidants and antimicrobials are among the most suitable candidate to serve this purpose. Besides, various food processors are progressively using various food additives to ascertain the sensory appeal and reliability of the end product. Emulsifiers have been extensively used to enhance the texture of bread, ice-creams, and dressings. Stabilizers and thickeners are also employed to impart appeal, consistency, and texture to the finished product. Sorbitol is one additive that can serve the function of humectant and sweetener. The use of artificial sweeteners could also be employed to formulate products with fewer or no calories [17]. The leaning trend toward the healthy eating has recently compelled the customers to look for novel food products of natural origin, with an intensifying attention for food products incorporated with all-natural constituents or additives [18]. In view of this, various plant-based constituents such as EOs [18], antimicrobials [19], and antioxidants [20] have been added to numerous food products in recent times.

12.2.3 Mechanism of Action of Food Preservatives

Preservatives are among the most extensively employed food additive to assure the wholesomeness and safety of the finished food product. They mainly work by various modes of actions that primarily aim at the destruction or inhibition of microbes in food products. Among various conventional preservatives, organic acids like lactic, acetic, citric, benzoic, and sorbic acid are most extensively employed. They work by establishing a pH-dependent equilibrium among dissociated and undissociated phases in an aqueous state. They generally have optimal activity at lower pH levels as it favors the undissociated phase to enter freely in the cell. Whereas, higher pH favors the discharge of charged anions and protons which are unable to freely pass the membrane and enter the cell. The preservative agent diffuses the cell till equilibrium is reached in agreement with the pH gradient through the membrane and results in the gathering of anions and protons inside the cell. The major mechanism of action of weak acids includes membrane disruption, disturbance in intracellular homeostasis, and inhibits vital metabolic processes, and gathering of toxic anions inside the cell. Hydrogen peroxide (H_2O_2) also exhibits vital biocidal properties. Under suitable conditions, a short span lactoperoxidase system is generated. The further reaction leads to the generation of singlet oxygen species which exhibit extensive biocidal property. Also, the partial reduction of molecular oxygen gives superoxide radicals. Then, the H_2O_2 together with superoxide radicals and transition metal ions [e.g., Fe (II)] enters the Fenton reaction and forms an extremely biocidal hydroxyl radical. It acts antagonistic to both gram-positive and gram-negative bacteria [21].

12.3 Regulatory Aspects of Food Additives and Preservatives

Safety regulation of food additives is the key parameter concerning public health. The regulation of food additives is controlled by different bodies, namely, Food and Drug Administration (FDA), Food and Agricultural Organization (FAO), European Union (EU), and World Health Organization (WHO) [15]. The regulations developed for food additives generally rely on the following principles: (1) determination of safe intake levels through vigilantly designed animal studies, (2) establish safe intake levels based on maximum dietary intake of additive causing no severe health implication, (3) the intake should be lower than the amount producing no severe effects in animals, and (4) reliable judgment should be made by scientists that promote effective determination of safe levels. This is universally accepted that for extensive use of a novel food additive, clinical trials should be conducted vigilantly to report any potential carcinogenic, mutagenic, chronic, and reproductive toxicity.

12.3.1 Generally Recognized as Safe

GRAS is a system for evaluation and approval of ingredients for addition to food [22]. It represents the group of additives that are declared safe and do not show any harmful effect on basis of their extensive records or through scientific studies considered "Generally Recognized as Safe" and comes under GRAS status. The intentional addition of a new substance falls beneath the category of food additive that must be approved by the FDA by rigorous testing to prove their safety before addition. The food additive petitions (FAPs) have to be submitted to the FDA. These petitions should contain adequate safety information to permit the organization to meet its criteria for their positive consent. The new substance added either follows the GRAS status or follows the status of regulated food additive. Unintentional additives that come in food through processing, packaging, or various other sources and do not come under GRAS status or follow regulated additive status will be regulated through special regulations provided that they will remain at a negligible amount and the manufacture will have to follow GMP to avoid their presence. For getting approval under GRAS status, the additive must have to validate its safety through various toxicological tests and scientific data.

12.3.2 FSSAI Regulations on Permissible Limits of Food Additives

The Food Safety and Standards Act, 2006 was aimed to enhance the global food safety compliance of food and facilitate fair trade practices inside and outside the nation. The FSSA has associated accountability to assure food safety solely on the Food Safety and Standards Authority of India (FSSAI) [23]. It is the key regulatory body that is solely responsible for all food safety regulations in India. FSSAI has provided regulatory standards for food additives, "Food Safety and Standards (Food Products Standards and Food Additives) Regulation" in 2011. FSSAI has also released a manual entitled "Manual for Analysis of Food Additives" to assist in the testing of various food additives in processed products in 2012.

12.4 Health Concerns of Conventional Food Additives

Escalating utilization of various food additives in various processed food formulations has aroused a situation of great concern due to its potential health implications associated with frequent intake if taken in excess [1].

The effects have been shown to exert a range of effects that may vary with age, frequency, and immune response. They may cause several allergic conditions such as hyperactivity and attention deficit sickness in individuals with higher sensitivity toward certain additives. The common reactions that may occur as a result of food additive allergic reactions are vomiting, rashes, headache, tight chest, etc. [15]. Apart from allergies, food additives can lead to breathing problems, stomach aches, etc. Bromates can result in nausea and diarrhea. Saccharin may lead to disorders of the gastrointestinal tract and heart as well as tumors. High intake of sodium chloride can also result in hypertension, heart attack, stroke, etc. Nitrates and nitrites have been known to induce carcinogenesis and hence have been banned by various countries. The butylated hydroxyanisole (BHA) as an antioxidant had shown to exert a carcinogenic effect in long term. In addition, the hostile effects associated with the consumption of food color tartrazine, the emulsifier sorbitol, the flavoring compound MSG, and the artificial sweeteners aspartame and saccharin have also been reported.

12.5 Technological Advancements in Food Additives and Preservatives

12.5.1 Novel Food Additives

The rising demand of consumers for health-promoting food products has aroused the need for the replacement of artificial or synthetic additives with green or natural ingredients exhibiting the same technological advantage [24]. Given the increasing concern regarding their thorough evaluation of its adverse health impacts, there is a slanting trend toward the use of natural components (like EOs and phytochemicals) and greener approaches (like green antimicrobial particles) to alleviate additive related health implications [25]. Therefore, food processors are continuously incorporating natural ingredients and are indicated on the label as the named ingredient without "E-number" providing a "clean label" to the product. Various food ingredients have been reported recently to provide a clean label status like lecithin as an emulsifier, olive leaf extract, lycopene, anthocyanin, etc. Some bacterial cultures named "protective cultures" have been known to retard bacterial and fungal growth. They act by producing acids like acetic acid and lactic acid and help manufacture claim clean labels [24].

12.5.1.1 Essential Oils/Phytochemicals

Food systems containing natural antioxidants and preservatives for enhancement of shelf life and oxidative stability of products are a booming trend among food processors and researchers. They are considered more effective, efficient, and harmless than their synthetic counterparts. Natural antioxidants and antimicrobials are ubiquitous in spices, herbs, vegetables, fruits, seeds, oils, peels, etc. Herbs and spices are a rich source of various phytochemicals and EOs exhibiting strong antimicrobial and antioxidant properties [26]. EOs represent a diverse class of volatile organic compounds having a low molecular weight and are extracted from different parts of the plant. They contain various groups of active compounds such as terpenoids and terpenes [27]. They are usually extracted using steam or hydrodistillation, expression, and supercritical fluid extraction. They contain components that exhibit strong antioxidants and anti-microbial and insecticidal properties [28]. The usage of plant EO as a natural food additive has been the emphasis of numerous investigations for their proficient use of synthetic food additives. They can also be added into food packages for the creation of an active packaging system that releases their compound to food on time. EOs extracted from cardamom seeds [*Elettaria cardamomum (L.) Maton*], basil (*Ocimum basilicum L.*), and rosemary (*Rosmarinus officinalis L.*) have been reported to be incorporated into food packaging materials as antioxidant and antimicrobial with GRAS consideration [29]. Similarly, phytochemicals are the secondary metabolites derived from plant and plant parts exhibiting good antioxidant and antimicrobial activities. Some of the common examples of antimicrobials from the plant are curcumin from turmeric, eugenol from cloves, and allicin from garlic. They have been shown to suppress the growth of various microorganisms. For instance, thyme oil has been reported to retard the growth of *Clostridium perfringens*. These phytochemicals can be preferentially derived from essentials and have antimicrobial mechanism different from the conventional antimicrobials. Phytochemicals are mainly categorized into flavonoids, alkaloids, phenolic acids, carotenoids, isoflavones, and monoterpenes [28]. Further, various agri-food by-products have provided an economical source of various secondary metabolites such as polyphenols, peptides, and functional pigments with good antioxidant and antimicrobial potential and hence promote a circular economic model. They serve as a multifunctional food additive with multiple technological frontiers [6].

12.5.1.2 Metallic Nanoparticles as Antimicrobial (Green Route)

The exponential expansion of green nanoscience holds massive potential for the creation of nanoparticles using biological entities. Both microorganisms and plants have been utilized for the proficient conversion of metal ions to nano dimension particles [30]. These nanoparticles due to their smaller size exhibit potent antimicrobial activity. Green nanotechnology has been effectively employed for the creation of nanoparticles of gold, silver, palladium, iron, and copper using non-toxic solvent and reducing agents [31]. Utilization of green synthesized metallic nanoparticles as antimicrobial agents against a broad range of microbes for the creation of antimicrobial active packaging systems has gained remarkable potential in shelf life extension of foods. These antimicrobial nanoparticles are safe, cheap, and environment-friendly. Green creation of nanoparticles generally follows a bottom-up approach where smaller particles give rise to larger size particles [32]. Silver nanoparticles fabricated using green technology have been reported to be the most widely used antimicrobial nanoparticles among others. They have shown to exhibit strong antimicrobial activity [33]. These metallic nanoparticles exert their antimicrobial effect through various mechanisms such as cell wall disruption, oxidative stress, disturbance of electron transport system, and the free radical generation that results in microbial cell destruction [34]. The use of nano-dimension materials may lead to certain safety uncertainties and hence demands the implementation of practices that critically take into consideration human health and their safety aspect. The use of nanomaterials as antimicrobials is widely adopted in the food packaging domain but is still accompanied by numerous safety and regulatory issues that must be addressed vigilantly [35].

12.6 Novel Technological Approaches for Enhanced Functionality

12.6.1 Nanoencapsulation

Oxidative degradation and microbial contamination are the two key issues of concerns that pose serious threat to public health. The current research has been focused on natural antioxidants and antimicrobials for their effective incorporation into food systems to alleviate the negative impact of synthetic compounds. Bioactive compounds, phytochemicals from fruits and vegetables, whole grains, and agro-industrial waste serve as an abundant source of these antimicrobial and antioxidant compounds and are not accompanied by the toxicity associated with synthetic substances. Bio-preservation is intensively utilizing the natural antimicrobial from sources like plants, microbes, and animals for their wide spectrum action against foodborne pathogens. The most probable antimicrobial mechanism of action includes (i) membrane disruption, (ii) inhibition of ATPase activity, (iii) essential biomolecules leaching from the cell, (iv) enzyme inactivation, and (v) interruption of proton motive force. The major concern associated with their effective incorporation characterizes bioavailability of bioactive compounds as they cannot be simply expected as their concentration in the usual media, does not essentially symbolizes the same concentration in the target tissue [36]. This has created the need for targeted release. Encapsulation has come up as a speedily emerging technology to offer tremendous frontiers in this regard. In this technique, the core material is entrapped in an outer matrix that acts as a protective layer for the core material from Surroundings. Various bioactive compounds such as phenols, peptides, probiotics, enzymes, and nutrients have been effectively subjected to this technique for enhanced stability and controlled targeted release of these bioactive. Nanoencapsulation is most extensively employed currently in various areas that result in bio packing of bioactive in nano-dimension. It facilitates the control and targeted release of bioactive with improved bioavailability [37]. It has been widely employed to incorporate plant-derived antimicrobials and nutraceuticals for their distinctive functional attribute. The use of this technology for encapsulation of various natural antimicrobials has been shown to boost their inhibitory effect on microbes [25].

12.6.1.1 Fundamentals and Techniques

Encapsulation is a process of entrapping one substance within another, to give particles with a size range of few micrometers to nanometers. From a technical perspective, an efficient encapsulation system for conveyance of bioactive constituents for addition into food must meet the following requirements: (i) must be obtained from natural ingredients without any use of solvent, (2) should be food grade, (3) should exhibit high stability on incorporation of bioactive ingredient, (4) exert minimal effect on organoleptic properties, (5) should maximize the uptake of bioactive compound, (6) should provide protection to encapsulated compound from external factors like pH, temperature, and light, and (7) should be simply scalable for commercial production. In view of wide spectrum of the requirement to be achieved by delivery systems, it has elicited the need of various encapsulation techniques like coacervation, spray drying, nanoprecipitation, inclusion complexation, freeze drying, extrusion technologies, high pressure homogenization, electrospinning, and electrospraying [38, 39].

12.6.1.2 Types of Encapsulating Material

Various edible materials like polysaccharides-based, protein-based, and lipid-based have been effectively involved in development of bio-compatible micro and nano encapsulation systems for food applications. These delivery systems offer a wide variety of wall materials for encapsulation [40]. Carbohydrates-based carrier systems are most suitable for food application due to its abundance, low cost, biodegradability, and good potential for modification to exhibit a desired functionality. Various starches, gums, and oligosaccharides have been effectively employed for encapsulation [41]. Protein-based encapsulation systems have also attained great potential for encapsulation due to their (i) good adsorption properties that prevents aggregation of particles and (ii) ability to gather into assemblies like spheres, tubes, or fiber [42]. Among proteins derived delivery systems, materials like whey, casein, gelatin, chitosan, zein, soy, or gliadin have been extensively explored for their ability to entrap bioactive components. Lipid-based nano-delivery systems are among the most favorable encapsulation system that offers incomparable advantages over other two systems due to its capacity to capture materials with varying solubilities, enhanced targeted delivery and ability to guard a constituent from pH, temperature, and free radicals and enzymes [43]. Liposomes, nanoemulsions, and solid-lipid nanoparticles have been expansively evaluated for their encapsulating ability. Liposomes are the self-accumulated colloidal structure that is spherical in shape and encloses both aqueous and lipid phases. Its formation is mainly attributed to lipophobic-lipophilic interaction among water-soluble substances and phospholipids. Owing to their amphipathic nature, they can be employed to encapsulate both hydrophilic and hydrophobic constituents [44]. Nanoemulsions are made by dispersing to immiscible liquids together with the help of emulsifiers. Ultra-high pressure homogenizers are used for the fabrication of nanoemulsions with nano-dimension droplets. They exhibit improved stability, appearance, texture, and activity. Solid lipid nanoparticles are biodegradable and biocompatible colloidal delivery systems that remain solid at ambient conditions. They are extensively utilized as an entrapment matrix for chemically unsteady hydrophobic compounds. Solid-lipid nanoparticles could significantly increase the stability and bioavailability of hydrophobic antimicrobials like EOs and other bioactive in comparison to conventional liquid emulsion systems [25]. Nano encapsulated EOs and plant bioactive exhibit robust antimicrobial potential over synthetic alternatives [45]. These bioactive acts interacting conversely with biochemical and structural components leading to disturbance in metabolic reactions and membrane integrity that ultimately leads to cellular destruction [46]. Nanoencapsulation of plant-derived antimicrobials such as phytochemicals and EOs and antioxidants have shown to lessen the chances of microbial deterioration of food with much improved organoleptic attributes. An array of products have been developed based on this nanotechnological approach to offer inventive frontiers in numerous fields of food such as packaging, packaging, preservation, safety, quality, and nutraceuticals [47]. Insight of tremendous offerings provided by nanotechnology, they can play a significant contribution to the development of programmable food that could be an extraordinary futuristic concept [48].

12.7 Methods for Food Additives Determination

12.7.1 Analytical Methods

Analytical methods are employed to detect and measure the quantity of various food additives in food formulations. These techniques for effective implementation should provide enhanced reliable, sensitive, selective, and rapid determination in an environment-friendly and cost-effective manner [4].

12.7.1.1 Spectroscopy Techniques

The spectroscopic techniques utilize the electromagnetic radiations to study its interaction with matter. The heat energy or electric charge generated promotes quick excitation of electrons to a higher energy state. When it starts losing its energy, it goes back to its native state. If the light added is greater than the emitted light then it is known as emission spectroscopy and if the light added is higher than the light emitted then it is called absorption spectroscopy. The different functional groups present in the food additives permit absorption and adsorption of light at a known wavelength in different manner. This can be used to quantify and identify the chemical structures of various food additives. Among the spectroscopy techniques the ultraviolet/visible radiation (UV/Vis) spectroscopy and infrared spectroscopy (IR) are the most widely employed conventional techniques. The use of UV/Vis for food quality control is indispensible in view of its operational simplicity, efficiency, low cost, non-destructive, and environmentally benign methodology. But due to complex composition of food stuff, this method faces undesirable selectively and need to be coupled with another method such as artificial neural network to model multicomponent absorbance analysis. Other extensively employed newer techniques are NIR (near-infrared) and FTIR (Fourier transform infrared) spectroscopy. They have been widely used for the analysis of products such as biscuits, butter, and nutritional bars. These techniques provide more reliable data with non-destructive procedures. Another important technique based on the scattering of light is called as Raman spectroscopy. The Raman spectra are obtained as a result of an inelastic collision between incident single-wavelength light and the sample. It is used for both qualitative and quantitative estimation.

12.7.1.2 *Chromatographic Techniques*

Chromatographic techniques depict an array of methods that are used to detect, separate, and quantify organic and inorganic compounds on the basis of sample distribution between a stationary and mobile phase. They are mainly classified based on physicochemical principles employed in the separation. Some of the common chromatographic techniques are adsorption, partition, and ion-exchange chromatography. The other advance chromatographic techniques that are widely employed for food analysis are gas chromatography (GC) and liquid chromatography (LC). GC separates and detects complex volatile mixtures of volatile compounds using a gaseous mobile phase and an immobilized liquid or solid packed in the close tube as a stationary medium. LC or especially HPLC (high-pressure liquid chromatography) is a highly appropriate technique that comprises of a solid stationary phase and a liquid mobile phase. Separation of organic/inorganic compounds is based on their differentia affinity toward mobile and stationary phases. Several detectors are used for this technique and their selection depends upon the detection criteria. Coupling of these technologies with mass spectroscopy (MS) enhances its identification accuracy and enables more proficient compound identification.

Apart from the analytical techniques employed for analysis, the satisfactory assessment of the sample requires efficient sampling and pretreatment phases and better interpretation of results stands accountable for successful analysis of compounds [4].

12.7.1.3 Electroanalytical Techniques

In last few years, there is a growing popularity of electro analytical techniques in food quality control in view of its less cost, simplicity, and easier sample preparation steps. These techniques are based on electrical properties such as current, potential, charge, resistance, conductance, impedance and conductivity, to detect and measure compounds as the electrical properties are proportional to the target compounds. For food analysis, the electroanalytical techniques make use of potential to encourage an electron transfer reaction following a current measurement, known as voltammetric techniques. It is widely employed to detect and measure additives level in composite food samples. In voltammetric techniques, the potential is changed to encourage propagation of a reaction of electron transfer and the resultant current is noted as a function of the potential applied, giving a voltammogram for detection and quantification of an electroactive target compound. Differential pulse voltammetry (DPV) and square wave voltammetry (SWV) are classically employed for analytical measurements as the responses are based on the superior dismissal of the capacitive/background current, encouraging a sensitivity equivalent to chromatographic methods [4].

12.8 Future Prospects

The extensive health concerns associated with the consumption of certain food additives have affected consumer behavior toward processed food. There is a leaning trend in the market toward food products free of additives or containing natural additives. Hence, more attention should be given to the development and application of natural food additives with a thorough assessment of their safe intake limits. Further, the toxicology and safety assessments of newer food additives demand novel strategies for determining potential risks not depending only on the apical endpoints in animal models but also on the mechanism of toxicity. Also, there is a need to develop suitable newer methodologies for data acquisition. The QSAR (quantitative structure-activity relationship) models, 3-D cell cultures, bioconversion and kinetics, and cultures of human stem cells are essential for this. Additionally, the development of concentration-dependent effects and chronic exposure assessment of low concentration stands obligatory. Also, appropriate assessment of complex food ingredients must be conducted individually in the case of infant formula. The integration of toxicity data with biological approaches such as mode of action, kinetics, and outcome pathway increases the reliability of toxicity and safety assessment techniques. The adoption of the "fit-for-purpose" approach provides flexibility in establishing the results of safety assessment. Other novel approaches offer organ-specific features like metabolism and absorption patterns by mimicking human organs and provide newer outcomes about protein, gene, and metabolite level. The choice of appropriate assessment methodology relies on the sensitivity, significance, and validity of the system [49].

12.9 Conclusion

Food additives are substances added to perform specific functions and provide significant technological benefits in terms of maintaining uniform quality, enhancing appeal, ensuring nutritional value, and providing consumer satisfaction. The permitted usage levels of any food additive must be declared after evaluating its maximum adequate level and ADI. These values should be clear before their large scale production. Sometimes, these additives give rise to unwanted or unhealthy effects. Both immunological and non-immunological reactions can be caused due to the usage of some food additives. These direct toxicological and potential unhealthy effects of food additives are of utmost concern about food safety. FDA approves only those food additives that are proved safe for human consumption after rigorous assessments. Before approval, various toxicological tests, studies on sensitive animals and humans, data assessment, and proper monitoring will be conducted. FDA sets usage limits of food additives for long term use. Also before launching a new product in the market, food producers have to prove the safety of their product to the FDA. To determine the safety of margins, the regulatory authority conducts toxicological tests including studies on lifetime feeding effects and reaction of the body toward additive. ADI can be expressed as the approximate quantity of food additive taken daily for a lifelong period on a bodyweight basis without any negative health consequences. No-observed-adverse-effect level (NOAEL) is a key factor in the determination of ADI. NOAEL is determined by conducting studies in the most sensitive animal species. Any novel food additive prior to approval needs to undergo rigorous and judicious toxicity assessment, together with acute and chronic testing including biochemical estimation, teratogenic effects, and reproductive studies apart from estimation of LD50 dose. Further, there are leaning trends toward various green or natural additives in food formulations. Phytochemicals and essentials are the two most extensively used natural preservative and antioxidant agents in recent times. Additionally, the use of green routes created nanoparticles for their potential antimicrobial activity is also a booming approach. Technologically advanced techniques of nanoencapsulation have been extensively employed to boost the antimicrobial and antioxidant effect of these novel food additives in various food formulations and food systems. The thorough assessment of toxicity and safety of these ingredients stands obligatory to establish their critical intake limits.

References

1. Jain, A. and Mathur, P., Intake of processed foods and selected food additives among teenagers (13-19 years old) of Delhi, India. *Asian J. Multidiscip. Stud.*, 2, 2, 64–77, 2014.
2. Awuchi, C.G., Twinomuhwezi, H., Igwe, V.S., Amagwula, I.O., Food Additives and food preservatives for domestic and industrial food applications. *J. Anim. Health*, 2, 1, 1–16, 2020.
3. FAO/WHO, *Food additive functional classes*, 2017, http://www.fao.org/gsfa online/reference/techfuncs.html. Accessed 5 April 2018.
4. Martins, F.C., Sentanin, M.A., De Souza, D., Analytical methods in food additives determination: compounds with functional applications. *Food Chem.*, 272, 732–750, 2019.
5. Inetianbor, J.E., Yakubu, J.M., Ezeonu, S.C., Effects of food additives and preservatives on man-a review. *Asian J. Inf. Technol.*, 6, 2, 1118–1135, 2015.
6. Faustino, M., Veiga, M., Sousa, P., Costa, E.M., Silva, S., Pintado, M., Agro-food byproducts as a new source of natural food additives. *Molecules*, 24, 6, 1056, 2019.
7. Maryam, I., Huzaifa, U., Hindatu, H., Zubaida, S., Nanoencapsulation of essential oils with enhanced antimicrobial activity: A new way of combating antimicrobial Resistance. *Int. J. Pharmacogn. Phytochem.*, 4, 3, 165, 2015.
8. Bazana, M.T., Codevilla, C.F., de Menezes, C.R., Nanoencapsulation of bioactive compounds: challenges and perspectives. *Curr. Opin. Food Sci.*, 26, 47–56, 2019.
9. Ethiraj, A.S., Jayanthi, S., Ramalingam, C., Banerjee, C., Control of size and antimicrobial activity of green synthesized silver nanoparticles. *Mater. Lett.*, 185, 526–529, 2016.
10. Bindhu, M.R. and Umadevi, M., Antibacterial activities of green synthesized gold nanoparticles. *Mater. Lett.*, 120, 122–125, 2014.
11. Amer, M.W. and Awwad, A.M., Green synthesis of copper nanoparticles by *Citrus limon* fruits extract, characterization and antibacterial activity. *Chem. Int.*, 7, 1–8, 2020.
12. Msagati, T.A., *The chemistry of food additives and preservatives*, Wiley-Blackwell, 2013.

13. Sun, B. and Wang, J., Food additives, in: *Food Safety in China: Science, Technology, Management and Regulation*, pp. 186–200, 2017.
14. Laganà, P., Avventuroso, E., Romano, G., Gioffré, M.E., Patanè, P., Parisi, S., Delia, S., Classification and technological purposes of food additives: the European point of view, in: *Chemistry and Hygiene of Food Additives*, pp. 1–21, Springer, Cham, 2017.
15. Gherezgihier, B.A., Mahmud, A., Admassu, H., Shui, X.W., Fang, Y., Tsighe, N., Mohammed, J.K., Food additives: Functions, effects, regulations, approval and safety evaluation. *J. Acad. Ind. Res.*, 6, 62–68, 2017.
16. Amit, S.K., Uddin, M.M., Rahman, R., Islam, S.R., Khan, M.S., A review on mechanisms and commercial aspects of food preservation and processing. *Agric. Food Secur.*, 6, 1, 51, 2017.
17. Somogyi, L.P., Food additives. *Kirk-Othmer Encyclopedia of Chemical Technology*, pp. 1–59, 2000.
18. Fathi, M., Vinceković, M., Jurić, S., Viskić, M., Režek Jambrak, A., Donsì, F., Food-grade colloidal systems for the delivery of essential oils. *Food Rev. Int.*, 37, 1, 1–45, 2021.
19. Ravishankar, S., Plant-based antimicrobials for clean and green approaches to food safety, in: *Natural and Bio-Based Antimicrobials for Food Applications*, pp. 45–61, American Chemical Society, Washington, DC, 2018.
20. Makris, D.P. and Boskou, D., Plant-derived antioxidants as food additives, in: *Plants as a Source of Natural Antioxidants*, vol. 398, pp. 169–190, 2014.
21. Brul, S. and Coote, P., Preservative agents in foods: mode of action and microbial resistance mechanisms. *Int. J. Food Microbiol.*, 50, 1–2, 1–17, 1999.
22. Burdock, G.A. and Carabin, I.G., Generally recognized as safe (GRAS): history and description. *Toxicol. Lett.*, 150, 1, 3–18, 2004.
23. Yang, H. Food Safety in India: Status and Challenges. *Gates Open Res.*, 3, 1043, 2019, 2017.
24. Saltmarsh, M. and Saltmarsh, M. (Eds.), *Essential guide to food additives*, Royal Society of Chemistry, Cambridge, UK, 2013.
25. Prakash, B., Kujur, A., Yadav, A., Kumar, A., Singh, P.P., Dubey, N.K., Nanoencapsulation: An efficient technology to boost the antimicrobial potential of plant essential oils in food system. *Food Control*, 89, 1–11, 2018.
26. Inanç, T. and Maskan, M., The potential application of plant essential oils/extracts as natural preservatives in oils during processing: a review. *J. Food Sci. Eng.*, 2, 1, 1, 2012.
27. Hyldgaard, M., Mygind, T., Meyer, R.L., Essential oils in food preservation: mode of action, synergies, and interactions with food matrix components. *Front. Microbiol.*, 3, 12, 2012.
28. Seow, Y.X., Yeo, C.R., Chung, H.L., Yuk, H.G., Plant essential oils as active antimicrobial agents. *Crit. Rev. Food Sci.*, 54, 5, 625–644, 2014.
29. Ribeiro-Santos, R., Andrade, M., de Melo, N.R., Sanches-Silva, A., Use of essential oils in active food packaging: Recent advances and future trends. *Trends Food Sci. Technol.*, 61, 132–140, 2017.

30. Rolim, W.R., Pelegrino, M.T., de Araújo Lima, B., Ferraz, L.S., Costa, F.N., Bernardes, J.S., Rodigues, T., Brocchi, M., Seabra, A.B., Green tea extract mediated biogenic synthesis of silver nanoparticles: characterization, cytotoxicity evaluation and antibacterial activity. *Appl. Surf. Sci.*, 463, 66–74, 2019.
31. Saha, S.K., Chowdhury, P., Saini, P., Babu, S.P.S., Ultrasound-assisted green synthesis of poly (vinyl alcohol) capped silver nanoparticles for the study of its antifilarial efficacy. *Appl. Surf. Sci.*, 288, 625–6, 2014.
32. Shah, M., Fawcett, D., Sharma, S., Tripathy, S.K., Poinern, G.E.J., Green synthesis of metallic nanoparticles via biological entities. *Materials*, 8, 11, 7278–7308, 2015.
33. Manosalva, N., Tortella, G., Diez, M.C., Schalchli, H., Seabra, A.B., Durán, N., Rubilar, O., Green synthesis of silver nanoparticles: effect of synthesis reaction parameters on antimicrobial activity. *World J. Microbiol. Biotechnol.*, 35, 6, 88, 2019.
34. Gopinath, K., Shanmugam, V.K., Gowri, S., Senthilkumar, V., Kumaresan., S., Arumugam, A., Antibacterial activity of ruthenium nanoparticles synthesized using *Gloriosa superba L.* leaf extract. *J. Nanostruct. Chem.*, 4, 83, 2014.
35. Rai, M., Ingle, A.P., Gupta, I., Pandit, R., Paralikar, P., Gade, A., Chaud, M.V., dos Santos, C.A., Smart nanopackaging for the enhancement of food shelf life. *Environ. Chem. Lett.*, 17, 1, 277–290, 2019.
36. Pisoschi, A.M., Pop, A., Cimpeanu, C., Turcuș, V., Predoi, G., Iordache, F., Nanoencapsulation techniques for compounds and products with antioxidant and antimicrobial activity-A critical view. *Eur. J. Med. Chem.*, 157, 1326–1345, 2018.
37. Ezhilarasi, P.N., Karthik, P., Chhanwal, N., Anandharamakrishnan, C., Nanoencapsulation techniques for food bioactive components: a review. *Food Bioprocess Technol.*, 6, 3, 628–647, 2013.
38. Đorđević, V., Balanč, B., Belščak-Cvitanović, A., Lević, S., Trifković, K., Kalušević, A., Nedović, V., Trends in encapsulation technologies for delivery of food bioactive compounds. *Food Eng. Rev.*, 7, 4, 452–490, 2015.
39. Burgos, N., Mellinas, A.C., García-Serna, E., Jiménez, A., Nanoencapsulation of flavor and aromas in food packaging, in: *Food packaging*, pp. 567–601, Academic Press, 2017.
40. de Souza Simões, L., Madalena, D.A., Pinheiro, A.C., Teixeira, J.A., Vicente, A.A., Ramos, Ó. L., Micro-and nano bio-based delivery systems for food applications: *In vitro* behavior. *Adv. Colloid Interface Sci.*, 243, 23–45, 2017.
41. Fathi, M., Martin, A., McClements, D.J., Nanoencapsulation of food ingredients using carbohydrate based delivery systems. *Trends Food Sci. Technol.*, 39, 1, 18–39, 2014.
42 Samaranayaka, A. G., Li-Chan, E. C., Food-derived peptidic antioxidants: A review of their production, assessment, and potential applications. *J. Funct. Foods*, 3, 4, 229–254, 2011.

43. Mozafari, M.R., Flanagan, J., Matia-Merino, L., Awati, A., Omri, A., Suntres, Z.E., Singh, H., Recent trends in the lipid-based nanoencapsulation of antioxidants and `their role in foods. *J. Sci. Food Agric.*, 86, 13, 2038–2045, 2006.
44. Bozzuto, G. and Molinari, A., Liposomes as nanomedical devices. *Int. J. Nanomedicine*, 10, 975, 2015.
45. Donsì, F., Annunziata, M., Vincensi, M., Ferrari, G., Design of nanoemulsion-based delivery systems of natural antimicrobials: effect of the emulsifier. *J. Biotechnol.*, 159, 4, 342–350, 2012.
46. Kujur, A., Kiran, S., Dubey, N.K., Prakash, B., Microencapsulation of *Gaultheria procumbens* essential oil using chitosan-cinnamic acid microgel: improvement of antimicrobial activity, stability and mode of action. *LWT-Food Sci. Technol.*, 86, 132–138, 2017.
47. Chellaram, C., Murugaboopathi, G., John, A.A., Sivakumar, R., Ganesan, S., Krithika, S., Priya, G., Significance of nanotechnology in food industry. *APCBEE Proc.*, 8, 109–113, 2014.
48. Ravichandran, R., Nanotechnology applications in food and food processing: innovative green approaches, opportunities and uncertainties for global market. *Int. J. Green Nanotechnol. :Phys. Chem.*, 1, 2, 72–96, 2010.
49. Pressman, P., Clemens, R., Hayes, W., Reddy, C., Food additive safety: A review of toxicologic and regulatory issues. *Toxicol. Res. Appl.*, 1, 1–22, 2017.

13

Sensors for Non-Destructive Quality Evaluation of Food

Krishna Gopalakrishnan[1], Arun Sharma[1,2,3*], Neela Emanuel[1], Pramod K. Prabhakar[1] and Ritesh Kumar[2,3]

[1]National Institute of Food Technology Entrepreneurship and Management (NIFTEM), Kundli, Sonepat, Haryana, India
[2]Academy of Scientific and Innovative Research (AcSIR), Ghaziabad, India
[3]Council of Scientific and Industrial Research–Central Scientific Instruments Organisation (CSIR–CSIO), Chandigarh, India

Abstract

Food safety and quality are of utmost importance for consumers, suppliers, and regulators. The methodologies for determining the food quality are majorly classified as destructive and non-destructive techniques. The destructive techniques used for determining the food quality are labor-intensive, time-consuming, cost-intensive, and biased, while non-destructive techniques are gaining popularity due to ease of operation, reliability, and real-time results. Non-destructive methods play an important role in food industries due to its inherent nature. Both external and internal quality can be determined by non-destructive methods and rapidly help in sorting the superior quality of food products. Currently, different non-destructive methods such as electromagnetic, optical, mechanical, and dynamic are gaining popularity due to ease of operation, reliability, and faster turnover. The major drawback of non-destructive methods is the high equipment cost and the use of various instruments to analyze various parameters. But even then, these methods help to ensure customer satisfaction of products as it helps in providing good quality products without rupturing it. This chapter gives broader concept of non-destructive method used in quality assessment of food products and their applications for food evaluation and quality.

---

*Corresponding author: arunsharma1712@gmail.com; arun.sharma@niftem.ac.in

Mousumi Sen (ed.) Food Chemistry: The Role of Additives, Preservatives and Adulteration, (397–450) © 2022 Scrivener Publishing LLC

Keywords: Quality, safety, sensors, non-destructive methods, analyze, rapid, food, technology

13.1 Introduction

There have been growing cases of food warnings over the past decade, such as artificial milk, argemone oil in mustard oil, and pesticides in soft drinks, which created a confidence crisis in consumers [1]. This has led to public anxiety recently in the availability of good quality and safe agro-food. Food safety refers to the food which contains permissible levels of various components, with absence of toxins and contaminants harmful to human health. Food quality, however, includes good appearance such as food texture, odor, and beneficial attributes preferred by consumers with high nutritional value [2].

Quality is a term that indicates the level of high standard, excellence, or value of a product [3]. Quality is human construct and it is perceived differently depending on the customer's intentions and needs. Quality produce comprises of sensory properties (texture, appearance, aroma, and taste), mechanical properties, nutritive values (chemical constituents), defects, and functional properties [1]. Table 13.1 enlists some of the quality parameters evaluated in food. Food quality includes both sensory attributes that human senses readily perceive and hidden attributes that need to be measured by sophisticated instrumentation, such as safety and nutrition.

Table 13.1 Quality parameters in vegetables and fruits [2].

External quality factors	Internal quality factors
Size Volume, dimension, weight	**Flavor** Sourness, sweetness, aroma, astringency
Shape Diameter/depth ratio	**Texture** Crispness, juiciness, firmness
Color Uniformity intensity	**Nutrition** Carbohydrates, vitamins, proteins
Defect Stab, bruise, spot	**Defect** Water core, frost damage, internal cavity, rotten

Thus, quality assessment and monitoring are of supreme importance for the management of food products in the supply chain and post-harvest handling.

Depending on the orientation of the food, different methods are used in evaluating the quality parameters. It is broadly classified into objective or analytical methods (product characteristics) and sensory or subjective methods (consumer-centered) [1]. Safety aspects can be determined by either analytical or objective methods. The objective methods can be further divided into destructive and non-destructive methods. Traditional food quality assessment is primarily associated with appearance characteristics, tactile features, and internal quality (IQ) characteristics and destructive methods for detecting internal defects and disorders [4]. Most of the techniques used in assessment of fruit quality are destructive, labor intensive, time consuming, cost intensive, and biased [3]. Even the capacity of human visual perception is too limited, making it impossible to determine quality factors which include nutrient quantities, texture, and internal injuries [5].

Non-destructive sensing methods can be widely adopted in food industry for testing the food quality [5]. This analysis refers mainly to surface testing, without any intrusive techniques which affect the appearance and quality of the food. These methods are defined briefly in agricultural products and processed foods as qualitative and quantitative measurements which have been investigated without causing any physical, chemical, thermal, and mechanical damage [5]. Non-destructive techniques are better than traditional approaches, because they are focused largely on physical properties that are associated with other food quality parameters which cannot be detected by human senses [3].

With the growth of the technology, non-destructive methods have developed accurate measuring instruments. Non-destructive methods are suitable for detecting external defects (deformation and discoloration) and internal defects (freezing damage, internal bruises, internal browning, presence of insects in the core, etc.) that may degrade fruit and vegetable quality [3]. However, it is difficult to study chemical and physical parameters using a single non-destructive method. Therefore, the external and IQ assessment requires different methods. These methods help ensure customer satisfaction as it helps to deliver good quality products without rupturing the food product. This chapter includes details on the latest applications of various non-destructive technologies for food product quality evaluation.

13.2 Different Types of Non-Destructive Methods

The non-destructive methods can be categorized broadly as (i) dynamic methods, (ii) electromagnetic (EM) methods, (iii) mechanical methods, and (iv) optical methods. Table 13.2 briefly describes about various non-destructive methods used and various features it measures.

13.2.1 Mechanical Method

Mechanical methods consist of low-mass impact testing and microphone testing. It is used in fruit texture assessment by evaluating the elastic and stiffness properties [3]. The impact test has been shown to be a successful tool for the assessment of food quality. This technique is commonly used in measuring mechanical properties [6].

13.2.1.1 Mechanical Thumb Method

This method is mainly dependent on the reaction of fruits based on the force applied. The theory behind this method is by inserting a cylindrical

Table 13.2 Non-destructive methods for the analysis of food product quality characteristics [5].

Non-destructive method	Technologies used	Measured features
Optical	Image processing	Color, shape, size, external defects
	Spectroscopic transmission, reflectance and absorption, laser spectroscopy	Color, soluble solid content, sugar, acidity, stiffness, internal and external defects
Dynamic	Imaging X-ray, ultrasonic, CT	Degree of maturity, inner cavity structure
Mechanical	Mechanical thumb method	Stiffness, degree of maturity, viscoelasticity
Electromagnetic	MRI, NIR, and NMR	The inner cavity, sugar, moisture content
Chemical	E-tongue, E-nose	Sugar, acidity

head into the flesh of a peeled fruit and firmness of the fruit is measured to determine the maximum penetration power [3]. This method consists of contact head that had a penetration limited to 1.27 mm. It even analyzes fruit color by differences in the firmness of the fruit. Mizrach et al. experimented on the peel of tomatoes and oranges to determine the firmness of tomatoes and oranges using a small flat-head pin of 3-mm diameter by determining the deformation and force of the peel [3]. The "mechanical thumb" was used to design an online mechanical system to estimate firmness and thereby to compare the color of tomatoes [7].

13.2.1.2 Sinclair IQ™–Firmness Tester (SIQ-FT)

Sinclair International has developed SIQ-FT, a low-mass impact sensor [8]. The sensing factor at the tip of a bellow is used in online device firmness measurement. Using air pressure, the component hits the fruit and captures the responding impact signal, thereby helps in measuring the fruit IQ [9]. It is easy to apply the impact sensor used in SIQ-FT. These techniques have achieved better results in the firmness prediction of certain fruits, such as pears, peaches, and tropical fruits, but not in apples [9]. A bench top version of SIQ-FT in static form was used to measure internal quality (IQ) of apple [10].

13.2.1.3 Laser Air-Puff

This system (laser air puff) uses a short compressed air puff to deform the surface of the product by about 1 mm, and then, a laser displacement sensor provides a quick and precise deformation measurement [9]. Based on the distinct firmness range needed for the goods, the pressure is simply adjusted to the amount required. The adjustment does not cause the fruit to undergo injury or defects. While this method has high speed efficiency, it is possible to distinguish only very soft fruits accurately [9]. McGlone and Jordan used this method to calculate apricot and kiwi firmness and stated that on the basis of firmness, laser air-puff is ideal for sorting fruits into two classes [11].

13.2.2 Chemical Method

13.2.2.1 Electronic Nose

E-nose is an instrument that imitates the biological system's sense of smell [1]. The electronic nose or artificial nose consist of electronic and chemical sensors those try to stimulate olfactory system function [3]. The analytical instruments and human expert panel for characterization of aroma involve high cost. In the case of analytical methods, experienced and skilled individuals are required to operate these tools and to elaborate sample preparation and long analytical time, while the human expert panel is an expensive method because it needs qualified individuals who can work for a limited period of time. E-nose is a non-invasive, low-cost, and quick aroma measuring technique [1].

Working
E-nose mainly comprises of three elements which are [1] as follows:

(i) Sampling handling system
This system is used in incorporating the volatile compounds found in the headspace of the sample to the detection system.
(ii) Detection system
It helps in odor capture and the sensor technology is used in capturing.
(iii) Data processing system
The data acquisition card stores the sensor response in the PC and these data sets are processed in order to extract information.

Thus, E-nose consists of a detecting device, a variety of chemical gas sensors with different selectivity, and a computer with an appropriate pattern recognition algorithm capable of evaluating single or complex vapors, odors, or gases qualitatively and/or quantitatively [12]. This is an

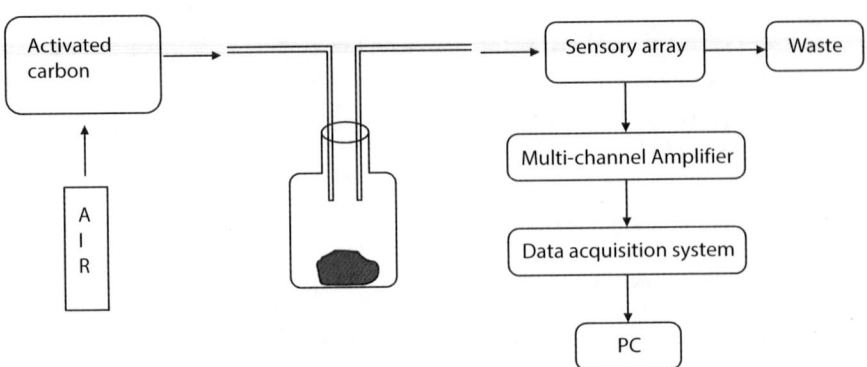

Figure 13.1 Schematic diagram of the electronic nose system [14].

instrument that can recognize, sort, and categorize the volatile fruit-emitted fingerprint patterns [13]. The schematic diagram of E-nose system is depicted in Figure 13.1 [14]. The sensors present in the E-nose interact in a non-selective manner with volatile compounds/odors to produce physical or chemical changes and thus produce signal and the computer reads the signals produced. The response to each sample produced by the sensors is known as "electronic finger print" of a mixture or compound [3]. Based on the reference of a standard finger print, sensors recognize the odor on the basis of the response pattern generated by sensors. Then, multivariate statistics are used to interpret the computed results.

Application in Food Industry
E-nose applications can be divided into five categories and they are as follows:

(i) Food Monitoring
E-noses have also been used for bioprocess surveillance where microbiological processes [15] include food processing, i.e., for screening the aroma generation of lactic acid bacteria strains in cheese and other fermented milk products [15]. E-noses have been used to classify rotten Iberian hams during the curing phase [16]. In the bioconversion process, it has also helped control the production of aroma from fermented grapes [16]. The black tea aroma and taste was attempted, and promising results were obtained using E-nose [17, 18].

(ii) Shelf life investigation
E-nose monitored fruit and vegetables ripeness states based on their aroma profile which further helped in evaluating its shelf life. It even includes shelf life investigation in milk, cheese, and oil samples [19–23].

(iii) Freshness evaluation
In the food industry, freshness is an essential quality trait. In the prediction of spoilage or freshness of different food raw materials and goods, electronic noses have been proved to be useful. The rapid degradation due to bacterial processes, such as oysters, fish, eggs, shrimps, soybean, meats, and curds can be found by the volatile releases in foods during storage using E-nose [24–30]. E-nose is also used in predicting the storage time and rancidity indices of peanuts using Multiple Linear Regression [31].

13.2.3 Electromagnetic Method

EM spectrum comprise of microwave, X-rays, ultraviolet (UV) rays, gamma rays, radio wave, visible light, and infrared rays. The Figure 13.2 shows the spectrum of electromagnetic radiation.

13.2.3.1 Nuclear Magnetic Resonance (NMR)

NMR technique is rapid, non-destructive and powerful technique which requires minimum sample preparation time. It is the study of molecules by recording the interaction between EM radio frequency (RF) radiation with molecular nuclei positioned in a strong magnetic field. NMR operates under the principle of absorption of magnetic energy when the nucleus is embedded in an alternating magnetic field. The energy is absorbed by the odd number of protons and neutrons in nuclei and can determine the amount of these protons present in the food [12]. NMR spectroscopy is used for structural and compositional aspects in food science, food analysis, authentication, and quality control of food products [32, 33]. This method can be applied in many foods for non-invasive quality assessment.

Working
NMR spectroscopy is an instrument that focuses on the magnetic properties of nuclei that have an even or odd mass number, but an odd atomic number. These nuclei have "s" nuclear spin, which is a form of "p" angular momentum and is distinguished by an "I" quantity of nuclear spin with different $2I + 1$ states. Both nuclei are rotating and formed by a magnetic field, μ, which is proportional to "p" and the gyromagnetic ratio γ ($\mu = \gamma p$). The interaction of external field B_0 and nuclear magnetic moment μ, formed

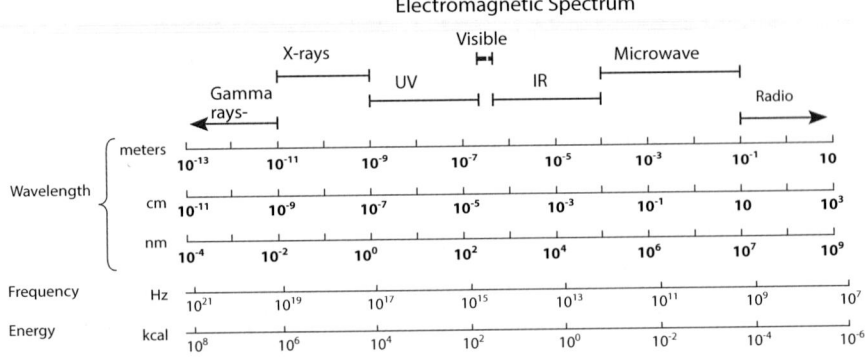

Figure 13.2 Electromagnetic spectrum.

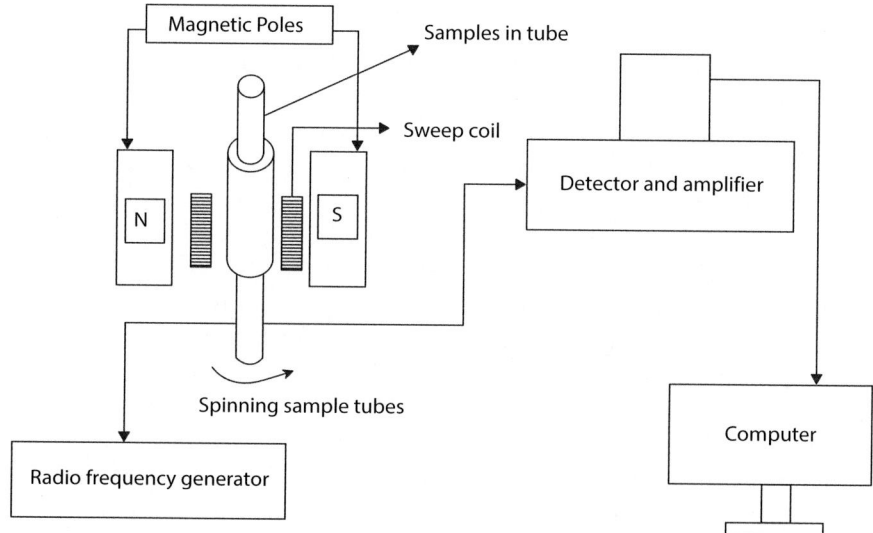

Figure 13.3 Schematic diagram of NMR spectrometer [32].

by the NMR tool in two separate ways to produce two energy states for absorption and emission of EM radiation, so that the nuclei can be studied using NMR. The most common nuclei used in food related applications in NMR are ^{13}C, ^{1}H, and ^{31}P, all of which have I = ½.

NMR instrument consist of superconducting magnets surrounded by two layers of jackets. The outer jacket is filled with liquid nitrogen, whereas inner one is filled with liquid helium to cool the magnets. The probe inside the magnet bore has small magnetic coils which receives the power from the NMR instrument [32]. It also has coils to receive and transmit RF energy. The sample is inserted at the top of the bore and lowered down to the magnetic probe. The probe is connected to an electronic chamber consisting of transmitter, receiver, and other systems to control NMR instrument. The transmitter produces the pulse at correct frequency to observe the nucleus. With the help of radio waves, the nuclei in the sample are excited into nuclear magnetic resonance to create NMR signals and these signals are detected by sensitive radio receivers. In a molecule, the resonance frequency of an atom is modified by the intra-molecular magnetic field that surrounds it. The schematic diagram of NMR spectrometer is demonstrated in Figure 13.3 [32].

Applications in Food Industry
NMR technique is applied to liquid and solid samples and helpful in the evaluation of multicomponent food such as fruits, vegetables, grains,

oils and lipids, meats, dairy products, and beverages without performing tedious process of separation and purification [34–39]. This technology is helpful in the detection of fruit and vegetable pits. NMR signals also help to determine the freeze damage caused by frozen fruit and vegetables stored. It is powerful technique helpful for food industries to meet the consumer's demand of safety and quality of food. Validation of food authenticity is the most common application of NMR sensors [40]. Oliveira *et al.* were able to differentiate adulterated honey from pure honey by adding high fructose glucose syrup and time domain NMR was suggested by Santos *et al.* as a rapid method for testing adulteration on milk samples [41]. A fully automated NMR screening method was proposed by Spraul *et al.* (2009) for quality control and authentication of various fruit juices [42]. NMR was also used to investigate the effect of high pressure processing on the properties and composition of food [42]. NMR spectroscopy method is also used in quantification of oil and water content in food samples. Since foods are abundant in protons, such as water, fat, carbohydrates, and proteins, and are therefore measured by the NMR technique [43]. In fish oil supplements, the quantification of lipids is accomplished by incorporating sufficient NMR signals into the applicable spectrum [44]. In a number of agro-food products, NMR has been devoted in quantification of fat and oil. NMR has shown the potential for successful application in quality monitoring in potatoes [45]. A recent research by Zehl *et al.* depicted the ability of NMR in controlling their specific bioactive compounds for the quality regulation of medicinal herbal products [46].

13.2.3.2 Magnetic Resonance Imaging

It is an instrument that generates three dimensional images which is applicable for food quality measurement and detects storage disorders developed over time. This approach helps to determine both the quantitative and the qualitative properties of biological materials [5]. It is on the basis of association of RF EM radiation with characteristic nuclei such as hydrogen and carbon in the presence of a strong magnetic field [5].

Working
Magnetic Resonance Imaging (MRI) technique is implemented through the following steps [1]:

1. Place the sample in the magnetic field
2. Radio waves are transmitted into material

3. Radio wave transmitter is turned off
4. Radio waves re-transmitted by subject is received
5. Measured RF data is converted to image

RF and magnetic fields waves are used in MRI. The magnetic moments of an atomic nucleus appear to align to an external magnetic field either as parallel or anti-parallel. The magnetic moment can be induced on the nucleus to flip their orientation by absorbing characteristic energy from an imposed EM radio wave. The emission of the absorbed energy brings back to ground state. This continuous absorption and emission of energy by nuclei either by RF or the external magnetic field can produce resonance which dependent on the target element and the strength of the magnetic field. Removal of RF source return the magnetic-moment vector to its ground state give rise to a characteristics signal depending on the location and orientation of the proton present in the food/biological samples. An array of sensitive receiver coils are located around the body with high-gain, low-noise amplifiers which serve as antennas to track the transmitted signal that is transformed into an image.

Application in Food Industry
MRI has been successfully used for IQ measurement [47]. MRI has been successfully used for internal fruit maturation condition measurement [47]. MRI helps to diagnose the determination of physical tissue injury (such as bruising) and is also used in the online sorting or identification of internal defects. For food processing processes, such as curing, brining and freezing, baking, fermentation, and internal consistency measurement, online MRI is used [48]. In a research study by MRI, the redistribution of water during the freezing and drying of apple tissue was evaluated [49]. MRI has been used by Hernandez-Sanchez *et al.* to distinguish healthy and freeze damaged oranges [50]. High and homogeneous signal strength was produced by undamaged oranges, while low intensity signals were produced by freeze damaged oranges. Water mobility and moisture migration in cereals and cookies such as rice kernel, corn flakes, or caramel candies have been investigated by MRI [51, 52]. MRI has been confirmed useful in distinguishing mealy from fresh fruits such as apple and peach [53]. It has been shown that different internal features of vegetables and fruits can be detected using rapid MRI techniques [54–56]. These features include physical, such as the foreign body's presence or physical/physiological damage, and the MRI signals may be associated with certain biochemical parameters, such as sugar levels. It is also used for assessing meat products and fish products [57–59].

Recent improvements have been taking place in MRI to assess sheep and goat carcass and meat quality [60].

13.2.4 Optical Method

The optical properties are characterized by transmission, absorption, reflection, or scattering. Spectral-optical methods can be used for analyzing starch acids, soluble solid content (SSC), and maturity index. This method includes visible/near infrared spectroscopy (VIR), image analysis (evaluation of food sample shape, size, color, and surface defects), reflectivity, transmittance, and laser and absorption spectroscopy [3].

13.2.4.1 NIR Spectroscopy

Visual spectroscopy and near-infrared (NIR) are rapid and effective non-destructive method for measuring the biological material composition. The NIR spectrum (700–2,500 nm) is sensitive to chemical components such as agricultural moisture, protein, oil, and food products [1]. The main gain is that NIR can usually penetrate far deeper into a sample and requires little or no preparation for the sample. It works in accordance with the Beer Lambert law which is based on the transmission, reflection, scattering, and/or absorption of light through food material [12]. The NIR radiation energy has the power to cause different molecular vibrations such as symmetrical and asymmetrical stretching and various bending vibrations. Different molecule bonds absorb vibrations and rotation at unique frequencies that determine the characteristics of the composition of the sample. The reflection or absorption of light in the defined wavelength range is measured and associated with different food material quality parameters. The light absorbed and transmitted from the sample reveals about chemical properties and surface structure of the sample. The light transmitted from fresh produce depicts the tissue microstructure, and the presence of chemical components in the sample is correlated with the absorption of light. Hence, all these phenomena are useful in quality assessment.

Working
NIR includes the use of light in the 780–2,500 nm wavelength range and the depth of light penetration, depending on the wavelength and sample characteristics, such as functional groups [6]. Studies conducted by Lammertyn *et al.* found that in the wavelength band 700–900 nm, penetration varied up to 4 mm and from 2 to 3 mm in the 900–1,900 nm range [61]. NIR spectroscopy is based on EM radiation absorption at wavelengths

ranging from 780 to 2,500 nm and depends on the mass of the atoms in the bond and strength of the bond. Quartz chamber is used for liquid samples and diffuse reflection with accessories is applicable for solid samples. The light from the tungsten-halogen lamp passing through the interferometer interacts with the sample and its transmittance and absorption are measured by the detector. Transmittance applies to the amount of light passing through the sample completely and striking the detector. Absorbance is a light measurement where the sample absorbs. The detector detects through the sample the light being emitted or absorbed and translates this data into a digital display. The working of NIR spectroscopy is illustrated in Figures 13.4 and 13.5.

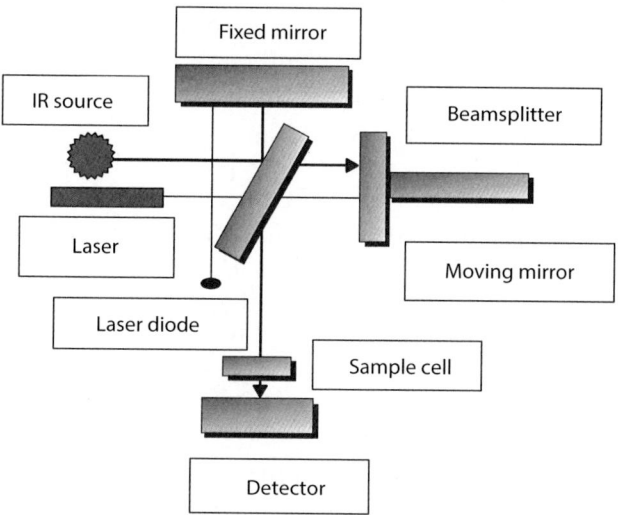

Figure 13.4 Schematic optical diagram of NIR spectrometer.

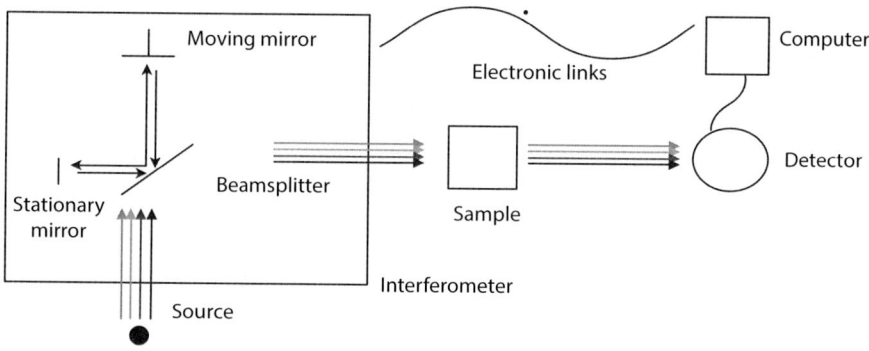

Figure 13.5 Schematic representation of NIR spectroscopy.

Application in Food Industry

The advantage of this technique for food safety and control is because of its inherent property such as rapid detection, ease of use, non-invasive, and minimal sample preparation. For oils, proteins, dry matter, firmness, and total soluble solids (TSSs), the non-destructive food quality assessment using the NIR spectroscopy approach has been widely applied. Wang et al. investigated use of VIS-NIR spectroscopy to measure vitamin C content in chillies [62]. Near infrared spectroscopy has shown the adulteration of fruit juice with high sugar solutions and fructose corn syrup [63]. Various apple juice samples adulterated with the addition of sugar were graded into adulterated and unadulterated samples [64]. Merzlyak et al. investigated the light reflectance on apple fruit using a spectrometer in 400- to 800-nm wavelength range [65]. A strong association between the reflectance and apple chlorophyll content was demonstrated by five different apples. McGlone et al. stated that for IQ evaluation of mandarin fruits, NIR spectroscopy with a 750- to 1,100-nm region was used, while Bergaz et al. evaluated NIR spectroscopy with a SSC and mandarin acidity range of 350–2500 nm [66, 67]. The ability to measure protein content, sprout damage, and alpha-amylase production in wheat and barley is shown by NIR-based methods, in addition to evaluation of the quality parameters in other cereals, such as maize, rice and oats, as well as prediction of fungal infections [68, 69]. NIR spectroscopy is efficient in predicting important chemical components such as fat, protein, and moisture but is not stable in predicting low-quantity chemicals in cheese products [70]. Applications of NIR in recent years are cereals and cereals products [69], meat and meat products [71–74], fruits and vegetables [75–77], fish and its products [78–81], dairy and its products [82, 83], irradiated sucrose [84], corn meal [85], soya bean meal [86], edible oils [87–89], coffee [90], honey [91], instant noodles [92], and beverages [93].

13.2.4.2 Image Analysis Techniques

This image system technique is used to test for shape, color, size, and surface texture and to detect surface defects in food samples, but not suitable for detecting chemical properties of a food product. This technique includes various methods and these are as follows.

13.2.4.2.1 Hyperspectral Imaging

Hyperspectral imaging (HSI) combines traditional imaging and spectroscopy to obtain spectral and spatial information from an object simultaneously. It was initially designed for remote sensing purposes; this technology has emerged as an influential tool for rapid and non-destructive analysis of food. HSI system comprises of both a spectrograph and a digital camera [1]. Each spatial location of the object are made up of hundreds of contiguous wavebands of hyperspectral images, also called hyper cubes. Consequently, each pixel contains the spectrum of that particular spot in a hyperspectral image. The hypercube allows the biochemical components of the sample to be visualized, segregated into different areas of the image, since the chemical composition of the regions of the sample with identical spectral properties are similar. It is an effective tool due to the ability of the target being scanned to gather data with both spectral and spatial characteristics [5]. It is also useful in providing data relating to various food systems constituents. HSI incorporates the benefits of computer vision and conventional spectroscopy to provide three-dimensional information [94]. Using different spectroscopic systems, internal defects were generally detected far more than external defects [45]. The schematic representation of HSI is illustrated in Figure 13.6.

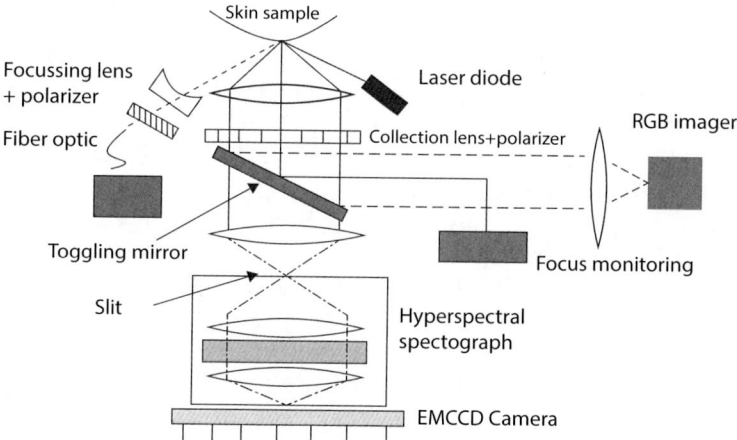

Figure 13.6 Schematic representation of HSI.

Working

HSI system comprises of the following components:

1. Illumination (light source): Illumination systems produce light that illuminates the target objects which are inspected; hence, the performance of illumination system success can have a huge impact on image quality and plays an important part in the overall reliability and precision of the devices.
2. Spectrograph (wavelength dispersion device): Three standard wavelength dispersion devices are Prism, Grating, and Filter. It is used to disperse broadband light over multiple wavelengths, using these optical instruments to project the scattered light onto the area detector.
3. Camera (area detector): It is used as image acquisition devices. Charging coupled devices (CCD) and complementary metal oxide semiconductors are the image sensors used to produce the image digitally (CMOS). Generally, CMOS image sensors are used in less challenging quality specifications, and the CCD image sensor is used to generate high-quality image data.
4. Transportation stage: This is specifically used to transfer the sample past the target's objective lens while a line of the illuminated sample is captured by the sensor.
5. Computer with corresponding software: The computer is used for data processing, acquisition and analysis of image and spectral data for specific applications in controlling the HSI system. It also provides hyperspectral image space for storage. Apart from that, computer helps in scanning the entire surface of sample and creates hyperspectral image which is displayed on computer.

Application in Food Industry

The HSI methodology was successfully assessed for food safety applications [95]. HSI systems were used to detect contamination in apples and damage caused by cucumber chilling [96, 97], and caps of white mushrooms (Agaricusbisporus) that can be used on the processing line for non-destructive surveillance of damaged mushrooms [98]. HSI device in the visible and near-infrared (400–1,000 nm) ranges is used, for the non-destructive evaluation of moisture content, TSSs, and strawberry acidity. In addition, an image texture analysis was also performed to classify strawberries based on the ripeness stage [99]. To detect the chilling injury to the cucumber, a HSI device was developed by Liu et al. [100].

Leivavalenzuela *et al.* investigated the use of HSI to estimate the solid soluble content of blueberries in the range of 500–1,000 nm [101]. HSI is also used for the classification of egg freshness [102]. Imaging and spectroscopy have shown positive results and potential applications for the quality assessment of potatoes and sweet potatoes [103]. HSI has the potential to assess the chemical and physical characteristics of cheese [104]. HSI was also evaluated to be useful for rapid non-destructive screening of single coffee beans [105]. HSI is also a promising approach to evaluate cocoa bean composition and thereby offering a valid tool for food inspection and quality control [106]. Su and Sun used HSI technique to detect adulteration of organic wheat flour [107].

13.2.4.2.2 Machine Vision

The machine vision system has been increasingly used to analyze fruit and vegetables [108]. The digital sensor is protected inside the specialized optical camera which acquire images and computer can do further processing of various parameter for meaningful results. Computer vision system is an integrated tool that automatically receives and interprets an image of a real scene [5]. Their functions include capturing, processing, and analyzing two-dimensional (2-D) images, and thereby helping to perceive and understand an image electronically. The schematic representation of an imaging system technique is depicted in Figure 13.7 [5].

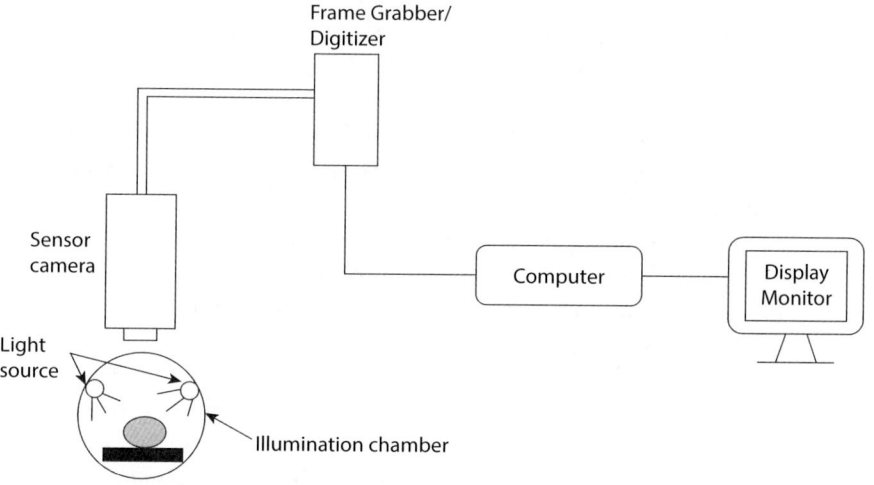

Figure 13.7 Schematic representation of an imaging system technique [5].

Working
The essential components of a typical computer vision system are as follows:

(i) Computer: corresponding to the human brain,
(ii) Sensor or camera: corresponding to the human eyes,
(iii) Illumination system: facilitates image capture,
(iv) Frame grabber/digitizer: digitizes the image information from the camera,
(v) Monitor: used in displaying the processed images.

A typical computer vision task involves the following steps:

a. Image acquisition: Appropriate light sources are required for image formation. It may consist of incandescent light, fiber-optic bundle, arc lamp, fluorescent tube, or strobe light. The laser beam is used for distance measurement in the triangulation system. To reduce glare or increase contrast, polarized or UV light is used. It is essential that the light source is correctly placed as it influences the image's contrast.
b. Image processing: Typically, a camera can create an image 30 times per second. At each time interval, the entire image has to be captured and frozen for processing by an image processor. Image preprocessing, segmentation, and feature extraction are the basic steps in image processing [1]. The objective of image preprocessing is to improve the accuracy of the image produced, which is also affected by noise and distortion in the electronic and optical systems of the input device.
c. Image analysis: Analysis is carried out by describing and measuring the properties of several image features which may belong to either regions of the image or the image as a whole. The image interpretation process begins by analyzing simple features and then adding more complicated features to fully define them. It therefore implies recognition and interpretation of images.
d. Image interpretation: Interpretation of images includes the identification of an object based on its image recognition. The comparison of results of the analysis with a pre-stored set of standard criteria and different conclusions is made.

Applications in Food Industry

Applications of computer vision technologies are widely used for shape characterization, defect identification, and assessment of quality. Due to advancements in machine vision and low-cost hardware and software availability, manual work on fruit categorization and grading has now been replaced by automated machine vision systems [109]. Machine vision devices are widely used in agricultural goods to detect external insect damage, but the detection of internal defects is not effective due to the difficulties involved in penetrating visible light into the fruit. An automatic machine vision system was developed to detect small insects in raspberry fruit holes by Okamoto et al., where insect contamination was hard to detect with human eyes [2]. Many studies have been conducted in recent years using computer vision to examine the quality attributes of cheese, and these studies have focused primarily on physical attributes [104].

13.2.4.3 *Time-Resolved Reflectance Spectrometry*

Time-resolved Reflectance Spectrometry (TRS) is applicable in highly diffusive media for optical characterization. Light penetration, through a diffusive medium, is depended on the distance between the detector and the source and the optical properties of the medium. The optical density (OD) is related to the μa coefficient of light absorption and absorption, which depends on the concentration of a given chromophore of interest. A fruit with a diffusive medium, in which the distribution of light is measured by the relationship between the phenomenon of scattering (due to fruit microstructure) and the absorption of light (due to the presence of chromophores and other chemical compounds).

Working

In TRS, a brief pulse of monochromatic light is inserted into the diffusive medium; if a photon strikes a dispersing center, then it changes its direction and begins to propagate in the medium until it inevitably re-emits through the boundary or is absorbed by an absorbing center [110]. It will delay, broaden, and attenuate the temporal propagation of the re-emitted photons at a distance from the injection point [110]. In order to simultaneously measure the absorption coefficient (μa) and the reduced dispersion coefficient (μs), it is possible to use the appropriate theoretical model. For TRS, a sample volume greater than 10 cm^3 is needed, but it could be carried out on strongly diffused samples with smaller dimensions. In most of the fruits and vegetables, light penetration obtained by TRS can be as high as 1–2 cm. It therefore offers information on the medium's internal properties

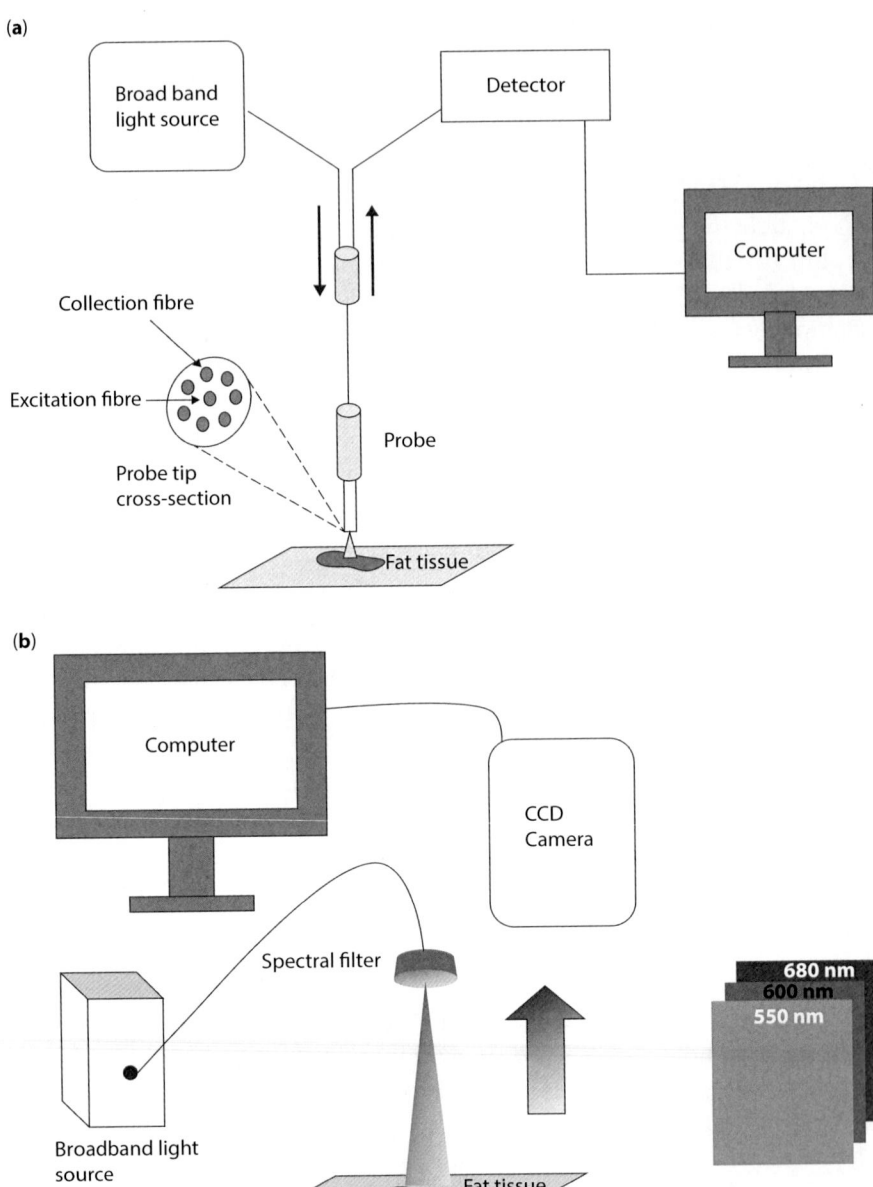

Figure 13.8 Schematic representation of the optical fiber–based setup of diffused reflectance spectroscopy (a) and multispectral imaging (b) [111].

and is independent of the surface features. It can be used for determining maturity, grading, analyzing shelf life and detecting internal defects [3]. The advantage of this method is that the ripening of the fruit can be determined, even if the fruit is coated with deep blush. In various fruits and vegetables, the reduced scattering coefficient and absorption coefficient are measured simultaneously using TRS. The amount of pigments present is determined by absorption, while scattering is due to local variations in the dielectric constant [89]. Constituents such as sugars, water, or pigments such as anthocyanins and chlorophyll are determined by absorption [5]. As it tests the bulk properties, it provides valuable information on internal fruit consistency. The drawback of this approach is that the required instrumentation is bulky and expensive. In addition, TRS measurements are not accurate when the dispersion is very low or the fruit/vegetable dimension is small [89]. The schematic representation of the optical fiber–based setup of diffused reflectance spectroscopy and multispectral imaging is shown in Figure 13.8 [110].

Applications in Food Industry

TRS applications were primarily based on vegetables and fruits for the non-destructive evaluation of bulk optical properties of foods [110]. For small yet highly diffusive samples such as cherries, the applicability of TRS depends on the diffusivity, providing a significant expansion of laser pulses, while TRS does not have stable results for samples such as gels, or in the presence of voids such as cereal flakes which has weak diffusion [112]. TRS can also be used to assess fruit maturity; the absorption coefficient calculated by TRS has been shown to decrease with maturation at a wavelength in the range of 630–690 nm, near the chlorophyll peak. The optical scattering and absorption properties measured by the TRS depict the fruit's textural characteristics. Braeburn apples were found to have near relationships between sensory texture profiles and mechanical characteristics of pulp with TRS scattering parameters in the range of 630–900 nm [113]. Vanoli *et al.* (2015) demonstrated that "Braeburn" apples with high scattering values were the apples with the highest stiffness and firmness values, and were the most crispy and juicy [114]. Various studies have confirmed that in the presence of internal disorder, TRS optical properties change in response [110]. The studies were conducted on "Braeburn" apples, pears, and potatoes having internal disorders using TRS [115, 116]. In the spectral range of 670–940 nm, the browned fruit displayed slightly greater μ_a values than the healthy ones [110]. In TRS, the absorption coefficient (μ_a 540) is used to measure the composition and content of the main antioxidants in the mango pulp [117]. It is also feasible tool for detecting internal defects in potato tubers of medium large size [118].

13.2.5 Dynamic Method

13.2.5.1 X-Rays

This method was used in various inspection applications within the food and agricultural industries. X-ray radiography producing 2-D images and X-ray computed tomography (CT) producing 3-D images are the most widely used methods [3]. The key issue associated with X-ray use is that biological cells can be ionized and destroyed by high-energy radioactive radiation, and thus, a shield is required. Therefore, equipment designed for radiography should meet both functional and radiation safety criteria.

Working
X-ray imaging consists of four basic elements:

i. X-ray source: Radio-active compounds and X-ray tubes are the sources of X-rays. X-rays are formed in X-ray tubes by association between the target's energetic electrons and atoms. Monochromatic X-rays can be released by radioactive materials, and X-ray tubes may create polychromatic beams.
ii. X-ray converter: This prohibits the penetration of X-rays into the imaging medium and generates a visible output proportional to the X-ray incident photons, e.g., phosphor screen.
iii. Imaging medium: The imaging is performed to record the image on photographic films. Through the object being studied, the X-ray is transmitted and a sensing film is used in forming the object image.
iv. Casing for imaging medium.

There is high energy in X-rays which can penetrate into several objects. In an X-ray pulse, photons travel through a biological substance and can be scattered, transmitted, or absorbed. Radiography attempts to identify the variation in the emitted X-ray beam in the form of a visible contrast in the signal, independent of the material difference. This comparison may be an indicator of a certain material's spatial and quantitative distribution depending on food composition. Based on changes in water content and density, physiological defects in tissue including internal browning or freezing damage or pit presence, presence of infested fruits, foreign particles, etc., can be detected.

Applications in Food Industry
X-ray inspection has become more widespread for processed, packaged foods, particularly those in cans, bottles, pouches, and jars because of its high penetration capacity. In food analysis, X-ray imaging can be used to identify physical contamination and to investigate the internal composition of food items for quality purposes. X-ray–based quality monitoring devices (X-ray imaging and CT) have recently been used successfully as possible detectors for examining the internal properties of agricultural commodities [119]. This technique has been used to analyze various internal defects and maturity in fruits and it has also shown promising results for some grains and nuts. Internal injuries caused by insects in pear, plum, cherry tomato, peach, and orange at different time intervals have been investigated by Yang *et al.* [120]. Vitreousness in durum wheat has been successfully classified by X-ray imaging [121]. Recently, dual-energy X-ray absorptiometry (DXA) has been used to test sheep and goat carcasses and meat consistency [60].

13.2.5.2 Computed Tomography

Another important tool from a scientific point of view is X-ray CT. The fundamental concept behind CT is that it is possible to reconstruct an object's internal structure from multiple object X-ray projections [1].

Working
Another important technique in X-ray CT is the rotation of the sample in the X-ray beam in these devices, while the detector and the X-ray source remain stationary. On the observed sample, an X-ray beam is oriented and a shadow image is recorded representing X-ray attenuation along the beam direction. The sample rotation provides successive images that are processed and then analyzed by computer-aided tomography or CAT scanning [122]. Conventional imaging approaches typically provide 2-D images of a sample's surface or cross-section, while X-ray CT can give 3-D images of foods that can be properly studied. Recently, X-ray CT technology has become a flexible tool for assessing the consistency of agricultural products, allowing sample composition, physicochemical characteristics, and internal structure to be better understood. In particular, a large variety of agricultural products such as cereals (wheat, maize, and rice), fruit (pear and apple) and meat (lamb and beef) could be studied using high-resolution 3-D and 2-D X-ray CT visualization technology [123].

Applications in Food Industry

- Dairy products: Quantitative assessment of eye shape in cheese, micro structural evolution analysis (samples of ice cream, mayonnaise and cheese), fat microstructure (yoghurt), and loose-packed and compacted milk powder microstructure [122].
- Bakery products: Impact of crumb morphology of water retention and migration on crispness, structural parameters and crystallization of starch (cake), pore structure of bread crumbs, effects of fat and sugar (sugar-snap cookies), study of distribution of bubble size, and growth and setting of gas bubbles (wheat flour dough) [122].
- Fruits and vegetables: Detecting water core disorder and "Braeburn" browning disorder (apples) characterization, internal structure quantification and characterization (pomegranate), and maturity determination (tomatoes) [122].
- Coffee beans and nuts: Post-harvest examination of internal degradation (chestnuts), roasting-induced micro structural changes (coffee beans), and insect activity (pecan nuts) [122].

13.2.5.3 Ultrasonic

Ultrasonic is a mechanical wave that propagates by particle vibration in the medium and penetrates through optically opaque surfaces to provide physical features, such as texture and shape, with internal or surface details. This transmits ultrasonic waves through a material and specifies the properties of the ultrasonic waves that are emitted and/or reflected. In the quality assessment of materials, low-intensity ultrasonic with a power range of up to 1 W/cm^2 was used, as high-intensity ultrasonic with a power spectrum above 1 W/cm^2 can lead to chemical/physical disturbances. The attenuation coefficient and ultrasonic velocity are the parameters that are most commonly used in ultrasonic measurements. Ultrasonic velocity for a substance in a given state is a constant quantity that depends on its physical properties. Attenuation is the amount of a wave's diminishing power as it passes through a material. Absorption and scattering are the key causes for attenuation. The attenuation value of the receiver's ultrasonic waves depends on the condition of the fruit [124]. Detailed knowledge of physicochemical properties is given by ultrasonic parameters. The strong ultrasonic reflection and refraction at the interfaces of the host tissue and foreign object can be identified accurately using ultrasonic techniques [1].

The non-contact ultrasonic parameters however are very susceptible to the unevenness of the sample surface and small surface defects.

Working
The elements of any ultrasonic system include the following:

1. Pulsing/receiving source
2. Transducer
3. Data visualization

The pulsing system produces an electrical pulse that drives the transducer to emit ultrasonic energy at a very high frequency. Sound waves pass through the sample to be tested for defects. Any anomalies or discontinuity in the material can partly reflect back any ultrasonic energy to the receiver. The receiver once again translates the reflected ultrasonic energy into an electrical signal and visualizes this data in one way or another.

Application in Food Industry
Ultrasound is an evolving technique that can be used for quality reservation and food safety security [125]. Mizrach *et al.* examined non-destructive control of physicochemical changes in the entire avocado during ultrasonic device maturation [126]. The propagation of surface waves is the most powerful tool for ultrasonic study of vegetables and fruits. It entails the transmission of energy into fruits and then the study of response energy. With expanded storage time, the researchers found the attenuation of ultrasonic waves in mangos to be greater. Although the mango flesh softens with the maturity of the fruit with increased storage time, when the firmness of the mango decreases, the attenuation would be greater [124]. Abeyratne and Morrison have successfully researched the firmness, density, and moisture content of oranges using ultrasonic waves [127]. It can also be used to monitor and assess cheese maturity on the basis of moisture and rheological parameters [128]. This technology can be used to monitor the consistency of fresh vegetables and fruit, pre and post-harvest bread, cheeses, industrial cooking oils, and cereal products [129].

13.2.5.4 Acoustic Techniques

Acoustic approaches (also referred to as sonic testing, vibration testing, mechanical impedance testing) focus on local effects on exciting vibrations in a specimen and then measure those vibration properties, such as resonant frequency, and decay rate. Acoustic analysis is a technique that is quick, accurate, economical, and non-destructive. The instrumental acoustic methods for are becoming more popular, mainly the audible frequency spectrum from 20 Hz to 20 kHz is used in food analysis [5, 130]. When an acoustic wave reaches agricultural products, the acoustic wave that is reflected or transmitted is based on the acoustic properties of agricultural products.

Working
The acoustic impulse resonance frequency system utilizes the fruit's intact natural frequency produced by capturing the sound generated by striking the fruit and then conducting out a Fourier transformation signal. This approach tests the resonance frequency with maturation, the resonance frequency of the fruit increases, and thereby defines the maturity of the fruit. The internal consistency and maturity of the fruit are demonstrated by the impact reaction to the application of force received from the fruit. The schematic representation of sound measurement technique using instrumental compression is illustrated in Figure 13.9 [5].

Applications in Food Industry
Acoustic vibration characteristics of agricultural products may be used for texture assessment, harvest maturity, ripeness classification, and defect detection [131]. Antihus *et al.* tested the response of acoustic signal's ability to track mandarin fruit firmness shifts during storage [132]. Fathizadeh, Z *et al.* used acoustic-vibration response method for assessment of the apple firmness [133]. The acoustic vibration approach is used in monitoring its elasticity index to assess the ripening speed of the melon cultivar "Miyabi-Haruaki" [134]. They suggested that the non-destructive measurement of the optimum maturity period with respect to the elasticity index is useful for calculating the shelf life of melons [135].

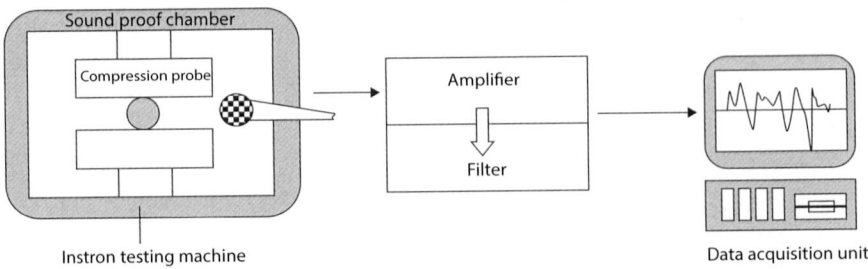

Figure 13.9 Sound measurement technique using instrumental compression [5].

The acoustic approach has proven that it is an effective method for grading and sorting of agricultural products [135]. A low-frequency acoustic technique to describe the kinetics of the phenomenon of bread dough fermentation [136] was proposed by Skaf *et al.* In the dairy sector, some researchers have used an acoustic approach to determine the structure of dairy products [137]. Acoustic methods used by Nassar *et al.* to monitor the quality of cheese [138].

13.2.6 Sensor Fusion

Sensor fusion is a collaborative approach with multi sensors, in order to enhance the quality evaluation of agricultural and food products. For applications in fruit quality evaluations, the sensor fusion approach has clear potential to assist human panels in decision-making [139].

Working

For non-destructive successful assessment of food samples, an acoustic sensor, an E-nose (EN), NIR spectrometer and online HSI are combined as illustrated in Figure 13.10 [6]. The strategy for sensor fusion is based on the signal fusion from multiple sensors (similar or different sensors) that will yield good results than a single sensor [140].

For the fusion process at various stages, different approaches are used (feature and decision level). Data fusion is processed at various levels: a feature level, a raw data level, a decision level, or a combination of these methods. For example, when a camera is one of the sensors, the raw data is the intensity values of each image pixel, the level of the feature indicates the features extracted from the image, and the decision-level indicates the

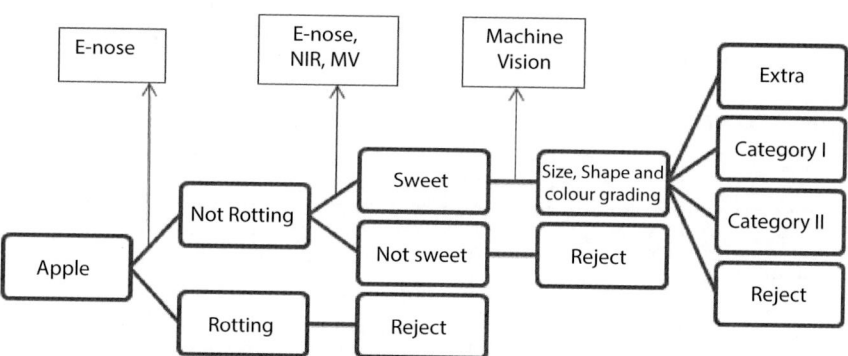

Figure 13.10 Schematic diagram of sensor fusion for assessment of final quality of apple [6].

decision-based image [140]. The fusion takes place using neural networks at the feature level. Artificial neural network versions, or adaptive neural networks, contain layers of components or nodes that can be interconnected in various ways [140]. The primary role of neural networks is to translate multi-sensor data into an entity's joint identity declaration. At the decision stage, fusion is made using the solution that is determined for each sensor, and with these decisions, the fusion is processed. Thus, based on the fusion of data at all the three levels, it helps in assessment of the quality of the sample.

Table 13.3 Summary of some non-destructive techniques [1].

Technique	Principle	Advantages	Disadvantages
Visual testing	Mechanical-optical	Minimum training, Low cost	Time consuming, repeatability, low resolution and high error
Radiographic testing	Penetrating radiation	Non-contact	Time consuming, hazardous
NMR	Magnetic field	Quick, can be used in routine analysis	Expensive
Ultrasonic testing	Sonic-ultrasonic	Non-contact	Single point measurement, limited to acoustics impedance
X-ray topography	Electromagnetic spectrum wavelength (<1 nm)	Non-contact	Time consuming, hazardous, not applicable to bulk flowing products
Hyperspectral imaging	Spectroscopy	Multi-constituent information, sensitivity to minor components	Requires training, cost intensive

Application in Food Industry

It is used in the evaluation of juice, fruit, tea, and other agro-food products [6]. Multi-sensor fusion has been applied to assess peach firmness by integrating two separate non-destructive instruments using acoustic resonance [141]. Multi-sensor fusion in melons with sensors measuring different fruit properties (color, sugar content, fineness, and aroma) was applied by V. Steinmetz *et al.* [142]. Al-Habaibeh *et al.* studied infrared and visual image processing sensor fusion to detect the consistency of laser sealed food containers. This approach can be used as effective quality control and assurance system [143]. The sensor fusion of e-nose and acoustic was studied by Ammar Zakaria *et al.* to categorize and evaluate the ripening and maturity stages of the mango samples. This study has proven that the classification of mango samples was enhanced by using this technology [139]. The Table 13.3 summarizes the various non-destructive techniques used in quality evaluation of food [1].

13.3 Non-Destructive Quality Testing in Various Food Commodities

13.3.1 Staple Foods

Staple foods are those forms of foods that influence the regular diet of humans by supplying essential oils, starch, proteins, vitamins, and minerals. The most important staple food types are cereals, tuber/root crops, and legumes [144]. Cereals include mainly wheat, maize, barley, millet, rice, oats, sorghum, and rye that are abundant in minerals, vitamins, oils, fat, carbohydrates, and protein and provide half the calories required human needs. Mycotoxin outbreaks and adulteration of both cereals and their products are the current threats associated with cereals. In order to solve such issues, both classical and emerging methods are commonly used [145]. The legumes include mainly soybean, bean, pea, lentil, tamarind, lupin, carob, peanut, and mesquite and carry protein content ranging from 17% to 40% [144]. Tuber/root crops are plants that contain tubers, corms, starchy roots, rhizomes, and stems, with high calcium, carbohydrate, and vitamin C content. It is distinguished into potatoes, cassavas, taro, yautia, sweet potatoes, and yams by Food and Agriculture Organization (FAO).

Nowadays, staple food quality is evaluated on the basis of chemical (starch, protein), adulteration (origin, species), sensory (color, external defect), parasitic infection, mycotoxin, and internal physiological (black heart, hollow heart) [144]. The traditional operations are generally

Table 13.4 Some of the research studies conducted on staple foods using NDT methods [2].

Staple foods	Technique	Parameter	Spectral range (nm)	Reference
Corn	HIS	Moisture content, oil content	750–1,090	Cogdill, R.P et al. (2004) [147]
Wheat	HIS	Detection of insect damage	960–1,700	Singh, C. et al. (2009) [148]
Cereal foods	AE	Water content		Roudaut, G. et al. (1998) [149]
Wheat	MV	Disease infection		Ruan, R. et al. (1997) [150]
Corn	MV	Size		Paulus, I. et al. (1997) [151]
Rice	MV	Grading		Wan, Y. et al. (2000) [152]

HIS, hyperspectral imaging; MV, machine vision; AE, acoustic emission.

time-consuming, laborious, manual and destructive. The Table 13.4 reports various studies regarding the non-destructive techniques implemented in staple food [2]. The various applications of spectroscopic techniques in quality evaluation of staple foods are discussed below.

13.3.1.1 Sensory Aspect

In terms of the sensory attributes of staple food, the important components of food quality involve shape, color, texture, hardness, and external defect. VIS (Visible) and NIR (Near Infrared) spectroscopy have determined the sensory attributes including texture and color [144]. Predicting these sensory attributes accurately and quickly is a primary concern in agriculture and the food industry. VIS and NIR spectroscopy have determined the sensory attributes including color and texture. As discussed in the preceding chapter, VIS/NIR work on the basis of transmission, reflection, scattering, and/or absorption of light in or through food material according to the Beer Lambert law [87]. The IR and Vis spectroscopy techniques are believed to have the ability to meet the sensory evaluation requirements with the enhancements in sensors and instruments. Even the sensory assessment of staple foods using HSI systems focuses mainly on determining color, hardness, and external defects.

13.3.1.2 Adulteration Aspects

Adulteration can decrease the nutritional value of cereal products and cause consumer health problems [146]. Most of the adulterated goods appear to be quite similar with the original commodities; thus, when mixed together, it is impossible to recognize them with the naked eye. Various spectroscopic techniques are used in determining the adulterations. It was found in various researches that Raman spectroscopy and NMR spectroscopy could accurately and quickly determine adulteration in the staple foods [144].

13.3.1.3 Chemical Aspects

The composition of the food is the basis for the evaluation of the overall acceptance and nutritional value, because the chemical components are factors which affect the quality of the food. The chemical composition of legumes and cereals is mainly contributed by carbohydrate followed by ash, moisture, crude fiber, fat and crude protein, whereas the highest humidity is found in root tubers. The techniques of VIS/IR spectroscopy have a tremendous potential in analyzing the chemical component in staple foods.

13.3.2 Fruits

The quality of the fruit is determined primarily by consumers on the basis of its characteristics which include aroma, visual appeal (blemishes, lack of color, texture, and size), flavor, and maturity [153]. Fruit quality is measured by flavor, aroma, textural and color, and characteristics that change regularly from pre-harvest through post-harvest periods as fruit grows and matures. Fruits contain a wide variety of volatile organic compounds that contribute distinctive flavor characteristics and convey distinct aromas. Conventionally, professional human panelists and graders were used to assess the quality of the fruit based on the characteristics of the visual and the aroma. One of the most valuable characteristics determining consumer

Table 13.5 Some of the research studies conducted on fruits using NDT methods [3].

NDT Method	Parameter	Fruit	Type/Variety	Observations	Reference
Acoustic impulse	Firmness	Apricot	Green mature, Half ripe, Ripe	With increase in ripening stage, acoustic firmness index decreased	Petrisor, C et al. (2010) [160]
VIS-NIR spectrophotometer	Sugar content and chlorophyll content	Banana	Unripe, Ripe, Very Ripe	Internal fruit sugar contents (sucrose, glucose, and fructose) were predicted with high accuracy. The chlorophyll content decreased with ripening	Zude, M (2003) [157]
Electronic nose	Maturity	Apple	Immature, Mature, Over Mature	Electronic nose responses were influenced by maturity of fruit. The apples were categorized into three classes of maturity.	Pathange, L. P et al. (2006) [156]
Ultrasonic	Chilling injury	Tomato		The attenuation of ultrasonic waves increased with chilling injury	Verlinden, B.E et al. (2004) [159]
VIS-NIR spectrophotometer	Soluble solids content	Orange		Mean value of 13.17° Brix was obtained. Both Destructive (conventional) and Non-destructive methods exhibited similar values.	Liu, Y et al. (2008) [158]
Electronic nose	Freshness prediction	Peaches	Firmness and total soluble solids	Firmness decreased and Total soluble solids increased with decreasing freshness	Guohua, H et al. (2012) [155]

choice and fruit quality is fruit aroma, since aroma is the best fruit flavor indicator [153]. The Table 13.5 reports various studies regarding the non-destructive techniques implemented in fruits [3].

13.3.2.1 Fruit Quality Inspection Using Electronic Nose

E-noses contain a sensor array that analyses the chemical components found in the aroma mixture (as a whole sample) and covers all the sensors in the array's electrical output signals (through a transducer) and assembles them together to create a distinctive digital series, known as an Electronic Aroma Signature Pattern (EASP), which is precise and extremely unique to the specific gas mixture. With the introduction of electronic nose instruments, grading perishable foods and fruits have more accurate quantitative and qualitative measurement of aromatic characteristics that prevent highly uncertain subjective judgments of human graders [153]. Various artificial prototypes of E-nose has been developed and designed to differentiate between complex odor mixtures, which comprises of varying volatile organic compounds [154]. Using multivariate statistics, the volatile data generated helps in discriminating different harvest maturities and maturing stages. E-nose was used in freshness prediction in peaches having the known parameters of TSSs and firmness [155]. E-nose was also able to predict the freshness accurately. Another study conducted by Pathange, L. P *et al.* depending on E-nose response; the apples were categorized into three classes of maturity [156].

13.3.2.2 Fruit Quality Inspection Using UV-VIS-NIR Spectroscopy

Depending on their intrinsic features, such as acidity and starch, the fruit has various reflecting and absorption properties in the light of varying wavelength bands. The optical properties suggest the fruit's response to the NIR band (700–2,500 nm) and visible light band (400–700 nm) [154]. When a beam of light falls on an object, it is either reflected back on the surface or transmitted through the surface. The transmitted radiation can either be scattered or absorbed, and the absorbed portion of radiation may be converted into various energy parts such as luminescence, chemical changes, and heat. NIR and VIS spectrum is mainly used for internal fruit quality inspection, whereas UV band is being used for fruit and beverage disinfection from various bacteria, viruses, etc. [154]. Compared with the visible radiation, NIR radiation is extremely penetrable, so it may be applied directly to the sample without any preparation. Sugar and chlorophyll content was determined in different ripening stages of banana using VIS-NIR spectrophotometer [157]. The internal fruit sugar content (glucose, sucrose, and fructose) was estimated to be highly precise and the chlorophyll content decreased with maturation. Liu Y et al. had determined soluble solids content in oranges using VIS-NIR spectrophotometer [158].

13.3.2.3 Fruit Quality Inspection Using Ultrasound Sensing Technique

Recently, there is a growing increase in the usage of an ultrasound method for non-destructive assessment of food safety. The benefits of ultrasound sensing techniques are that they are extremely sensitive, adapt rapidly, robust and able to monitor high speeds. This technology is mainly used for the monitoring and control of various food quality applications. Different researchers have explored various types of ultrasonic sensing modes which are pitch catch, pulse echo, and transmission to measure the test sample [154]. The test sample is placed between the receiver and transmitter, and different waves are generated for test sample evaluation based on the sensing mode. Ultrasonic sensing mode selection depends on the transmitter and receiver frequency range as well as the nature of application [154]. Frequency level for the generation of an acoustic signal is based on the nature of the application and those are quality assurance, product modification, safety or others. Verlinden, B.E et al. conducted a research study on chilling injury in tomatoes using ultrasonic technology [159].

13.3.2.4 Fruit Quality Inspection Using Machine Vision Sensing Technique

Automated machine vision can reduce a range of time-consuming testing tasks that require a long time to execute and involve sophisticated apparatus. An automatic fruit quality inspection machine vision sensing system consists of imaging process strategies that also incorporate mechanical and instrumental tools to eliminate human interference. The experimental setup of computer vision typically consists of a camera capturing the sample image, a light source with a directional waveguide for suitable color filters, illumination and connectivity to a personal computer or an embedded image acquisition, processing, and storage device [154]. For grading on the basis of shape, surface defects, color and height, counting the number of ripened fruits on the tree, detecting and recognizing diseases, etc., a machine vision–based system was used.

13.3.2.5 *Fruit Quality Inspection Using Acoustic Impulse Technique*

This method is mainly used in the evaluation of fruit texture. This method measures the resonance frequency generated by fruits, and with ripening, the resonance frequency of fruits changes. The impact response of the fruit on the application of force is measured and used to evaluate the internal maturity and quality of the fruit. In shelf life tests, this methodology also gives valuable knowledge on fruit consistency. Petrisor, C et al. performed a study on apricot firmness using the process of acoustic impulse [160].

13.3.3 Vegetables

Vegetables are the most valuable group of foods which plays a vital role in maintaining human health by preventing diseases and repairing the body via maintaining alkaline reserve (FSSAI, 2012). Different vegetables are consumed in different forms such as roots, leaves, stems, seeds and fruits and contribute to a healthy diet (USDA, 2012). Therefore, the main concern is the maintenance of quality of vegetables [161]. Vegetable quality is evaluated by different physicochemical parameters like TSSs, dry matter, acidity, and pH which require laboratory techniques that are destructive in nature [161]. Hence, non-destructive techniques are used in evaluation as it is fast, accurate, cheaper, and reliable.

Different non-destructive techniques are used in IQ assessment and sorting of vegetables and by applying these techniques. The scientific principle in evaluation of vegetable is by applying non-destructive technique and estimate vegetable quality by measuring the change in energy, applied in the target [161]. Figure 13.11 depicts the principle of non-destructive quality estimation for vegetables [161].

Different non-destructive methods used in assessing the quality of vegetables are as follows:

13.3.3.1 *Spectroscopic Techniques*

13.3.3.1.1 Visual Spectroscopy

The food contains various chemical components which absorb light energy at specific wavelength, and thus, compositional information of the food can be obtained using spectrophotometers. The major light absorbing component of vegetables are the pigments such as anthocyanin, chlorophyll, and other colored compounds in the visible wavelength range [161]. The wavelengths influence the perceived color which is dependent on both

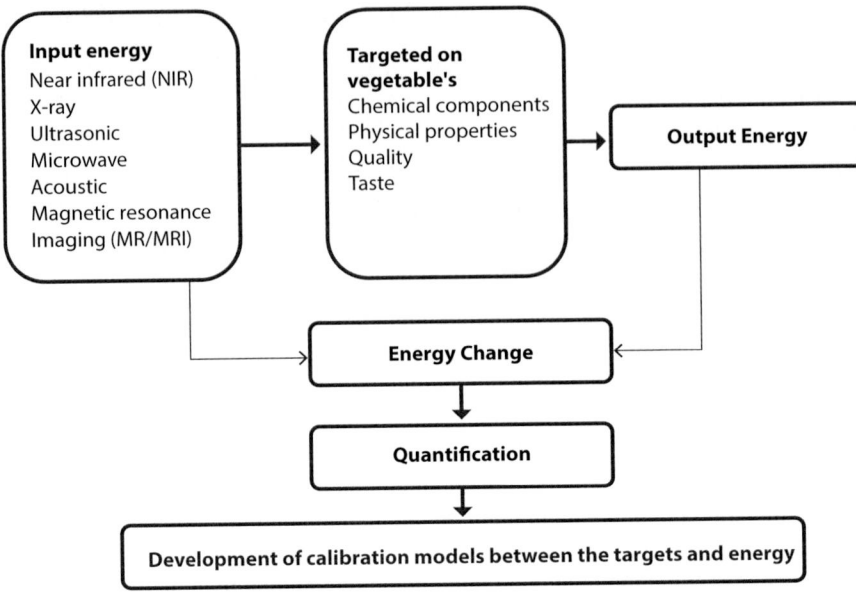

Figure 13.11 Scientific principle of non-destructive quality estimation for vegetables [161].

the absorption and light source by the object. Color is generally regarded as the appearing property that is attributed to the spectral light distribution. There are a number of factors which influence the radiation and exact color that one perceives. Jha et al. had developed a spectral radiometer that is used in determining the freshness of the eggplant based on the surface gloss and the weight at the time of storage [162]. The Table 13.6 reports various studies regarding the non-destructive techniques implemented in vegetables [79].

13.3.3.1.2 Near-Infrared Spectroscopy

This is efficient, fast, and reliable technique for evaluating biological materials' qualitative and quantitative properties. This process is easily used to determine the quality of fruits and vegetables and to rapidly evaluate carbohydrate, protein, fat, ash content, etc. [161]. The quality measurement is focused primarily on the absorption and scattering of light, i.e., the light scattered from fresh produce exposes the tissue microstructure, and the absorption of light is correlated with the chemical composition in the research sample. The freshness of agricultural products is usually evaluated based on moisture content. Hence, the NIR spectrum of fresh agriculture product is vastly controlled by water content as water has a high absorption in NIR region of light [161].

Table 13.6 Some of the research studies conducted on vegetables using NDT methods [79].

Vegetables	Technique	Parameters	Spectral range (nm)	Reference
Spinach	HIS	Escherichia coli detection	400–1,000	Siripatrawan, U *et al.* (2011) [173]
Onion	HIS	Prediction of cooking time	400–1,000	Do Trong, N.N. *et al.* (2011) [174]
Cucumbers	HIS	Chilling injury	447–951	Cheng, X *et al.* (2004) [175]
Cabbage	HIS	Bacterial contamination	700–1,100	Suthiluk, P *et al.* (2008) [176]
Potato	HIS	Prediction of cooking time	400–1,000	Do Trong, N.N. *et al.* (2011) [174]
Onion	MV	Bacterial infection detection		Wang, W *et al.* (2012) [167]
Broccoli	MV	Mature		Ramirez, R.A. (2006) [177]
Tomato	AE	Ripening stages		Baltazar, A *et al.* (2007) [178]
Eggplant	Spectral radiometer	Freshness of the eggplant		Jha *et al.* (2012) [162]

HIS, hyperspectral imaging; MV, machine vision; AE, acoustic emission.

13.3.3.1.3 Microwave Dielectric Spectroscopy

The molecular structure determines the physical and chemical properties of samples. The dielectric properties will therefore uniquely identify different molecules which constitute a given material. A tested material can successfully diversify the physical and chemical properties [161]. The specific method used depends on the type of target material and the frequency range. The water content, permittivity measurements and SSC of watermelon were evaluated and they were used as a quality factor in correlation with the dielectric properties [163].

13.3.3.1.4 X-Ray and Computerized Tomography

This is a method used specifically for the fast identification of strongly attenuating materials. Recently, techniques based on CT and 2-D X-ray imaging have been examined and used for examination of IQ of food [161]. X-ray is short wave radiation (0.01–10nm) with high energy ($1.92 \times 10^{-17} - 1.92 \times 10^{-14}$J) and can quickly penetrate into matter. They are produced by a metallic anode bombardment of electrons (X-ray tube). This technique is gaining tremendous attraction with the improvement in high-performance computers.

13.3.3.2 *Sound Waves Techniques*

Different waves, which are acoustic sound waves and ultrasonic waves, are used as non-destructive measurements to assess the quality of fresh vegetables. When the sound waves pass through the object, their sound characteristics demonstrate the quality aspect of fruit and vegetables.

13.3.3.2.1 Acoustics

This can be used in the prediction of maturity, ripening level, internal consistency, and other related parameters. This approach is largely based on the response to vibration and sounds when the source is softly tapped [161]. Similar agricultural products have different acoustic features dependent on internal tissue structures [164, 165]. The transmitted acoustic wave produced when an acoustic wave passes through the agricultural products, is based on the acoustic characteristics of the agricultural products.

13.3.3.2.2 Ultrasound

Ultrasound technique is used in evaluating and testing the biological and food materials [161]. Ultrasonic vibration is above the range of audible frequencies that is > 20 kHz. It can be used as a high frequency and low power evaluation tool and can assist with very high power processing. A piezoelectric element that serves as a receiver and transforms back to electrical energy and detects the ultrasonic signal emanating from the test sample [161]. Mizrach *et al.* stated that the non-destructive ultrasonic measurement method was used to evaluate the same transmission parameters with which maturation, growth, firmness, and other fruits and vegetables could have a quantitative relationship [166].

13.3.3.3 Imaging Analysis Techniques

A typical method of extracting spatial information from samples of monochromatic forms or color images is the imaging system technique. The imaging system is mostly used to assess the shape, color, size, and surface texture of food products and to identify surface defects in food samples, but it is not possible to identify or diagnose the chemical properties of food products [43].

13.3.3.3.1 Hyperspectral Imaging

This is an accurate method capable of gathering data for both spectral and spatial properties of the object being scanned [161]. Wang *et al.* developed the NIR reflectance HSI method for detecting sour-skin in Vidalia onions [167]. Itoh *et al.* have used this approach to measure the distribution of nitrate content in a vegetable leaf [168]. Polder *et al.* measured the carotene and chlorophyll surface distribution at spectral range of 400–700 nm in ripening tomatoes with 1-nm resolution [169]. The line-scan HSI system with three sensing modes (interactance, transmittance, and reflectance) used to predict the quality of onions in the spectral range of 400–1000 nm and TSSs, carotenoid, ascorbic acid, and total chlorophyll content during bell pepper maturity at spectral range of 550–850 nm [161].

13.3.3.3.2 Machine Vision

Recently, the machine vision technology has been growingly used for fruit and vegetable inspection purposes, in particular in applications for quality inspection and defect sorting. This system involves capturing, processing, and analyzing 2-D images, with a further note aimed at replicating the effect of human vision by analyzing and electronically perceiving an image. This system's core operations are image analysis and image processing with numerous methods and algorithms convenient to achieve the required measurement and classification [161]. Machine vision system was developed for guiding robot arm to pick the ripe tomato during harvest by acquiring images from tomato plant [170].

13.3.3.3.3 Magnetic Resonance Imaging

This method is mainly used for the non-destructive analysis of internal food structure. This non-destructive method is used to assess the qualitative as well as the quantitative properties of biological materials [161]. It is used to evaluate physical tissue damage assessment (as bruising), food processing (as drying), and others for online sorting or detection of internal defects. The physiological changes in tomatoes at different stages of maturity have been visualized using MRI [171, 172]. The thawing process for frozen and boiled edible vegetables such as broad beans, okra, green soybeans, asparagus, and taro was traced using MRI [161].

13.4 Conclusion

In India, the majority of the population's survival is based on agricultural products. Any business and organization making, displaying, transporting, or preparing food for sale will have to check the quality of the food. There are two methodologies, which are destructive and non-destructive. Non-destructive sensing methods can be widely adopted in food industry for testing the food quality, as this analysis mainly involves the surface testing without any invasive approach that influences the quality of the food. With the growth of the technology, non-destructive methods have developed accurate measuring instruments. This chapter provides information about the recent applications of various non-destructive technologies for assessing the quality of agricultural and food products. Imaging methods like X-ray and MRI are helpful in IQ evaluation but do not provide information on the composition. CT is another powerful technique that provides a 3-D image of food and thereby helps in analyzing food in a better manner. Acoustic techniques can be used to study important attributes of quality in food. Detailed knowledge about physicochemical properties can be given by the ultrasonic NDT process. It also refers to the identification of traces of foreign objects such as glass, bone, or metal particles in or on food items. Biosensors such as electronic nose–based quality determination are also the potential method used prominently due to their application in food process monitoring, shelf life studies, and freshness evaluation. Grading perishable foods provide more dependable quantitative and qualitative measures of aroma features with the advent of electronic nose devices. The sensor fusion technique promises for overall attributed quality measurement in the food industry. It is a collaborative approach with multi-sensors to enhance the quality assessment of agro-food products. The major drawback of non-destructive methods is the high cost of the equipment and the use of various instruments to analyze the samples. But even then, these methods help to ensure customer satisfaction of products as it helps in providing good quality products without rupturing the food product.

References

1. Jha, S.N., Near infrared spectroscopy, in: *Nondestructive Evaluation of Food Quality*, pp. 141–212, Springer, Berlin, Heidelberg, 2010.
2. El-Mesery, H.S., Mao, H., Abomohra, A.E.F., Applications of non-destructive technologies for agricultural and food products quality inspection. *Sensors*, 19, 4, 846, 2019.
3. Chauhan, O.P., Lakshmi, S., Pandey, A.K., Ravi, N., Gopalan, N., Sharma, R.K., Non-destructive quality monitoring of fresh fruits and vegetables. *Def. Life Sci. J.*, 20, 2, 103, 2017.
4. Sarig, Y., Potential applications of artificial olfactory sensing for quality evaluation of fresh produce. *J. Agric. Eng. Res.*, 77, 3, 239–258, 2000.
5. Aboonajmi, M. and Faridi, H., Nondestructive quality assessment of Agro-food products, in: *Proceedings of the 3rd Iranian international NDT conference*, 2016.
6. Abasi, S., Minaei, S., Jamshidi, B., Fathi, D., Dedicated non-destructive devices for food quality measurement: A review. *Trends Food Sci. Technol.*, 78, 197–205, 2018.
7. Mizrach, A., Nahir, D., Ronen, B., Mechanical Thumb sensor for fruit and vegetable sorting. *Trans. ASABE*, 35, 247–250, 1992.
8. Howarth, M.S. and Ioannides, Y., Sinclair IQ-firmness tester, in: *Proc Intnl Conf Agricultural Engineering*, Budapest, June, 2002.
9. Khalifa, S., Komarizadeh, M.H., Tousi, B., Usage of fruit response to both force and forced vibration applied to assess fruit firmness-a review. *Aust. J. Crop Sci.*, 5, 5, 516, 2011.
10. Shmulevich, I., Galili, N., Howarth, M.S., Non destructive impact and acoustic testing for quality assessment of apples. *In Proc Intnl Conf Agricultural Engineering.* Budapest. June, 2002.
11. McGlone, V.A. and Jordan, R.B., Kiwifruit and apricot firmness measurement by the non-contact laser air-puff method. *Postharvest Biol. Technol.*, 19, 1, 47–54, 2000.

12. Narsaiah, K. and Jha, S.N., Nondestructive methods for quality evaluation of livestock products. *J. Food Sci. Technol.*, 49, 3, 342–348, 2012.
13. Brezmes, J. and Llobet, E., Electronic noses for monitoring the quality of fruit, in: *Electronic Noses and Tongues in Food Science*, pp. 49–58, Academic Press, USA, 2016.
14. Tian, X.Y., Cai, Q., Zhang, Y.M., Rapid classification of hairtail fish and pork freshness using an electronic nose based on the PCA method. *Sensors*, 12, 1, 260–277, 2012.
15. Marilley, L., Ampuero, S., Zesiger, T. *et al.*, Screening of aroma-producing lactic acid bacteria with an electronic nose. *Int. Dairy J.*, 14, 849, 2004.
16. García, M., Aleixandre, M., Horrillo, M.C., Electronic nose for the identification of spoiled Iberian hams. *Spanish on Conference Electron Devices*, Tarragona, pp. 537–540, 2005.
17. Kawakami, M., Sarma, S., Himizu, K. *et al.*, Aroma characteristics of Darjeeling tea, in: *Proceedings of International Conference O-CHA (Tea) Culture Science*, Shizuoka, Japan, pp. 110–116, 2004.
18. Bhattacharyya, N., Seth, S., Tudu, B. *et al.*, Detection of optimum fermentation time for black tea manufacturing using electronic nose. *Sens. Actuators B Chem.*, 122, 627–634, 2007.
19. Benedetti, S., Sinelli, N., Buratti, S. *et al.*, Shelf life of crescenza cheese as measured by electronic nose. *J. Dairy Sci.*, 88, 3044, 2005.
20. Riva, M. and Mannino, S., Shelf-life monitoring and modeling by e-nose and image-analysis. *Ital. Food Beverage Technol.*, 42, 11, 2005.
21. Labreche, S., Bazzo, S., Cade, S. *et al.*, Shelf life determination by electronic nose: application to milk. *Sens. Actuators B Chem.*, 106, 199, 2005.
22. Cosio, M.S., Ballabio, D., Benedetti, S. *et al.*, Evaluation of different storage conditions of extra virgin olive oils with an innovative recognition tool built by means of electronic nose and electronic tongue. *Food Chem.*, 101, 485, 2007.
23. Mildner-Szkudlarz, S., Jelen, H.H., Zawirska-Wojtasiak, R., The use of electronic and human nose for monitoring rapeseed oil autoxidation. *Eur. J. Technol.*, 110, 61, 2008.
24. Di Natale, C., Olafsdottir, G., Einarsson, S. *et al.*, Comparison and integration of different electronic noses for freshness evaluation of cod-fish fillets. *Sens. Actuators B Chem.*, 77, 572, 2001.
25. Du, W.X., Lin, C.M., Huang, T. *et al.*, Potential application of the electronic nose for quality assessment of salmon fillets under various storage conditions. *J. Food Sci.*, 67, 307, 2002.
26. Olafsdottir, G., Chanie, E., Westad, F. *et al.*, Prediction of microbial and sensory quality of cold smoked Atlantic salmon by electronic nose. *J. Food Sci.*, 70, S563, 2005.
27. Chantarachoti, J., Oliveira, A.C.M., Himelbloom, B.H., Portable electronic nose for detection of spoiling Alaska pink salmon (Oncorhynchus gorbuscha). *J. Food Sci.*, 71, S414, 2006.

28. Korel, F., Luzuriaga, D.A., Balaban, M.Ö., Objective quality assessment of raw tilapia (Oreochromis niloticus) fillets using electronic nose and machine vision. *J. Food Sci.*, 66, 1018, 2001.
29. Tokusoglu, O. and Balaban, M., Correlation of odor and color profiles of oysters (Crassostrea virginica) with electronic nose and color machine vision. *J. Shellfish Res.*, 23, 143, 2004.
30. Dutta, R., Hines, E.L., Gardner, J.W., Non-destructive egg freshness determination: an electronic nose based approach. *Meas. Sci. Technol.*, 14, 190, 2003.
31. El Barbri, N., Llobet, E., El Bari, N. *et al.*, Electronic nose based on metal oxide semiconductor sensors as an alternative technique for the spoilage classification of red meat. *Sensors*, 8, 142, 2008.
32. Mahato, D.K., Verma, D.K., Billoria, S., Kopari, M., Prabhakar, P.K., Kumar, A., Behera, S.M., Srivastav, P.P., Applications of nuclear magnetic resonance in food processing and packaging management, in: *Developing Technologies in Food Science*, pp. 109–142, Apple Academic Press, USA, 2017.
33. Capitani, D., Sobolev, A.P., Di Tullio, V., Mannina, L., Proietti, N., Portable NMR in food analysis. *Chem. Biol. Technol. Agric.*, 4, 1, 17, 2017.
34. Alonso-Salces, R.M., Holland, M.V., Guillou, C., 1H-NMR fingerprinting to evaluate the stability of olive oil. *Food Control*, 22, 12, 2041–2046, 2011.
35. Lee, J.-E., Lee, B.-J., Chung, J.-O., Kim, H.-N., Kim, E.-H., Jung, S., Hong, Y.-S., Metabolomic unveiling of a diverse range of green tea (Camellia sinensis) metabolites dependent on geography. *Food Chem.*, 174, 452–459, 2015.
36. Mannina, L., Marini, F., Antiochia, R., Cesa, S., Magrì, A., Capitani, D., Sobolev, A.P., Tracing the origin of beer samples by NMR and chemometrics: Trappist beers as a case study. *Electrophoresis*, 37, 20, 2710–2719, 2016.
37. Merkx, D.W.H., Westphal, Y., van, V.E.J.J., Thakoer, K.V., de, R.N., van, D.J.P.M., Quantification of food polysaccharide mixtures by ^{1}H NMR. *Carbohydr. Polym.*, 179, 379–385, 2018.
38. Schievano, E., Tonoli, M., Rastrelli, F., NMR quantification of carbohydrates in complex mixtures. A challenge on honey. *Anal. Chem.*, 89, 24, 13405–13414, 2017.
39. Sobolev, A.P., Testone, G., Santoro, F., Nicolodi, C., Iannelli, M.A., Amato, M.E., Mannina, L., Quality traits of conventional and transgenic lettuce (Lactuca sativa L.) at harvesting by NMR metabolic profiling. *J. Agric. Food Chem.*, 58, 11, 6928–6936, 2010.
40. Spraul, M., Schutz, B., Humpfer, E., Mortter, M., Schafer, H., Koswig, S., Rinke, P., Mixture analysis by NMR as applied to fruit juice quality control. *Magn. Reson. Chem.*, 47, S130–S137, 2009.
41. Kirtil, E., Cikrikci, S., McCarthy, M.J., Oztop, M.H., Recent advances in time domain NMR & MRI sensors and their food applications. *Curr. Opin. Food Sci.*, 17, 9–15, 2017.
42. Lou, X., Ye, Y., Wang, Y., Sun, Y., Pan, D., Cao, J., Effect of high-pressure treatment on taste and metabolite profiles of ducks with two different vinasse-curing processes. *Food Res. Int.*, 105, 703–712, 2018.

43. Marcone, M.F., Wang, S., Albabish, W., Nie, S., Somnarain, D., Hill, A., Diverse food-based applications of nuclear magnetic resonance (NMR) technology. *Food Res. Int.*, *51*, 2, 729–747, 2013.
44. Williamson, K. and Hatzakis, E., NMR spectroscopy as a robust tool for the rapid evaluation of the lipid profile of fish oil supplements. *JoVE (J. Visualized Exp.)*, 123, 55547, 2017.
45. Rady, A.M. and Guyer, D.E., Rapid and/or nondestructive quality evaluation methods for potatoes: A review. *Comput. Electron. Agric.*, 117, 31–48, 2015.
46. Zehl, M., Braunberger, C., Conrad, J., Crnogorac, M., Krasteva, S., Vogler, B., Krenn, L., Identification and quantification of flavonoids and ellagic acid derivatives in therapeutically important Drosera species by LC-DAD, LC-NMR, NMR, and LC-MS. *Anal. Bioanal. Chem.*, 400, 8, 2565–2576, 2011.
47. Hussain, A., Pu, H., Sun, D.W., Innovative nondestructive imaging techniques for ripening and maturity of fruits–a review of recent applications. *Trends Food Sci. Technol.*, 72, 144–152, 2018.
48. Ezeanaka, M. C., Nsor-Atindana, J., Zhang, M., Online low-field nuclear magnetic resonance (LF-NMR) and magnetic resonance imaging (MRI) for food quality optimization in food processing. *Food Bioproc. Tech.*, 9, 1435-1451, 2019.
49. Hills, B.P. and Remigereau, B., NMR studies of changes in subcellular water compartmentation in parenchyma apple tissue during drying and freezing. *Int. J. Food Sci. Technol.*, 32, 51–61, 1997.
50. Hernandez-Sanchez, N., Barreiro, P., Ruiz-Altisenta, M., Ruiz-Cabello, J., Fernandez-Valle, M.E., Detection of freeze injury in oranges by magnetic resonance imaging of moving samples. *Appl. Magn. Reson.*, 26, 431–445, 2004.
51. Hwang, S.S., Cheng, Y.C., Chang, C., Lur, H.S., Lin, T.T., Magnetic resonance imaging and analyses of tempering processes in rice kernels. *J. Cereal Sci.*, 50, 1, 36–42, 2009.
52. Miquel, M.E. and Hall, L.D., Measurement by MRI of storage changes in commercial chocolate confectionery products. *Food Res. Int.*, 35, 10, 993–998, 2002.
53. Barreiro, P., Ortiz, C., Ruiz-Altisent, M., Ruiz-Cabello, J., Fernández-Valle, M.E., Recasens, I., Asensio, M., Mealiness assessment in apples and peaches using MRI techniques. *Magn. Reson. Imaging*, 18, 9, 1175–1181, 2000.
54. Haishi, T., Koizumi, H., Arai, T., Koizumi, M., Kano, H., Rapid detection of infestation of apple fruits by the peach fruit moth, Carposina sasakii Matsumura, larvae using a 0.2-T dedicated magnetic resonance imaging apparatus. *Appl. Magn. Reson.*, 41, 1, 1, 2011.
55. Milczarek, R.R. and McCarthy, M.J., Low-field MR sensors for fruit inspection, in: *Magnetic Resonance Microscopy: Spatially Resolved NMR Techniques and Applications*, 2009.

56. Morsy, N. and Sun, D.-W., Robust linear and non-linear models of NIR spectroscopy for detection and quantification of adulterants in fresh and frozen-thawed minced beef. *Meat Sci.*, 93, 2, 292–302, 2012.
57. Bouhrara, M., Lehallier, B., Clerjon, S., Damez, J.L., Bonny, J.M., Mapping of muscle deformation during heating: in situ dynamic MRI and nonlinear registration. *Magn. Reson. Imaging*, 30, 3, 422–430, 2012.
58. Kremer, P.V., Förster, M., Scholz, A.M., Use of magnetic resonance imaging to predict the body composition of pigs in vivo. *Animal*, 7, 6, 879–884, 2013.
59. Wu, J.-L., Zhang, J.-L., Du, X.-X., Shen, Y.-J., Lao, X., Zhang, M.-L., Chen, L.-Q., Du, Z.-Y., Evaluation of the distribution of adipose tissues in fish using magnetic resonance imaging (MRI). *Aquaculture*, 448, 112–122, 2015.
60. Silva, S., Guedes, C., Rodrigues, S., Teixeira, A., Non-Destructive Imaging and Spectroscopic Techniques for Assessment of Carcass and Meat Quality in Sheep and Goats: A Review. *Foods*, 9, 1074, 2020.
61. Lammertyn, J., Peirs, A., De Baerdemaeker, J., Nicolai, B., Light penetration properties of NIR radiation in fruit with respect to non-destructive quality assessment. *Postharvest Biol. Technol.*, 18, 2, 121–132, 2000.
62. Wang, X., Xue, L., He, X., Liu, M., Vitamin C content estimation of chilies using Vis/NIR spectroscopy, in: *Proceedings of the 2011 International Conference on Electric Information and Control Engineering (ICEICE)*, Wuhan, China, IEEE, New York, NY, USA, pp. 1894–1897, 2011.
63. Leon, L., Kelly, J.D., Downey, G., Detection of apple juice adulteration using near-infrared ' transflectance spectroscopy. *Appl. Spectrosc.*, 59, 5, 593–599, 2005.
64. Kelly, J.F.D. and Downey, G., Detection of sugar adulterants in apple juice using Fourier transform infrared spectroscopy and chemometrics. *J. Agric. Food Chem.*, 53, 3281–3286, 2005.
65. Merzlyak, M.N., Solovchenko, A.E., Gitelson, A.A., Reflectance spectral features and non-destructive estimation of chlorophyll, carotenoid and anthocyanin content in apple fruit. *Postharvest Biol. Technol.*, 27, 197–211, 2003.
66. McGlone, V.A., Fraser, D.G., Jordan, R.B., Künnemeyer, R., Internal quality assessment of mandarin fruit by vis/NIR spectroscopy. *J. Near Infrared Spectrosc.*, 11, 323–332, 2003.
67. Bergaz, L.P., Ruiz, G.R., Gracia, L.M.N., Guimaraes, A.C., Gil, J.G., Bakery products quality control using computer vision: Napolitalas case, in: *Proceedings of the CIGR Workshop on Image Analysis in Agriculture*, Budapest, Hungary, pp. 26–27, 2010.
68. Caporaso, N., Whitworth, M.B., Fisk, I.D., Near-Infrared spectroscopy and hyperspectral imaging for non-destructive quality assessment of cereal grains. *Appl. Spectrosc. Rev.*, 53, 8, 667–687, 2018.
69. Dachoupakan Sirisomboon, C., Putthang, R., Sirisomboon, P., Application of near infrared spectroscopy to detect aflatoxigenic fungal contamination in rice. *Food Control*, 33, 1, 207–214, 2013.

70. Lei, T. and Sun, D.W., Developments of nondestructive techniques for evaluating quality attributes of cheeses: A review. *Trends Food Sci. Technol.*, 88, 527–542, 2019.
71. Alamprese, C., Casale, M., Sinelli, N., Lanteri, S., Casiraghi, E., Detection of minced beef adulteration with turkey meat by UV–vis, NIR and MIR spectroscopy. *LWT-Food Sci. Technol.*, 53, 1, 225–232, 2013.
72. Barbin, D.F., ElMasry, G., Sun, D.-W., Allen, P., Morsy, N., Non-destructive assessment of microbial contamination in porcine meat using NIR hyperspectral imaging. *Innov. Food Sci. Emerg. Technol.*, 17, 180–191, 2013.
73. Morsy, N. and Sun, D.-W., Robust linear and non-linear models of NIR spectroscopy for detection and quantification of adulterants in fresh and frozen-thawed minced beef. *Meat Sci.*, 93, 2, 292–302, 2012.
74. Zamora-Rojas, E., Perez-Marın, D., De Pedro-Sanz, E., Guerrero-Ginel, J.E., Garrido-Varo, A., In-situ Iberian pig carcass classification using a microelectro-mechanical system (MEMS)-based near infrared (NIR) spectrometer. *Meat Sci.*, 90, 3, 636–642, 2012.
75. Agelet, L.E., Armstrong, P.R., Tallada, J.G., Hurburgh Jr., C.R., Differences between conventional and glyphosate tolerant soybeans and moisture effect in their discrimination by near infrared spectroscopy. *Food Chem.*, 141, 3, 1895–1901, 2013.
76. Salguero-Chaparro, L., Gaitán-Jurado, A.J., Ortiz-Somovilla, V., Peña-Rodríguez, F., Feasibility of using NIR spectroscopy to detect herbicide residues in intact olives. *Food Control*, 30, 2, 504–509, 2013.
77. Xue, L., Cai, J., Li, J., Liu, M., Application of particle swarm optimization (PSO) algorithm to determine dichlorvos residue on the surface of navel range with Vis-NIR spectroscopy. *Proc. Eng.*, 29, 4124–4128, 2012.
78. Fasolato, L., Balzan, S., Riovanto, R., Berzaghi, P., Mirisola, M., Ferlito, J.C., Serva, L., Benozzo, F., Passera, R., Tepedino, V., Novelli, E., Comparison of visible and near-infrared reflectance spectroscopy to authenticate fresh and frozen-thawed Swordfish (Xiphias gladius L). *J. Aquat. Food Prod. Technol.*, 21, 5, 493–507, 2012.
79. Kimiya, T., Sivertsen, A.H., Heia, K., VIS/NIR spectroscopy for non-destructive freshness assessment of Atlantic salmon (Salmo salar L.) fillets. *J. Food Eng.*, 116, 3, 758–764, 2013.
80. Ottavian, M., Fasolato, L., Facco, P., Barolo, M., Foodstuff authentication from spectral data: Toward a species-independent discrimination between fresh and frozen-thawed fish samples. *J. Food Eng.*, 19, 4, 765–775, 2013.
81. Tito, N.B., Rodemann, T., Powell, S.M., Use of near-infra red spectroscopy to predict microbial numbers on Atlantic salmon. *Food Microbiol.*, 32, 2, 431–436, 2012.
82. Balabin, R.M. and Smirnov, S.V., Melamine detection by mid-and near-infrared (MIR/NIR) spectroscopy: A quick and sensitive method for dairy products analysis including liquid milk, infant formula, and milk powder. *Talanta*, 85, 1, 562–568, 2011.

83. Hsieh, C.L., Hung, C.Y., Kuo, C.Y., Quantization of adulteration ratio of raw cow milk by least squares support vector machines (LS-SVM) and visible/near infrared spectroscopy. *Eng. Appl. Neural Netw.*, 363, 130–139, 2011.

84. Gong, A., Qiu, Z., He, Y., Wang, Z., A non-destructive method for quantification the irradiation doses of irradiated sucrose using Vis/NIR spectroscopy. *Spectrochim. Acta Part A: Mol. Biomol. Spectrosc.*, 99, 7–11, 2012.

85. Gaspardo, B., Del Zotto, S., Torelli, E., Cividino, S.R., Firrao, G., Della Riccia, G., Stefanon, B., A rapid method for detection of fumonisins B1 and B2 in corn meal using Fourier transform near infrared spectroscopy (FT-NIR) implemented with integrating sphere. *Food Chem.*, 135, 3, 1608–1612, 2012.

86. Haughey, S.A., Graham, S.F., Cancouet, E., Elliott, C.T., The application of Near-Infrared Reflectance Spectroscopy (NIRS) to detect melamine adulteration of soya bean meal. *Food Chem.*, 136, 3, 1557–1561, 2013.

87. Kuligowski, J., Carrion, D., Quintas, G., Garrigues, S., de la Guardia, M., Direct determination of polymerised triacylglycerides in deep-frying vegetable oil by near infrared spectroscopy using Partial Least Squares regression. *Food Chem.*, 131, 1, 353–359, 2012.

88. Luna, A.S., da Silva, A.P., Pinho, J.S., Ferre, J., Boque, R., Rapid characterization of transgenic and non-transgenic soybean oils by chemometric methods using NIR spectroscopy. *Spectrochim. Acta A*, 100, 115–119, 2013.

89. Sarkar, M., Gupta, N., Assaad, M., Nondestructive Food Quality Monitoring Using Phase Information in Time-Resolved Reflectance Spectroscopy. *IEEE Trans. Instrum. Meas.*, 69, 10, 7787–7795, 2020.

90. Ebrahimi-Najafabadi, H., Leardi, R., Oliveri, P., Chiara Casolino, M., JalaliHeravi, M., Lanteri, S., Detection of addition of barley to coffee using near infrared spectroscopy and chemometric techniques. *Talanta*, 99, 175–179, 2012.

91. Chen, L., Xue, X., Ye, Z., Zhou, J., Chen, F., Zhao, J., Determination of Chinese honey adulterated with high fructose corn syrup by near infrared spectroscopy. *Food Chem.*, 128, 4, 1110–1114, 2011.

92. Liu, F. and He, Y., Classification of brands of instant noodles using Vis/NIR spectroscopy and chemometrics. *Food Res. Int.*, 41, 5, 562–567, 2008.

93. Pontes, M.J.C., Santos, S.R.B., Araujo, M.C.U., Almeida, L.F., Lima, R.A.C., Gaiao, E.N., Souto, U.T.C.P., Classification of distilled alcoholic beverages and verification of adulteration by near infrared spectrometry. *Food Res. Int.*, 39, 2, 182–189, 2006.

94. Liu, Y., Pu, H., Sun, D.W., Hyperspectral imaging technique for evaluating food quality and safety during various processes: A review of recent applications. *Trends Food Sci. Technol.*, 69, 25–35, 2017.

95. Kim, M.S., Chen, Y.R., Mehl, P.M., Hyperspectral reflectance and fluorescence imaging system for food quality and safety. *Trans. ASAE*, 44, 721–729, 2001.

96. Kim, M.S., Lefcourt, A.M., Chao, K., Chen, Y.R., Kim, I., Chan, D.E., Multispectral detection of fecal contamination on apples based on

hyperspectral imagery. Part I. Application of visible and near-infrared reflectance imaging. *Trans. ASAE*, 45, 2027–2037, 2002.
97. Cheng, X., Chen, Y.R., Tao, Y., Wang, C.Y., Kim, M.S., Lefcourt, A.M., A novel integrated PCA and FLD method on hyperspectral image feature extraction for cucumber chilling damage inspection. *Trans. ASAE*, 47, 1313–1320, 2004.
98. Gowen, A.A., O'Donnell, C.P., Taghizadeh, M. *et al.*, Hyperspectral imaging combined with principal component analysis for bruise damage detection on white mushrooms (Agaricus bisporus). *J. Chemom.*, 22, 259–267, 2008.
99. ElMasry, G., Wang, N., ElSayed, A., Hyperspectral imaging for nondestructive determination of some quality attributes for strawberry. *J. Food Eng.*, 81, 98–107, 2007.
100. Liu, Y., Chen, Y.R., Wang, C.Y., Chan, D.E., Kim, M.S., Development of hyperspectral imaging technique for the detection of chilling injury in cucumbers; spectral and image analysis. *Appl. Eng. Agric.*, 22, 1, 101–111, 2006.
101. Leiva-Valenzuela, G.A., Lu, R., Aguilera, J.M., Prediction of firmness and soluble solids content of blueberries using hyperspectral reflectance imaging. *J. Food Eng.*, 115, 1, 91–98, 2013.
102. Suktanarak, S. and Teerachaichayut, S., Non-destructive quality assessment of hens' eggs using hyperspectral images. *J. Food Eng.*, 215, 97–103, 2017.
103. Sanchez, P.D.C., Hashim, N., Shamsudin, R., Nor, M.Z.M., Applications of imaging and spectroscopy techniques for non-destructive quality evaluation of potatoes and sweet potatoes: A review. *Trends Food Sci. Technol.*, 96, 208–221, 2020.
104. Lei, T. and Sun, D.W., Developments of nondestructive techniques for evaluating quality attributes of cheeses: A review. *Trends Food Sci. Technol.*, 88, 527–542, 2019.
105. Caporaso, N., Whitworth, M.B., Grebby, S., Fisk, I.D., Non-destructive analysis of sucrose, caffeine and trigonelline on single green coffee beans by hyperspectral imaging. *Food Res. Int.*, 106, 193–203, 2018.
106. Caporaso, N., Whitworth, M.B., Fowler, M.S., Fisk, I.D., Hyperspectral imaging for non-destructive prediction of fermentation index, polyphenol content and antioxidant activity in single cocoa beans. *Food Chem.*, 258, 343–351, 2018.
107. Su, W.H. and Sun, D.W., Evaluation of spectral imaging for inspection of adulterants in terms of common wheat flour, cassava flour and corn flour in organic Avatar wheat (Triticum spp.) flour. *J. Food Eng.*, 200, 59–69, 2017.
108. Blasco, J., Munera, S., Aleixos, N., Cubero, S., Molto, E., Machine vision-based measurement systems for fruit and vegetable quality control in postharvest, in: *Measurement, Modeling and Automation in Advanced Food Processing*, pp. 71–91, Springer, Cham, 2017.
109. Naik, S. and Patel, B., Machine vision based fruit classification and grading-a review. *Int. J. Comput. Appl.*, 170, 9, 22–34, 2017.
110. Lu, R. (Ed.), *Light scattering technology for food property, quality and safety assessment*, Crc Press, Boca Raton, 2017.

111. Dinish, U.S., Wong, C.L., Sriram, S., Ong, W.K., Balasundaram, G., Sugii, S., Olivo, M., Diffuse optical spectroscopy and imaging to detect and quantify adipose tissue browning. *Sci. Rep.*, 7, 1, 1–11, 2017.
112. Torricelli, A., Spinelli, L., Vanoli, M. *et al.*, Optical coherence tomography (OCT), space resolved reflectance spectroscopy (SRS) and time-resolved reflectance spectroscopy (TRS): Principles and applications to food microstructures, in: *Food Microstructures: Microscopy, Measurement and Modelling*, V.J. Morris, and K. Groves (Eds.), pp. 132–162, Woodhead Publishing Limited, Cambridge, 2013.
113. Rizzolo, A., Vanoli, M., Bianchi, G. *et al.*, Relationship between texture sensory profiles and optical properties measured by time-resolved reflectance spectroscopy during post storage shelf life of "Braeburn" apples. *J. Hortic. Res.*, 22, 113–121, 2014.
114. Vanoli, M., Rizzolo, A., Grassi, M., Spinelli, L., Zanella, A., Torricelli., A., Characterizing apple texture during storage through mechanical, sensory and optical properties. *Acta Hortic.*, 1079, 383–390, 2015.
115. Eccher Zerbini, P., Grassi, M., Cubeddu, R., Pifferi, A., Torricelli, A., Nondestructive detection of brown heart in pears by time-resolved reflectance spectroscopy. *Postharvest Biol. Technol.*, 25, 87–97, 2002.
116. Vanoli, M., Rizzolo, A., Grassi, M., Spinelli, L., Verlinden, B.E., Torricelli, A., Studies on classification models to discriminate "Braeburn" apples affected by internal browning using the optical properties measured by time-resolved reflectance spectroscopy. *Postharvest Biol. Technol.*, 91, 112–121, 2014.
117. Vanoli, M., Grassi, M., Spinelli, L., Torricelli, A., Rizzolo, A., Quality and nutraceutical properties of mango fruit: influence of cultivar and biological age assessed by Time-resolved Reflectance Spectroscopy. *Adv. Hortic. Sci.*, 32, 3, 407–420, 2018.
118. Ibrahim, A., Grassi, M., Lovati, F., Parisi, B., Spinelli, L., Torricelli, A., Vanoli, M., Non-destructive detection of potato tubers internal defects: critical insight on the use of time-resolved spectroscopy. *Adv. Hortic. Sci.*, 34, 1S, 43–51, 2020.
119. Ahmed, M.R., Yasmin, J., Lee, W.H., Mo, C., Cho, B.K., Imaging technologies for nondestructive measurement of internal properties of agricultural products: A review. *J. Biosyst. Eng.*, 42, 3, 199–216, 2017.
120. Yang, E.C., Yang, M.M., Liao, L.H., Wu, W.Y., Chen, T.W., Chen, T.M., Jiang, J.A., Non-destructive quarantine technique-potential application of using x-ray images to detect early infestations caused by oriental fruit fly (Bactrocera dorsalis)(Diptera: Tephritidae) in fruit. *Formos. Entomol.*, 26, 171–186, 2006.
121. Neethirajan, S., Jayas, D.S., White, N.D.G., Detection of sprouted wheat kernel using soft X-ray image analysis. *J. Food Eng.*, 81, 3, 509–513, 2007.
122. Vidhya, M., Varadharaju, N., Kennedy, Z.J., Amirtham, D., Jesudas, D.M., Applications of X-ray computed tomography in food processing. *J. Food Process. Technol.*, 8, 5, 673, 2017.

123. Du, Z., Hu, Y., Ali Buttar, N., Mahmood, A., X-ray computed tomography for quality inspection of agricultural products: A review. *Food Sci. Nutr.*, 7, 10, 3146–3160, 2019.
124. Abd Shaib, M.F., Rahim, R.A., Muji, S.Z.M., Ahmad, A., Investigating Maturity State and Internal Properties of Fruits Using Non-Destructive Techniques-A Review. *Telkomnika*, 15, 4, 1574–1584, 2017.
125. Knorr, D., Froehling, A., Jaeger, H., Reineke, K., Schlueter, O., Schoessler, K., Emerging technologies in food processing. *Annu. Rev. Food Sci. Technol.*, 2, 203–235, 2011.
126. Mizrach, A. and Flitsanov, U., Nondestructive ultrasonic determination of avocado softening process. *J. Food Eng.*, 40, 139–144, 1999.
127. Morrison, D.S. and Abeyratne, U.R., Ultrasonic technique for non-destructive quality evaluation of oranges. *J. Food Eng.*, 141, 107–112, 2014.
128. Lee, H.O., Luan, H., Daut, D.G., Use of an ultrasonic technique to evaluate the rheological properties of cheese and dough. *J. Food Eng.*, 16, 127–150, 1992.
129. Arvanitoyannis, I.S., Kotsanopoulos, K.V., Savva, A.G., Use of ultrasounds in the food industry–Methods and effects on quality, safety, and organoleptic characteristics of foods: A review. *Crit. Rev. Food Sci. Nutr.*, 57, 1, 109–128, 2017.
130. Ali, M.M., Hashim, N., Bejo, S.K., Shamsudin, R., Rapid and nondestructive techniques for internal and external quality evaluation of watermelons: A review. *Sci. Hortic.*, 225, 689–699, 2017.
131. Zhang, W., Lv, Z., Xiong, S., Nondestructive quality evaluation of agro-products using acoustic vibration methods—A review. *Crit. Rev. Food Sci. Nutr.*, 58, 14, 2386–2397, 2018.
132. Wang, J., Gomez, A.H., Pereira, A.G., Acoustic impulse response for measuring the firmness of mandarin during storage. *J. Food Qual.*, 29, 4, 392–404, 2006.
133. Fathizadeh, Z., Aboonajmi, M., Beygi, S.R.H., Nondestructive firmness prediction of apple fruit using acoustic vibration response. *Sci. Hortic.*, 262, 109073, 2020.
134. Taniwaki, M., Tohro, M., Sakurai, N., Measurement of ripening speed and determination of the optimum ripeness of melons by a nondestructive acoustic vibration method. *Postharvest Biol. Technol.*, 56, 1, 101–103, 2010.
135. Aboonajmi, M., Jahangiri, M., Hassan-Beygi, S.R., A Review on Application of Acoustic Analysis in Quality Evaluation of Agro-food Products. *J. Food Process. Preserv.*, 39, 6, 3175–3188, 2015.
136. Skaf, A., Nassar, G., Lefebvre, F., Nongaillard, B., A new acoustic technique to monitor bread dough during the fermentation phase. *J. Food Eng.*, 93, 3, 365–378, 2009.
137. Wallhauber, E., Walid, B., Mohamed, A., Hinrichs, J., Becker, T.M., On the usage of acoustic properties combined with an artificial neural network – A new approach of determining presence of dairy fouling. *J. Food Eng.*, 103, 449–456, 2011.

138. Nassar, G., Lefbvre, F., Skaf, A., Carlier, J., Nongaillard, B., Noêl, Y., Ultrasonic and acoustic investigation of cheese matrix at the beginning and the end of ripening period. *J. Food Eng.*, 96, 1–13, 2010.
139. Zakaria, A., Shakaff, A.Y.M., Masnan, M.J., Saad, F.S.A., Adom, A.H., Ahmad, M.N., Kamarudin, L.M., Improved maturity and ripeness classifications of magnifera indica cv. harumanis mangoes through sensor fusion of an electronic nose and acoustic sensor. *Sensors*, 12, 5, 6023–6048, 2012.
140. Steinmetz, V. and Bellon, V., Sensor Fusion for Quality Control of Agricultural Products. *IFAC Proc. Volumes*, 28, 4, 237–243, 1995.
141. Armstrong P, R., Stone M, L., Brusewitz, G.H., Peach firmness determination using two di!erent nondestructive vibrational sensing instruments. *Trans. ASAE*, 40, 3, 699–703, 1997.
142. Steinmetz, V., Sevila, F., Bellon-Maurel, V., A methodology for sensor fusion design: application to fruit quality assessment. *J. Agric. Eng. Res.*, 74, 1, 21–31, 1999.
143. Al-Habaibeh, A., Shi, F., Brown, N., Kerr, D., Jackson, M., Parkin, R.M., A novel approach for quality control system using sensor fusion of infrared and visual image processing for laser sealing of food containers. *Meas. Sci. Technol.*, 15, 10, 1995, 2004.
144. Su, W.H., He, H.J., Sun, D.W., Non-Destructive and rapid evaluation of staple foods quality by using spectroscopic techniques: A review. *Crit. Rev. Food Sci. Nutr.*, 57, 5, 1039–1051, 2017.
145. Hussain, N., Sun, D.W., Pu, H., Classical and emerging non-destructive technologies for safety and quality evaluation of cereals: A review of recent applications. *Trends Food Sci. Technol.*, 91, 598–608, 2019.
146. Bansal, S., Singh, A., Mangal, M., Mangal, A.K., Kumar, S., Food adulteration: Sources, health risks, and detection methods. *Crit. Rev. Food Sci. Nutr.*, 57, 6, 1174–1189, 2017.
147. Cogdill, R.P., Hurburgh, C., Rippke, G.R., Bajic, S.J., Jones, R.W., McClelland, J.F., Jensen, T.C., Liu, J., Single-kernel maize analysis by near-infrared hyperspectral imaging. *Trans. ASAE*, 47, 311, 2004.
148. Singh, C., Jayas, D., Paliwal, J., White, N., Detection of insect-damaged wheat kernels using near-infrared hyperspectral imaging. *J. Stored Prod. Res.*, 45, 151–158, 2009.
149. Roudaut, G., Dacremont, C., Meste, M.L., Influence of water on the crispness of cereal-based foods: Acoustic, mechanical, and sensory studies. *J. Texture Stud.*, 29, 199–213, 1998.
150. Ruan, R., Ning, S., Ning, A., Jones, R., Chen, P., Estimation of scabby wheat incident rate using machine vision and neural network, in: *1997 ASAE Annual International Meeting Technical Papers*, ASAE, St. Joseph, MN, USA, pp. 49085–9659, 1997.
151. Paulus, I., De Busscher, R., Schrevens, E., Use of image analysis to investigate human quality classification of apples. *J. Agric. Eng. Res.*, 68, 341–353, 1997.

152. Wan, Y., Lin, C., Chiou, J., Adaptive classification method for an automatic grain quality inspection system using machine vision and neural network, in: *2000 ASAE Annual International Meeting Technical Papers*, ASAE, St. Joseph, MN, USA, pp. 1–19, 2000.
153. Baietto, M. and Wilson, A., Electronic-nose applications for fruit identification, ripeness and quality grading. *Sensors*, 15, 1, 899–931, 2015.
154. Srivastava, S. and Sadistap, S., Non-destructive sensing methods for quality assessment of on-tree fruits: a review. *J. Food Meas. Charact.*, 12, 1, 497–526, 2018.
155. Guohua, H., Yuling, W., Dandan, Y., Wenwen, D., Linshan, Z., Lvye, W., Study of peach freshness predictive method based on electronic nose. *Food Control*, 28, 1, 25–32, 2012.
156. Pathange, L.P., Mallikarjunan, P., Marini, R.P., OKeefe, S., Vaughan, D., Non-destructive evaluation of apple maturity using an electronic nose system. *J. Food Eng.*, 77, 1018–1023, 2006.
157. Zude, M., Non-destructive prediction of banana fruit quality using VIS/NIR spectroscopy. *Fruits*, 58, 3, 135–142, 2003.
158. Liu, Y., Sun, X., Ouyang, A., Nondestructive measurement of soluble solid content of navel orange fruit by visible–NIR spectrometric technique with PLSR and PCA-BPNN. *LWT-Food Sci. Technol.*, 43, 4, 602–607, 2010.
159. Verlinden, B.E., De Smedt, V., Nicolai, B.M., Evaluation of ultrasonic wave propagation to measure chilling injury in tomatoes. *Postharvest Biol. Technol.*, 32, 109–113, 2004.
160. Petrisor, C., Lucian-Radu, G., Balan, V., Campeanu, G., Rapid and non-destructive analytical techniques for measurement of apricot quality. *Rom. Biotechnol. Lett.*, 15, 5213–5216, 2010.
161. Singh, B. and Singh, S. (Eds.), *Advances in Postharvest Technologies of Vegetable Crops*, CRC Press, Waretown, NJ, 2018.
162. Jha, S.N., Matsuoka, T., Miyauchi, K., Surface Gloss and Weight of Eggplant during Storage. *Biosyst. Eng.*, 81, 407–412, 2002.
163. Nelson, S.O., Guo, W.C., Trabelsi, S., Kays, S.J., Dielectric Spectroscopy of Watermelons for Quality Sensing. *Meas. Sci. Technol.*, 18, 7, 1887, 2007.
164. Sugiyama, J., Otobe, K., Hayashi, S., Usui, S., Firmness Measurement of Muskmelons by Acoustic Impulse Transmission. *Trans. ASABE*, 37, 1234–1241, 1994.
165. Trnka, J., Stoklasová, P., Strnková, J., Nedomová, Š., Buchar, J., Vibration Properties of the Ostrich Eggshell at Impact. *Acta Acta Univ. Agric. Silvic. Mendelianae Brun.*, 61, 1873–1880, 2013.
166. Mizrach, A., Flitsanov, U., Akerman, M., Zauberman, G., Monitoring Avocado Softening in Low-Temperature Storage Using Ultrasonic Measurements. *Comput. Electron. Agric.*, 26, 199–207, 2000.
167. Wang, W., Thai, C., Li, C., Gitaitis, R., Tollner, E.W., Yoon, S.-C., Detection of Sour Skin Diseases in Vidalia Sweet Onions Using Near-Infrared

Hyperspectral Imaging. *ASABE Annual International Meeting*, Paper No: 096364, 2009.
168. Itoh, H., Kanda, S., Matsuura, H., Sakai, K., Sasao, A., Measurement of Nitrate Concentration Distribution in Vegetables by Near-Infrared Hyperspectral Imaging. *Environ. Control Biol.*, 48, 31–43, 2010.
169. Polder, G., van der Heijdena, G.W.A.M., van der Voeta, H., Young, I.T., Measuring Surface Distribution of Carotenes and Chlorophyll in Ripening Tomatoes Using Imaging Spectrometry. *Postharvest Biol. Technol.*, 34, 117–129, 2004.
170. Arefi, A., Motlagh, A.M., Mollazade, K., Teimourlou, R.F., Recognition and Localization of Ripen Tomato Based on Machine Vision. Austra. *J. Crop Sci.*, 5, 1144–1149, 2011.
171. Saltveit, M.E., Jr, Determining Tomato Fruit Maturity with Non-Destructive in vivo Nuclear Magnetic Resonance Imaging. *Postharvest Biol. Technol.*, 1, 153–159, 1992.
172. Zhanga, L. and McCarthy, M.J., Measurement and Evaluation of Tomato Maturity Using Magnetic Resonance Imaging. *Postharvest Biol. Technol.*, 67, 37–43, 2012.
173. Siripatrawan, U., Makino, Y., Kawagoe, Y., Oshita, S., Rapid detection of Escherichia coli contamination in packaged fresh spinach using hyperspectral imaging. *Talanta*, 85, 1, 276–281, 2011.
174. Do Trong, N.N., Tsuta, M., Nicolaï, B.M., De Baerdemaeker, J., Saeys, W., Prediction of optimal cooking time for boiled potatoes by hyperspectral imaging. *J. Food Eng.*, 105, 4, 617–624, 2011.
175. Cheng, X., Chen, Y.R., Tao, Y., Wang, C.Y., Kim, M.S., Lefcourt, A.M., A novel integrated PCA and FLD method on hyperspectral image feature extraction for cucumber chilling damage inspection. *Trans. ASAE*, 47, 4, 1313, 2004.
176. Suthiluk, P., Saranwong, S., Kawano, S., Numthuam, S., Satake, T., Possibility of using near infrared spectroscopy for evaluation of bacterial contamination in shredded cabbage. *Int. J. Food Sci. Technol.*, 43, 1, 160–165, 2008.
177. Ramirez, R.A., *Computer vision based analysis of broccoli for application in a selective autonomous harvester*, Doctoral dissertation, Virginia Tech, USA, 2006.
178. Baltazar, A., Espina-Lucero, J., Ramos-Torres, I., González-Aguilar, G., Effect of methyl jasmonate on properties of intact tomato fruit monitored with destructive and nondestructive tests. *J. Food Eng.*, 80, 4, 1086–1095, 2007.

Index

β-glucosidase, 230, 237

Acceptable daily intakes (ADIs), 376, 378
Acidity regulators, 9, 380
Acoustic techniques,
 application, 422
 working, 422
Active packaging, 82–83
Additive, 97–115, 322, 328
 natural additive, 328, 329, 331
 synthetic additive, 328, 330, 331
Additive manufacturing, 267
Adulterants, 172–173
Adulteration, 25–26, 118–120
 accidental, 120
 edible oil and fats, 176
 grains and pulses, 175
 herbs and spices, 175
 honey, 176
 incidental, 193
 incidental adulteration, 28
 intentional, 118, 193
 intentional adulteration, 27
 juice, 175
 meat, fish, and seafood, 177
 metallic, 120
 metallic adulteration, 28
 microbial, 120
 milk and dairy products, 172
 tea and coffee, 177
 types, 192
 unintentional, 118
Adulteration in foods,
 additives, 196
 adequate daily intake, 196
 adulteration, 200
 alkaloids, 196
 annatto, 197
 argemone, 198
 barium, 199
 brazil nut, 198
 calcium carbonate, 197
 cassava, 198
 curcumin, 197
 cuyanuric acid, 199
 erythrosine orange, 199
 ethylene, 197
 european union, 197
 fraud, 197
 fungicide, 196
 hazel nut, 198
 malpractices, 196
 melamine, 199
 metanil yellow, 197
 methyl orange, 199
 morpholine, 196
 oleomargarine, 196
 poisonous, 196
 rhodomine, 197
 rodium sulphate, 199
 supply chain, 197
 tartrazine, 199
 tatrazine, 197
 urea, 200

Advanced biomaterials, 260
Advanced robotics, 260
Aglycones, 223
Alflatoxins, 238
Alginate, 89
Alkaloid, 221, 227–228, 233
  glycoalkaloids (GA), 228
  proto-alkaloids, 227
  pseudo-alkaloids, 227
  pyrrolizidine alkaloids (PA), 228
  true-alkaloids, 227
  tryptamine alkaloids, 228
Alternative sources for protein, 260
Amino acid availability, 223
Amino acids, 363, 365
Analgesic, 228
Animal husbandry, 251
Anthocyanin, 355–357, 359, 364
Anti-amylase, 223
Antibacterial, 227
Anti-caking agents, 380
Anticancer, 228
Anticarcenogenic, 222, 224
Antidepressant, 227
Antimicrobial, 375, 378–379, 381, 384–389, 393
Antinutrient factors (ANF), 219–220, 227, 233–239
Anti-nutritional effects, 222
Antioxidants, 9, 97, 99, 110, 222, 227, 342, 346, 348, 375, 378–381, 385–386, 389
Apple, 362
Artificial intelligence, 260
Artificial sweeteners, 379–381, 384
Aseptic packaging, 84–85
Aspartame, 384
Awareness, 183

Binder jetting, 269
Bioactive, 378, 386–389
Bioactive compounds, 342, 344, 346–347, 355, 360

Biocompatible, 388
Biodegradable, 388
Biodiversity, 260
Biopreservation, 261, 387
Bioterrorism, 168
Biotic stresses, 263
Blanching, 303, 312, 320, 321
Blueberry, 355–356
Butylated hydroxyl anisole (BHA), 376, 380, 384
Butylated hydroxytoluene (BHT), 376, 380

Caesium, 327
Caramelizing, 237
Carbonyl group, 222
Cardboard, 80
Cardioprotective, 222
Cardiovascular disease, 224
Catechin, 222
Categories of food additive, 98, 110, 111, 115
Cavitations technologies, 261
C-glycosol flavones, 232
Chelators, 222
Chemicals, 97, 98, 99, 105
Chemotrypsin, 223
Child stunting, 252
Chilling, 318, 321
Choline chloride, 353, 359–360, 363–365
Chromatographic techniques, 390, 391
  gas chromatography (GC), 376, 390
  liquid chromatography (LC), 376, 390
  mass spectroscopy (MS), 376, 390
Citric acid, 353, 355, 359–360, 364–365
Cobalt, 327
CODEX, 97, 102
Codex alimentarius, 377
Codex Alimentarius Commission, 270
Cold plasma (CP), 288, 298–300, 323
Cold plasma technology, 261

Colors, 97, 98, 102, 103
Communication, 78
Co-morbidities, 252
Complex food architecture, 267
Computed tomography,
   application, 420
   working, 419
Containment, 76
Cooking, 234, 238–239
Cradle-to-cradle, 260
CRISPR, 262
Crystallization, 295
Crystals, 229–230
Culinary, 267
Cultured meat, 266
Customized diet, 260, 266
Cyanide toxicity, 230–231
Cynogenic glycosides, 221, 230–231, 233

Data carriers, 84
De-branning, 234
Dehydration, 303, 308, 316, 318
Deiodinization, 232
Delivery systems, 387–388
Designer foods, 266
Developing countries, 165, 166
Diet diversification, 271
Direct additive, 98, 101, 108
DNA bar-coding, 262
Drying, 308, 313, 316–318

Early life nutrition, 252
Ecofriendly, 262
Effects of adulteration,
   aluminium flouride, 202
   anaphylaxis, 202
   asthma, 202
   benzaldehyde, 202
   benzoic acid, 202
   biogenic amines, 201
   carcinogenesis, 201
   dropsy, 201
   dysentery, 202
   EMA, 201
   erythrosine, 201
   nitric acid, 201
   nutrition, 202
   respiratory distress, 201
   sanguinarine, 202
   sodium cyclamate, 201
   ulceration, 201
EFSA, 111
Electroanalytical techniques, 391
   differential pulse voltammetry (DPV), 391
   square wave voltammetry (SWV), 391
Electrode, 296–299, 304, 307, 314, 327
Electronic nose,
   application, 403
   working, 402
Emulsifiers, 10, 103, 109, 110, 115, 375, 380, 381, 384, 388
Encapsulating material, 388
Encapsulation, 375, 378, 386–389, 393
   coacervation, 388
   electrospraying, 388
   electrospinning, 388
   extrusion technologies, 388
   freeze drying, 388
   high pressure homogenization, 388
   inclusion complexation, 388
   nanoprecipitation, 388
   spray drying, 388
Enteric hyperoxaluria, 229
E-number, 384
E-numbering, 13
Environment conditions, 84
Enzyme inhibitors, 234
Enzyme preparation, 99, 100
Epicatechin, 222
Epicatechin-3-gallate, 222
Epigallocatechin-3-gallate, 222
Essential oils, 375, 376, 378, 385
   expression, 385
   hydro distillation, 385
   supercritical fluid extraction, 385

Estimated daily intake, 108–109
European Union (EU), 376, 379, 382
European Union legislation, 110
Evaporation, 303, 316, 323
Extraction, 289, 295, 304, 305
Extrusion, 234, 237–238, 295, 309, 318, 319, 354
  extrudate, 238
  extruder barrel, 238
  feed rate, 238
  high-temperature short-time (HTST), 237
  hopper, 238
  single screw, 319
  twin-screw, 319

Farm to fork, 250, 252, 259
Fecal nitrogen, 223
Fermentation, 230, 234–236, 282, 295, 304
Flavanols, 222
Flavonoids, 222, 232, 237, 342, 350, 355, 360
Flavor enhancer, 380
Flavoring agent, 98, 99, 100
Flavors, 99
Food additive, 97–101
Food additive petitions (FAPS), 376, 383
Food additives, 219, 237, 375–385, 389–393
Food adulteration, 165–166, 169
Food and agricultural organization (FAO), 376, 378, 382
Food and drug administration (FDA), 376, 382, 383, 392
Food color,
  artificial colors, 2
  natural colors, 2
  nature-identical colors, 2
Food contaminants, 219, 240
  biological contaminants, 219
  non-biological contaminants, 219
Food contamination, 218, 219, 238
Food for all, 251
Food for future, 250
Food fraud, 166, 179
Food printing, 266
Food processing, 250, 260
Food safety and standards act, 30–32
Food security, 252, 255
Food sensing technologies, 260
Food supply chain, 219, 251
Food toxicity, 219
Food wastage, 250
Foodborne disease (FBD), 218–219
Fortificants, 99
Fortifying agents, 11
Freeze drying, 354–355
Freezing, 282, 289, 307, 308, 315, 316, 318, 321, 323
  ice crystal, 308, 315, 316, 323
  latent heat, 315
  sensible heat, 315
  thermal conductivity, 303, 316, 318
Frying, 317, 323
FSSAI, 99, 375, 376, 383
Fused deposition manufacturing, 268

Gallic acid, 350, 357–358
Gamma radiations, 239
Gas expanded liquids, 345, 347
Gastric hemorrhage, 230
Gelling, 379
Generally recognized as safe (GRAS), 376, 378, 383, 385
Genome editing, 263
Genomic technology, 239
Gingerol, 356
Glass, 80
Global food industry, 250
Glycerol, 352, 354, 356–361, 363, 365
Glycoslated gallic acid, 222
GMOs, 251, 262
GMP, 376, 383
Goitrogens, 221, 231–233
  glucosinolates (GSL), 232
  goiter, 231, 233
  goitrin, 232

INDEX   455

Government and regulatory agency, 177
Gram-negative, 278, 297, 382
Gram-positive, 297, 382
Grape, 355, 357, 359, 363–364
GRAS, 100, 103
Green chemistry, 343, 352
Green drying, 261
Green food technologies, 259
Green nanotechnology, 386
Greenhouse gases, 251
GRIN technologies, 251

HDR, 263
Health effect of food additives and preservatives, 66–70
Heat and mass transfer, 237
Hemagglutinin, 220, 225–226, 233
  plant hemagglutinin, 226
Hematuria, 230
Hemolytic, 223–224
Herbicide resistance, 264
High hydrostatic pressure, 282, 292
High-pressure processing (HPP), 234, 285, 286, 292–295, 321
Homogenization, 295
Hot melt extrusion, 268
Humectants, 380
Hunger, 250
Hurdle technology, 22, 282, 320, 322, 323
  microbiological hurdles, 321
  physical hurdle, 321
  physical non-thermal hurdle, 321
  physicochemical hurdle, 321
Hydrogen bond acceptor, 352, 359, 363
Hydrogen bond donor, 352, 359, 363
Hydrogen cyanide (HCN), 230
Hydroxyl group, 222
Hypocholsterolemic, 224–225, 233

IBD, 253
Ideal diet, 257
Image analysis,
  hyperspectral imaging, 411–413, 435
  machine vision, 413–415, 430, 435
  time resolved reflectance, 415–417
Immunomodulatory, 222
Impact of adulteration on health, 120–121
  aflatoxin, 121
  anemia, 121
  appendicitis, 120
  epilepsy, 120
  metallic lead, 120
  metanil yellow, 120
  neurotoxicity, 120
  rhodamine, 121
  sudan dyes, 121, 125, 126
Indirect additive, 108
Infrared heating (IRH), 306, 311–313, 317
INS, 114
Intelligent packaging material, 83–84
Intentional adulteration, 169–171
International numbering system (INS), 376, 379
Iodine absorption, 232, 233
Ionic liquids, 341, 345, 348, 350–351
Iron deficiency anemia (IDA), 222
Irradiation, 275, 282, 291, 295, 321, 322, 324, 325, 327
  electron beam (β), 321, 327
  gamma (γ) rays, 321, 327
  ionizing radiation (IR), 321, 324–327
  kilo gray (kGy), 321
  x-ray, 311, 321, 326, 327
Isoflavones, 232, 237, 360, 363
  daidzein, 232
  genistein, 232
Isothiocyanates, 232–233

Joint expert committee on food additives (JECFA), 101, 376, 377

Kaempferol, 350, 361
Kidney stones, 229, 233
Konzo, 231

Labeling regulations, 114
Lab-on-a chip, 262
Lactic acid bacteria (LAB), 235
   Bifido bacteria, 236
   Lactobacillus spp., 236
   Streptococcus spp., 236
Lactic acid fermentation, 235, 236
Lacto peroxidase system, 382
Laser air-puff, 401
Layer-by-layer, 268
LD50, 376, 392
Leaching, 234, 235
Leaky gut, 225, 233
Lectins, 220, 225–227, 233, 237
   legere, 225
   phytohemagglutinins (PHA), 226–227
Life cycle assessment, 86, 90–91
Linamarin, 230
Lipase, 223, 224
LOC, 105–109
Loopholes, 179

Magnetic resonance imaging,
   application, 407–408
   working, 406–407
Malnutrition, 250
Mass customization, 266
Measures to mitigate food adulteration,
   Bureau of Indian Standards, 207
   chemical residues, 207
   Codex, 206
   Codex Alimentarius Commission, 207
   comprehensive, 209
   exposure, 207
   fertilizers, 206
   food fraud, 209
   Food Safety Management System, 207
   FSSAI, 206
   GAP, 206
   GMP, 206

HACCP, 206
   legislation, 208
   legitimation, 208
   pesticides, 206
   policies, 208
   prevention, 206
   public health, 208
   public health concern, 209
   regulation, 208
   regulatory agencies, 208
   TACCP, 206
   US Food and Drug Administration, 207
   VACCP, 206
   vulnerable, 208
Mechanical thumb method, 400
Membrane filtration, 290, 301–303
   microfiltration (MF), 290, 301, 302
   nanofiltration (NF), 301–303
   reverse osmosis (RO), 290, 301–303
   ultrafiltration (UF), 290, 301, 302
Merits and demerits of food additives and preservatives, 47–48
Metal, 79
Metallic nanoparticles, 375, 378, 386
Microbial enzyme activity, 223
Microbial infestation, 219
Micronutrient deficiencies, 250
Microwave (MW), 282, 306, 308, 311, 312, 317, 321, 323
   dielectric, 291, 299, 311–314, 317
   electromagnetic, 300, 311–314, 317
Microwave heating, 234
Milling, 234, 237
Minerals, 98–99
MSG, 384
Myoglobin, 278, 287
Myricetin, 350, 361
Myrosinase, 232

Nanoemulsions, 376, 378, 388
Nanoencapsulation, 375, 378, 386, 387, 389, 393
Nanoliposome, 378

Nanotechnology, 89, 260
Near infrared spectroscopy,
  application, 410
  working, 408–409
Neuro sensory deafness, 231
Neuroprotective, 227
NHE, 263
Nitrile group, 230
Nitriles, 232
Non destructive methods,
  chemical, 401–403
  dynamic, 418–424
  electromagnetic, 404–407
  mechanical, 400, 401
  optical, 408–417
Non destructive testing in,
  fruits, 427–430
  staple foods, 425–427
  vegetables, 431–436
Non-thermal, 276, 277, 282–290, 292, 295, 296, 298, 314, 320, 321, 332
No-observed-adverse-effect level (NOAEL), 376, 392
Novel, 182
Novel-thermal, 276, 277, 303, 305–310, 314, 319, 332
Nuclear magnetic resonance,
  application, 405–406
  working, 404–405
Nutrients, 223–225, 233–234, 238
Nutrigenomics, 260
Nutritional genomics, 254
Nutritional issues, 250
Nutritional transition, 251, 254, 256, 257
Nutrotherapeutic, 267, 269

Ohmic heating (OH), 261, 296, 303–305, 311
  cold spot, 304, 311
  electro conductive heating, 303
  electro heating, 303
  Joule heating, 303
Oils and spices, 21

Omnivorous, 254
Omnivorous diet, 257
Onion, 350, 358, 360–363
Optical atrophy, 231
Organic acids, 382
Organic farming, 260
Ornithine-derived alkaloids, 228
Oscillating magnetic field (OMF), 289, 300, 301
Oxalic acid, 228–230
  oxalates, 228–230, 233, 234, 236, 237, 239
Oxalobacter formigens, 229
Oxidative rancidity, 380
Ozone (O_3), 282, 292, 300
Ozone processing, 261
Ozonization, 261

Packaging material, 89
Packaging technology, 277, 278, 293, 296
  active packaging, 277
  biaxially oriented polypropylene (BOPP), 293
  carbonated gable, 281
  ethylene–vinyl alcohol (EVOH), 293
  high-density polyethylene (HDPE), 293
  low-density polyethylene (LDPE), 293
  modified atmospheric packaging (MAP), 277–284, 323
  nylon, 293
  polyester, 293
  polyethylene (PE), 293
  polypropylene (PP), 293
  respiration rate, 278, 283
  scavenger, 278
  thermoformed package, 279
Paper, 78–81
Particle size reduction, 237
Pasteurization, 292, 304, 312, 320, 321
  cold pasteurization, 292
Personalized nutrition, 266

Physical authentication techniques, 122–124
  fluorescence, 123, 124, 151
  FTIR, 124, 127, 128, 132
  GCMS, 124, 127
  HPLC, 126, 129
  microscopic, 122
  molecular, 136
    AFLP, 137, 139
    DNA, 128, 136, 137–140, 142–150
    DNA markers, 136
    hybridization, 136, 147–148
    isothermal, 142, 152
    ISSR, 136, 139
    NGS (next generation sequencing), 147–151
    polymerase chain reaction (PCR), 128, 136–148
    RAPD, 137, 138, 140
    RFLP, 136, 137
    RFLP markers, 137
    RNA, 136, 141, 145, 146
    RT PCR, 142–146
    sequencing based, 128, 146
    SNP, 137
    SSR, 137
    STS, 137
  NMR, 127–128, 131
  SEM, 122, 123
  starch, 123, 125, 128
  TEM, 122
  tetrahertz, 127
  UV, 124
Phytase, 234–236, 238
Phytates, 219, 220, 233–237, 239
Phytochemicals, 375, 378, 384–386, 389, 392
  alkaloids, 385
  carotenoids, 385
  flavonoids, 385
  isoflavone, 385
  monoterpenes, 385
  phenolic acids, 385
Plant secondary metabolites, 262

Plasma cholesterol, 224, 225
Plastic, 80
Plasticizers, 380
Polyphenolic compounds, 221
Polyphenols, 342, 355–357
Pomace, 355, 357–358, 364
Post-storage and processing condition, 219
Poverty, 250
Powder bed jetting, 269
Precision farming, 251
Preservation, 77
Preservative, 101–102
Preservatives, 375–382, 384–385
  influence of pH, 16
  properties of preservatives, 16
Primary hyperoxaluria, 229
Proanthocyanidin, 222
Processing, 76
Propyl gallate, 380
Protein digestibility, 223, 225, 233
Pulsed electric field (PEF), 282, 287, 288, 296–298
  capacitor, 296, 300
  electrolysis, 296, 291
  electroporation, 296, 304
  oscilloscope, 296

QSAR, 376, 391
Quality indicator, 84
Quercetin, 350, 357, 360–363

Radio frequency (RF) heating, 306
  dielectric, 307, 313, 314
Raising agents, 380
Recyclability, 86
Red book, 105
Remedy strategies, 177
Restricted provisions, 114
Resveratrol, 232
RNA, 239
Roasting, 234, 238, 239
Röntgen radiation, 327
Rutin, 349, 351, 361, 363

Saccharification, 236
Saccharin, 384
Safety assessment, 99
Safety of food additives and
    preservatives, 25
Saponin, 220, 223–225, 233, 234, 236
    sapogenin, 223
SDGs, 259
Secondary metabolites, 378, 385
Selective sintering technology, 268
Sensor fusion,
    application, 425
    working, 423–424
Sensory gait ataxia, 231
Shelf life, 81
Soaking, 230, 234–235, 239
Solid lipid nanoparticles, 376, 378, 388
Sorbitol, 381, 384
Spastic paraparesis, 231
Spectroscopic techniques, 389
    FTIR, 390
    infrared spectroscopy (IR), 389
    NIR, 376, 390
    Raman spectroscopy, 390
    ultraviolet/visible radiation (UV/Vis) spectroscopy, 376, 389
SSNs, 263
Stabilizers, 105, 379, 380, 381
Steroidal glycosides, 223
Subcritical water, 346
Sulforaphane, 232
Super critical carbon dioxide, 342, 347
Super critical fluid, 346
Surface active agents, 380
Sustainability, 86, 250, 254
Sustainable diet, 257
Sweeteners, 11
Synthetic colorants,
    fat soluble synthetic colorants, 8
    lake colorants, 8
    water soluble synthetic colors, 7

Tailor made, 266
TALENs, 263

Tannins, 220–223, 233–239
    condensed tannins, 222
    hydrolyzable tannins, 222
Taxiphyllin, 230
Teratogenic, 392
Texturizing, 237
Thickening, 379
Thiocyanates, 232
Three-dimensional (3-D) printing, 309, 310, 319, 320
    deformation, 319
Threshold regulations, 109
Thyroid gland, 231–233
    T3, 232
    T4, 232
    thyroid hormones, 231–233
    thyroid peroxidase (TPO), 232, 233
TOR, 108
Traditional foods, 113–114
Triterpene, 233
Triterpenoid, 224–225
Tropane, 228
Tropical ataxic neuropathy (TAN), 231
Trypsin, 220, 223–224, 233, 236, 238–239
Types of adulterants,
    adulteration, 193
    afflatoxins, 204
    alimentary toxic aleukia, 204
    ascariasis, 204
    cadmium, 205
    cancer, 203
    carcinogenous, 204
    contaminants, 193
    diarrhoea, 203
    flourosis, 203
    gastritis, 198
    hydrocarbons, 203
    incidental, 193
    insomnia, 205
    intentional, 193
    irradiation, 195
    mercury, 205
    metallic, 193
    microbial, 194

myocardial failure, 205
paralysis, 205
parasitosis, 204
peristalisis, 198
pesticides, 203
poisoning, 203
salmonellosis, 204
shigellosis, 204
Types of food additives and preservatives, 48–65
coloring agent, 51–53
flavoring agent, 51
miscellaneous additives, 54–65
nutritional additives, 51
preservatives, 48–50
texturizing agent, 53–54
Tyrosine, 232

Ultrasonic,
application, 421
working, 421

Ultrasound, 282, 286, 287, 295, 296, 322, 323
cavitation, 296
Ultraviolet (UV), 291, 311
Unintentional adulteration, 170
Urea, 352–358, 361, 365
USFDA, 106–108
Utility, 78

Value addition, 271
Vitamins, 108

Water activity, 376, 377
World Health Organization (WHO), 102, 375, 376, 378, 382

X-rays,
application, 419
working, 418

ZFNs, 263

Printed in the USA/Agawam, MA
January 28, 2022

788406.001